Modeling
Black Hole
Evaporation

Modeling Black Hole Evaporation

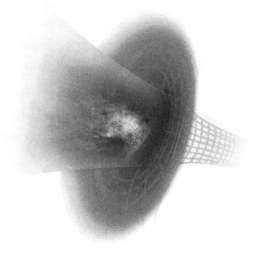

Alessandro Fabbri
Università di Bologna, Spain

José Navarro-Salas
Universidad de Valencia, Spain

Imperial College Press

Published by

Imperial College Press
57 Shelton Street
Covent Garden
London WC2H 9HE

and

World Scientific Publishing Co. Pte. Ltd.
5 Toh Tuck Link, Singapore 596224
USA office: 27 Warren Street, Suite 401-402, Hackensack, NJ 07601
UK office: 57 Shelton Street, Covent Garden, London WC2H 9HE

British Library Cataloguing-in-Publication Data
A catalogue record for this book is available from the British Library.

MODELING BLACK HOLE EVAPORATION

Copyright © 2005 by Imperial College Press and World Scientific Publishing Co. Pte. Ltd.

All rights reserved. This book, or parts thereof, may not be reproduced in any form or by any means, electronic or mechanical, including photocopying, recording or any information storage and retrieval system now known or to be invented, without written permission from the Publisher.

For photocopying of material in this volume, please pay a copying fee through the Copyright Clearance Center, Inc., 222 Rosewood Drive, Danvers, MA 01923, USA. In this case permission to photocopy is not required from the publisher.

ISBN-13 978-1-86094-527-4
ISBN-10 1-86094-527-9

Printed in Singapore

To my family
Sandro

To Lola, Sergi and María Dolores
Pepe

Preface

The aim of this book is to offer a comprehensible monograph on black hole quantum radiance and black hole evaporation. Most of the existing books on quantum field theory in curved spacetimes describe with detail the Hawking effect. However none of them extensively treats the issue of the backreaction of the evaporation process. Since this is a rather involved topic we shall take a modest, but pedagogical line to approach the subject.

The history of research on black holes tells us that important physical insights have been gained through the study of simplified models of gravitational collapse. The paradigm is the Oppenheimer and Snyder model (1939). Despite its simplicity, forced by the difficulties of the technical treatment of more realistic situations, it turns out to produce a very accurate picture of the gravitational collapse and its final outcome. All the main ingredients were there: different descriptions for external and infalling observers, divergent redshift at the horizon and existence of the internal singularity. The assumption of perfect spherical symmetry, the main criticism of black hole opponents, was not, in the end, a real drawback to invalidate the full picture offered by the model. In recent years, a model inspired by string theory, and proposed by Callan, Giddings, Harvey and Strominger (1992), also offers a simplified scenario which allows to study analytically the process of black hole formation and subsequent evaporation, including semiclassical backreaction effects. The results and techniques generated by this model renewed the interest in black hole evaporation.

Motivated by these two paradigmatic models, we take this line of thought and try to present a pedagogical view of the subject of black hole radiance and black hole evaporation. This is the reason for calling "modeling" the approach we take in this book.

The style and presentation of the different aspects involved have been chosen to make them accessible to a broad audience. We want to stress that we assume a basic knowledge of general relativity and also of quantum field theory, at the level of introductory graduate courses. Therefore, a wide spectrum of physicists, ranging from particle physicists to astrophysicists, and from beginner graduate students to senior researchers, can follow all chapters. Even those who are not very familiar with either general relativity or quantum field theory can find, we hope, this monograph accessible. With this respect we want to remark that this is indeed a book on quantum aspects of black holes and not on "quantum field theory in curved spacetimes". Since the latter subject plays a fundamental role to address the backreaction problem, we approach it following a simple, although somewhat unconventional, way. We try to escape from the standard technicalities of regularization schemes to derive conformal anomalies, effective actions, etc. Rather we try to present and rederive the fundamental results on a physically motivated basis.

The most delicate parts of the text were written while we worked together at the Department of Theoretical Physics and IFIC of the University of Valencia, and also at the Department of Physics of the University of Bologna, the Schrödinger Institute in Vienna and the Institut d'Astrophysique de Paris. We wish to thank our colleagues and collaborators in these institutions, and in particular Roberto Balbinot for a critical reading of the draft of the book. We also wish to thank the students who attended a Ph.D. course, delivered by both authors at the University of Valencia, that was based on preliminary notes of the present book. We found very useful the comments and questions posed by the students and our young collaborators Sara Farese and Gonzalo Olmo to improve the pedagogical style of the text. A special thanks to all our past and present collaborators, from whom we have benefited a lot. Finally we acknowledge financial support from the Spanish Ministerios de Educación y Ciencia y Tecnología, the Generalitat Valenciana, and the collaborative program CICYT-INFN.

Alessandro Fabbri and José Navarro-Salas

Contents

Preface — vii

1. Introduction — 1

2. Classical Black Holes — 7
   2.1 Modeling the Gravitational Collapse — 7
      2.1.1 The Oppenheimer–Snyder model — 11
      2.1.2 Non-spherical collapse — 18
   2.2 The Schwarzschild Black Hole — 20
      2.2.1 Eddington–Finkelstein coordinates — 20
      2.2.2 Kruskal coordinates — 23
   2.3 Causal Structure and Penrose Diagrams — 26
      2.3.1 Minkowski space — 27
      2.3.2 Schwarzschild spacetime — 29
   2.4 Kruskal and Locally Inertial Coordinates at the Horizons — 32
      2.4.1 Redshift factors and surface gravity — 36
   2.5 Charged Black Holes — 39
   2.6 The Extremal Reissner–Nordström Black Hole — 46
   2.7 Rotating Black Holes — 50
      2.7.1 Energy extraction from rotating black holes: the Penrose process — 54
   2.8 Trapped Surfaces, Apparent and Event Horizons — 56
      2.8.1 The area law theorem — 59
   2.9 The Laws of Black Hole Mechanics — 62
   2.10 Stringy Black Holes — 66

3. The Hawking Effect — 73
   3.1 Canonical Quantization in Minkowski Space 73
   3.2 Quantization in Curved Spacetimes 75
      3.2.1 Bogolubov transformations and particle production . 77
   3.3 Hawking Radiation in Vaidya Spacetime 81
      3.3.1 Ingoing and outgoing modes 85
      3.3.2 Wave packets 89
      3.3.3 Wick rotation 93
      3.3.4 Planck spectrum 96
      3.3.5 Uncorrelated thermal radiation 98
      3.3.6 Where are the correlations? 101
      3.3.7 Thermal density matrix 106
   3.4 Including the Backscattering 107
      3.4.1 Waves in the Schwarzschild geometry 108
      3.4.2 Late-time basis to accommodate backscattering . . . 111
      3.4.3 Thermal radiation and grey-body factors 114
      3.4.4 Estimations for the luminosity 116
         3.4.4.1 Emission of massless and massive particles . 118
   3.5 Importance of the Backreaction 118
   3.6 Late-Time Independence on the Details of the Collapse . . 119
      3.6.1 The example of two shock waves 119
      3.6.2 Geometric optics approximation 122
      3.6.3 General collapse 125
      3.6.4 Adding angular momentum and charge 126
   3.7 Black Hole Thermodynamics 128
   3.8 Physical Implications of Black Hole Radiance and the Information Loss Paradox 131
      3.8.1 Black hole evaporation 131
      3.8.2 Breakdown of quantum predictability 132
      3.8.3 Alternatives to restore quantum predictability 137

4. Near-Horizon Approximation and Conformal Symmetry — 145
   4.1 Rindler Space 146
   4.2 Conformal Symmetry, Stress Tensor and Particle Number . 148
      4.2.1 The normal ordered stress "tensor" operator 153
      4.2.2 The SO(d,2) conformal group and Möbius transformations 154
         4.2.2.1 Infinitesimal transformations 156

		4.2.3	The particle number operator	157
	4.3	Radiation in Rindler Space: Hawking and Unruh Effects . .	159
		4.3.1	The Hawking effect	161
			4.3.1.1 Correlation functions and thermal radiation	163
		4.3.2	Radiation through the horizon	165
		4.3.3	The Unruh effect	166
			4.3.3.1 Unruh modes	169
			4.3.3.2 Möbius invariance of the vacuum state . . .	171
		4.3.4	Three different vacuum states	172
	4.4	Anti-de Sitter Space as a Near-Horizon Geometry	173
		4.4.1	Extremal black holes	174
		4.4.2	Near-extremal black holes	176
	4.5	Radiation in Anti-de Sitter Space: the Hawking Effect . . .	177
		4.5.1	Three vacuum states	183
	4.6	The Moving-Mirror Analogy for the Hawking Effect	183
		4.6.1	Exponential trajectory: thermal radiation	185
		4.6.2	Radiationless trajectories	187
			4.6.2.1 Hyperbolic trajectories	188
		4.6.3	Asymptotically inertial trajectories and unitarity . .	189
5.	Stress Tensor, Anomalies and Effective Actions			193
	5.1	Relating the Virasoro and Trace Anomalies via Locally Inertial Coordinates	195
	5.2	The Polyakov Effective Action	203
		5.2.1	The role of the Weyl-invariant effective action	206
	5.3	Choice of the Quantum State	208
		5.3.1	Boulware state	209
		5.3.2	Hartle–Hawking state	210
		5.3.3	The "in" and Unruh vacuum states: the Hawking flux	212
			5.3.3.1 "in" vacuum state	212
			5.3.3.2 Unruh state	214
	5.4	Including the Backscattering in the Stress Tensor: the s-wave	216
		5.4.1	Two-dimensional symmetries	218
		5.4.2	The normal ordered stress tensor	219
			5.4.2.1 Transformation laws	220
		5.4.3	The covariant quantum stress tensor	222
		5.4.4	Effective action	225
			5.4.4.1 Anomaly induced effective action	225
			5.4.4.2 The Weyl-invariant effective action	226

	5.4.5	Quantum states and Hawking flux	229
		5.4.5.1 Boulware state	230
		5.4.5.2 Hartle–Hawking state	231
		5.4.5.3 The "in" and Unruh vacuum states: Hawking flux with backscattering	232

5.5 Beyond the s-wave Approximation 234
5.6 The Problem of Backreaction 235
    5.6.1 Backreaction equations from spherical reduction . . . 235
    5.6.2 The problem of determining the state-dependent functions 238
    5.6.3 Returning to the near-horizon approximation 241

6. **Models for Evaporating Black Holes**   **243**

6.1 The Near Horizon Approximation 245
    6.1.1 Schwarzschild 245
    6.1.2 Near-extremal Reissner–Nordström black holes: the JT model 248
    6.1.3 Near-extremal dilaton black holes 249
        6.1.3.1 Einstein frame 250
        6.1.3.2 String frame: the CGHS model 252
6.2 The CGHS Model 253
    6.2.1 The CGHS black hole 253
        6.2.1.1 Free field 255
        6.2.1.2 Linear dilaton vacuum 256
        6.2.1.3 Black hole solutions 257
        6.2.1.4 CGHS and four-dimensional dilaton black holes 259
        6.2.1.5 CGHS and Schwarzschild black holes 260
    6.2.2 Including matter fields 261
        6.2.2.1 Dynamical black hole solutions 263
        6.2.2.2 Singularities, event and apparent horizons . 264
    6.2.3 Bogolubov coefficients and Hawking radiation 265
    6.2.4 Quantum states for the CGHS black hole 270
        6.2.4.1 Boulware state 271
        6.2.4.2 Hartle–Hawking state 272
        6.2.4.3 Unruh state 273
        6.2.4.4 "in" vacuum state 273
6.3 The Problem of Backreaction in the CGHS Model 274
    6.3.1 State-dependent functions for evaporating black holes 276
    6.3.2 The RST model 278

		6.3.2.1	Liouville fields	279
		6.3.2.2	General solution	280
		6.3.2.3	Semiclassical static solutions	281
		6.3.2.4	Backreaction in the Unruh state	283
	6.3.3	Black hole evaporation in the RST model		286
		6.3.3.1	Information loss in the RST model	290
		6.3.3.2	Is the picture of RST black hole evaporation generic?	295
6.4	The Semiclassical JT Model			296
	6.4.1	Extremal solution		298
	6.4.2	Semiclassical static solutions		299
	6.4.3	General solutions		301
	6.4.4	Black hole evaporation in the JT model		303
		6.4.4.1	First approximation for the state-dependent functions	303
		6.4.4.2	Exact treatment for the state-dependent functions	308
	6.4.5	Beyond the near-horizon approximation and Hawking radiation		313
	6.4.6	Information loss in the JT model		315
	6.4.7	Unitarity in the semiclassical approximation?		316

Bibliography   319

Index   331

Chapter 1
Introduction

Black holes are among the most fascinating predictions of Einstein's theory of general relativity. They are an exotic, but natural outcome of the very basic feature of the gravitational interaction. The universality of gravity, nicely expressed by the equivalence principle, allows to produce an accumulative effect that can result in a very strong gravitational field. Indeed, it can be so strong that it cannot be counterbalanced by other forces and continues to grow up until extreme situations. Since gravity affects the spacetime geometry, the gravitational field produced by matter could become so strong as to substantially modify the ordinary causal structure of spacetime, and produce a region where even light is trapped and no particle can escape to infinity.

Quantum mechanics enters immediately into this game and its first consequence is to prevent, partly, this extreme situation. Chandrasekhar showed that the quantum degeneracy pressure of electrons can balance the gravitational force and avoid a complete implosion. However, this requires a limiting mass for the star of about 1.4 times the solar mass. Above this limit the full gravitational collapse can be prevented only by the degeneracy pressure of neutrons, but this requires again an upper mass limit (2–3 solar masses). Therefore for sufficiently massive stars, that do not throw away enough matter or radiation to reach the neutron star limit, complete collapse is inevitable and a black hole forms.

Nevertheless, quantum mechanics seems to continue conspiring against the black hole, but in a different way. In a remarkable discovery Hawking showed how this takes place. Black holes are not as "black" as general relativity predicts, but rather radiate thermally all types of existing quanta, although for kinematical reasons light particles dominate the emission. This is a very small effect, at least initially, for black holes created from gravitational collapse. Hawking's discovery, which can be rederived from different perspectives, puts forward an intriguing and close relation between black holes and thermodynamics. This astonishing result is obtained using the approximation that the spacetime background is kept fixed at all times. This is an accurate approximation for the first period of the evaporation, where the effect is small, and, likely, a reasonable approximation until the black hole reaches Planck-size. Eventually, by extrapolation of the results, the black hole will disappear. The conspiracy of quantum mechanics is such that it turns, as a "boomerang" effect, against itself. Indeed, the resulting physical picture, as it was pointed out by Hawking himself almost thirty years ago, seems not to be compatible with the principles of quantum mechanics. The type of radiation emitted does not allow the recovery of the information about the star from which the black hole was created. Therefore, with the disappearance of the black hole this information will be lost forever. But this is forbidden by the basic principles of quantum mechanics itself.

However, there is no definitive picture of the full evaporation process and the reason comes out immediately. The fixed background approximation ignores the effects of the radiation on the spacetime geometry, in other words, the backreaction effects. They play an important role, even before reaching the Planck scale, when a still unknown quantum theory of gravity is expected to dominate the process. They need to be taken into account to have a detailed view of the evaporation process, and indeed they could serve as an inspiration to attack the deep problems or paradoxes of black hole evaporation mentioned above. The hope of many researchers is that, at the end, quantum mechanics will keep conspiring in such a way that a full quantum gravity approach will modify Hawking's original picture and allow information retrieval. Most remarkably, Hawking himself now seems to be converging towards this belief after almost thirty years of skepticism.

In general it is hard to try to go beyond the fixed background approximation to include the backreaction, even at the semiclassical level. This requires to solve the so called semiclassical Einstein equations

$$G_{\mu\nu} = 8\pi \langle \Psi | T_{\mu\nu}(g_{\alpha\beta}) | \Psi \rangle \ . \tag{1.1}$$

This is not at all a purely technical problem. First of all one needs to have an explicit expression for the expectation values $\langle \Psi | T_{\mu\nu}(g_{\alpha\beta}) | \Psi \rangle$ for a large family of metrics (including those that could be potentially the solution of the semiclassical equations). Moreover, these quantities also depend on the quantum state of the matter $|\Psi\rangle$, and this is a rather non-trivial issue.

To properly model the process of black hole evaporation one should bypass these difficulties somewhat. This is a remarkable open problem which should be addressed, in the appropriate limit, by any theory containing general relativity and quantum mechanics.

The organization of the material follows the above brief historical "tour". In Chapter 2 we start by briefly overviewing the Oppenheimer-Snyder model and the main features of the gravitational collapse. After this we introduce the basic ingredients of stationary (charged and rotating) black holes, using the simplest one, i.e., the Schwarzschild solution. To make the presentation accessible for a wide community of scientists we have based our arguments, as much as possible, on the equivalence principle to derive the basic results. The Kruskal coordinates, usually introduced for global purposes, are motivated here on a local basis in terms of locally inertial coordinates, where the intuition of readers more familiar with flat spacetime physics is more solid. Nevertheless the global aspects of the solutions are, of course, very important and we also pay special attention to this issue. In addition to the most popular notion of "event horizon" we also introduce the concept of "apparent horizon", which is of special relevance in the analysis of time dependent settings such as the evaporation process of Chapter 6. We also present the intriguing formal analogy between the "laws of black hole mechanics" and the laws of thermodynamics. We point out the loophole of the analogy, which will be filled in as a result of Hawking's discovery (to be presented in Chapter 3). Finally we also introduce black hole solutions in theories different from pure general relativity, motivated by the fact that they will be the inspiration of the model proposed by Callan, Giddings, Harvey and Strominger, the main theme of Chapter 6.

The whole Chapter 3 is devoted to the Hawking effect. We shall follow the original derivation of Hawking, together with the works of Parker and Wald, to explicitly show that the radiation emitted by a black hole at late times is exactly described by a thermal density matrix (modulated by a "grey-body" factor). We divide the derivation in two steps aiming to show in a clear way the skeleton of the argument and separate the technicalities from the main physical ideas. The simplest possible model, involving the matching of Minkowski and Schwarzschild spacetimes, contains all the ingredients that produce the Hawking radiation: the existence of an event horizon in a non-stationary spacetime. The intention is to introduce the Hawking radiation in an elementary way and make it more accessible to readers with a basic background on general relativity. The second step is to show that the late time thermal radiation obtained in this simplified spacetime, mimicking a gravitational collapse, is insensitive to the details of a realistic collapse. Apart from this, the derivation is quite standard. We based it on the properties of the late time Bogolubov transformations and the features of wave propagation in the Schwarzschild geometry. We finish the chapter presenting the physical implications of the Hawking effect. We shall discuss in detail the challenges that black hole evaporation poses to the interphase of quantum mechanics and general relativity. The possible breakdown of quantum predictability is the cornerstone of this more conceptual discussion. It will serve to warn the reader that not everything in this subject is well understood.

In Chapter 4 we approach the Hawking effect from a different perspective, aiming to get a better understanding. Since the basic feature is the existence of a horizon we simplify the Schwarzschild geometry working with its "near-horizon approximation" Rindler geometry, which is nothing else but a wedge of Minkowski space. We want to remark that it is quite common in the literature to discuss Rindler space and the Unruh effect before considering its curved spacetime "analog". However, we prefer to present things the other way around. The Unruh effect is even more shocking than the Hawking effect for readers that have learned "quantum field theory" with too much emphasis on "Poincaré symmetry" (see at this respect [Wald (1994)]). For this reason we rederive Hawking radiation from the near-horizon Rindler geometry and, only as a by-product, the Unruh effect. This result finds justification from the equivalence principle. All this discussion allows us to introduce into the game, in a natural way, one of the strongest spacetime symmetries in physics: the two-dimensional conformal symmetry. The theory of conformal fields is usually applied to analyse second-order phase transitions in condensed matter systems [Di Francesco et al. (1997)] and it is also widely used in the formulation of string theory [Polchinski (1998)]. We shall show how it also plays an important role in the Hawking effect. The thermal character of the emitted radiation can be nicely reobtained in terms of the conformal properties of the correlation functions of the effective matter theory around the horizon. This will offer a new perspective to grasp the deep physical meaning of the Hawking effect.

In Chapter 5 we approach the problem of determining an expression for the expectation value of the stress tensor for the matter fields in a curved background. This is the first obstacle one faces before attacking the backreaction problem. This is the fundamental problem of "quantum field theory in curved spacetime" and it is indeed very hard. No exact analytical expression is known even for the fixed Schwarzschild spacetime. Only for conformally flat geometries and for conformal fields one has a solution to the problem. We are far away from having such an expression in a generic four-dimensional spacetime. Reduction of the gravity-matter system under spherical symmetry leads to an effective two-dimensional theory that can serve as the starting point for the analysis. The advantage of doing so is that in two dimensions every metric is conformally flat, and therefore one expects to find an expression for the expectation value of the stress tensor. The near-horizon approximation, together with the ensuing conformal symmetry, allows to find an exact solution to this problem. This result is usually presented as a consequence of the fact that the trace anomaly determines univocally all the components of the stress tensor. An understanding of this, at least for a non-expert reader, will require a derivation of the trace anomaly itself. We want to avoid entering into the technicalities of regularization in curved spacetime and for this reason we present an unconventional derivation of the expectation values of the stress tensor, including the trace anomaly itself. Our derivation is based on the use of locally inertial coordinates and of the transformation law of the normal ordered stress tensor obtained in Chapter 4. In this way a non expert reader can easily follow the derivation. This approach also allows to determine an expression for the stress tensor when the spherically symmetric reduction is not restricted to the near-horizon geometry. We consider in detail the problem of selecting the appropriate quantum states relevant for quantum black hole physics. We stress that this is a highly non-trivial problem, especially when addressing the backreaction problem for the case of evaporating black holes. The last part of the chapter is devoted to this issue.

The first model which bypasses all these difficulties was proposed twelve years ago in [Callan *et al.* (1992)] and describes the near-horizon properties of near-extremal stringy black holes. Starting from it, the first analytic description of the process of black hole formation and evaporation was given in [Russo *et al.* (1992b)]. This is the central theme of Chapter 6. We also study the near-horizon dynamics of near-extremal Reissner–Nordström black holes. Also in this case the semiclassical equations are solvable, despite the fact that the selection of the appropriate quantum state to describe the evaporation is more involved. On the basis of these exact solutions we discuss the implications of this approach for what concerns the information loss paradox, but also its limitations.

Notation: throughout this book we shall follow the conventions for the metric and the curvature used in [Misner *et al.* (1973)]. The metric signature is $(-+++)$ and the definition of the curvature tensors are: $R^\mu{}_{\nu\alpha\beta} = \partial_\alpha \Gamma^\mu_{\nu\beta} - \cdots$, $R_{\mu\nu} = R^\alpha{}_{\mu\alpha\nu}$. In most of the book we use geometrized units $G = 1 = c$. However, to emphasize the quantum aspects we maintain explicitly Planck's constant ($\hbar$) in the formulae.

Chapter 2

Classical Black Holes

2.1 Modeling the Gravitational Collapse

A physically realistic treatment of the process of gravitational collapse is certainly a hard one. However, as it historically happened, the assumption of spherical symmetry allows to simplify considerably the analysis keeping the main physical features of the process, in particular the eventual emergence of a region where gravity is so strong that even light cannot escape to infinity and remains trapped. To visualize and simplify further the process we identify two regions in the spacetime of a collapsing star. The first is the interior one described by a well-behaved metric, with all the complications involved in describing the matter making up the star. The second is the exterior region, which is in general a portion of a spherically symmetric vacuum solution to Einstein's field equations. Due to Birkhoff's theorem it is given by a portion of the Schwarzschild spacetime (obviously, in a realistic collapse the exterior region is much more complicated than that and only "at late times" it relaxes down to a vacuum geometry). The dynamics involved in the process will make the radius of the star decrease with time. This means that as time goes on bigger and bigger portions of the exterior vacuum solution will be needed for the description of the physical spacetime. Depending on the total mass M of the collapsing object we know that there are different possible outcomes. In particular we know that there is a mass threshold of order 1.4 times the solar mass, called the Chandrasekhar limit [Chandrasekhar (1931)], below which the collapse will stop and the final state of the collapsed object will be a white dwarf. If the initial mass is bigger but below a few solar masses the outcome is instead a neutron star.

One can get an estimate of the Chandrasekhar limit in the following simple way [Carter (1972); Hawking and Ellis (1973)].[1] A high density cold star at the end of the collapse, after having exhausted all its nuclear fuel, will be supported by fermion degeneracy pressure either of the electrons (white dwarf) or of neutrons (neutron star). Also, because electrons are much lighter than neutrons ($m_e \ll m_n$) electron degeneracy is the first to take place. Assuming one electron in a cube of size of its Compton wavelength λ_e we have that the average electron momentum $p_e \sim \frac{\hbar}{\lambda_e}$ is given, due to the uncertainty principle, by

$$p_e \sim \hbar n^{1/3} , \qquad (2.1)$$

where n is the number density of electrons composing the star. For non-relativistic electrons the total energy of the star is given by

$$E \sim E_K - \frac{GM^2}{r} , \qquad (2.2)$$

where r is the radius of the star and the kinetic term E_K is obtained by summing over the kinetic energies of the electrons

$$E_K \sim nr^3 E_{electron} \sim nr^3 \frac{p_e^2}{m_e} \sim \frac{\hbar^2 n^{5/3} r^3}{m_e} . \qquad (2.3)$$

Observing that the mass of the star is due to protons and neutrons (with $m_p \sim m_n$)

$$M \sim n m_n r^3 \qquad (2.4)$$

then we can eliminate n in E_K and get

$$E_K \sim \frac{\hbar^2}{m_e} \left(\frac{M}{m_n}\right)^{5/3} \frac{1}{r^2} . \qquad (2.5)$$

The stable endpoint of the collapse is then the value of r minimizing E, namely

$$r_{min} \sim \frac{\hbar^2}{G m_e M^{1/3} m_n^{5/3}} . \qquad (2.6)$$

It is important to stress that the condition for the validity of the non-relativistic approximation $p_e/m_e \ll c$ implies that $n \ll (m_e c/\hbar)^3$ which, in

---
[1] See also [Shapiro and Teukolsky (1983); Padmanabhan (2001)] for a detailed study.

turn, using Eq. (2.4) with $r = r_{min}$ is equivalent to

$$M \ll M_C = \frac{1}{m_n^2}\left(\frac{\hbar c}{G}\right)^{3/2}. \tag{2.7}$$

In the relativistic case $E_{electron} \sim p_e c$ and therefore

$$E_K \sim \hbar c \left(\frac{M}{m_n}\right)^{4/3}\frac{1}{r}. \tag{2.8}$$

Since E_K and the gravitational energy $E_G \sim -\frac{GM^2}{r}$ are both of order $1/r$ then equilibrium is possible only if

$$\hbar c \left(\frac{M}{m_n}\right)^{4/3} = GM^2, \tag{2.9}$$

which in turn implies

$$M = M_C. \tag{2.10}$$

The radius of the star can then be estimated taking into account that, in the relativistic regime, $m_e c \leq p_e$. Therefore $n \geq (m_e c/\hbar)^3$ and then

$$r_C \leq \frac{1}{m_e m_n}\left(\frac{\hbar^3}{Gc}\right)^{1/2}. \tag{2.11}$$

The typical mass density in this situation is $\rho \sim 10^7 - 10^{11} \text{g/cm}^3$. The critical value for the mass so obtained is the so called Chandrasekhar limit [Chandrasekhar (1931)] and corresponds to ~ 1.4 solar masses. Stars of mass bigger than M_C cannot be maintained in equilibrium by the degeneracy pressure of electrons. Note that the existence of this limit is a direct consequence of the shift in the compressibility index of the Fermi gas, from $5/4$ to $4/3$, when the velocities approach the relativistic regime.[2]

In the above extreme situation the electrons can produce neutrons, via inverse β decay

$$e^- + p^+ \rightarrow n + \nu_e \tag{2.12}$$

once the electrons have reached the energy $(m_n - m_p)c^2$. The conversion of electrons into neutrons makes the star unstable until it is converted into a neutron star. Then one would be tempted to use the same ideal Fermi gas approximation to find equilibrium configurations due to neutron degeneracy

---

[2] In the non-relativistic situation the equation of state of the (cold) degenerate Fermi gas is $PV^{\frac{5}{3}} = const.$, while in the relativistic regime one has $PV^{\frac{4}{3}} = const.$

pressure. This can be easily done, since it amounts to replace m_e with m_n in the previous formulae. The critical mass M_C stays the same, but the estimation for the radius has changed to

$$r_C \sim \frac{1}{m_n^2}\left(\frac{\hbar^3}{Gc}\right)^{1/2} = \frac{GM_C}{c^2} \; . \tag{2.13}$$

On one hand this shows that neutron stars are much more compact than white dwarfs. However, since r_C is of the order of the Schwarzschild radius, general relativistic corrections can no more be neglected and must be included in the analysis.

More reasonable approximations for the matter inside the star have to be considered. This requires to solve Einstein's equations (from now on we shall use geometrized units $G = 1 = c$)

$$G_{\mu\nu} = 8\pi T_{\mu\nu} \tag{2.14}$$

for a matter stress tensor of the *perfect fluid* form[3]

$$T_{\mu\nu} = (\rho + p)u_\mu u_\nu + p g_{\mu\nu} \; , \tag{2.15}$$

where u^μ is the four-velocity of a matter point, ρ the energy density and p its pressure. Spherical symmetry leads to the Oppenheimer–Tolman–Volkoff equation [Oppenheimer and Volkoff (1939); Tolman (1939)] for hydrostatic equilibrium[4]

$$\frac{dp}{dr} = -(p + \rho)\frac{m(r) + 4\pi r^3 p}{r(r - 2m(r))} \; , \tag{2.16}$$

where

$$m(r) = 4\pi \int_0^r dr' \rho r'^2 \; . \tag{2.17}$$

This together with the equation of state $p = p(\rho)$ provides a system of differential equations which allows to study the equilibrium configurations of cold matter. [Oppenheimer and Volkoff (1939)] were able to find, using a crude model of neutron stars, an upper mass limit for equilibrium. A detailed series of investigations were initiated, after the second world war, by Wheeler and his group and they confirm that for very high densities the

---

[3] For an introductory account see [Schutz (1985)].
[4] In the Newtonian limit it reduces to the usual hydrostatic equation $\frac{dp}{dr} = -\frac{\rho m(r)}{r^2}$.

only possible equilibrium configurations are neutron stars,[5] with an upper bound for the mass to be of the order of 2–3 solar masses. This indicates that a dramatic scenario happens when the initial mass of the collapsing star is above the threshold for neutron star formation. In this case no known force is capable to counterbalance the gravitational force responsible for the collapse until complete implosion, i.e., the radius of the star will decrease without limit down to, ideally, zero size. In this case a black hole is said to be formed. In 1957, Wheeler summarized the situation as follows [Adams et al. (1958)]: "Of all the implications of general relativity for the structure and evolution of the universe, this question of the fate of great masses of matter is one of the most challenging ... Perhaps there is no final equilibrium state: this is the proposal of Oppenheimer and Snyder ..."

However, before the black hole idea was generally accepted, additional investigations were required. The main reason was that a black hole possesses "very uncomfortable and disturbing properties", as was first described by the model of Oppenheimer and Snyder.

2.1.1 The Oppenheimer–Snyder model

Motivated by the earlier striking result of the existence of an upper mass limit for neutron stars, [Oppendeimer and Snyder (1939)] studied the process of gravitational collapse ending with a black hole. They had the intuition that the internal pressure has no essential role since, in any case, no internal force is able to stop the gravitational contraction. For this reason, and for mathematical simplicity, they constructed a model neglecting the internal pressure and assumed a uniform density, see Fig. 2.1. This leads to model the interior of the star as a homogeneous and isotropic ball of dust described by the energy-momentum tensor

$$T_{\mu\nu} = \rho u_\mu u_\nu \ . \tag{2.18}$$

Homogeneity means that ρ can only be a function of time τ in comoving and synchronous coordinates, i.e.,

$$u^\mu = \delta^\mu_\tau$$
$$ds^2 = -d\tau^2 + g_{ij}dx^i dx^j \ . \tag{2.19}$$

---

[5]This result was obtained at the end of the fifties and is reported in [Harrison et al. (1965)].

Modeling Black Hole Evaporation

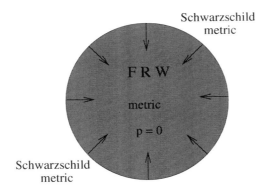

Fig. 2.1 A collapsing star in the Oppenheimer–Snyder model.

The solution to the field equations inside a spherical ball of dust is known and corresponds to a portion of a closed Friedmann–Robertson–Walker (FRW) spacetime of the form[6]

$$ds^2 = -d\tau^2 + R^2(\tau)d\Omega_{(3)}^2 , \qquad (2.20)$$

where $d\Omega_{(3)}^2$ is the metric on the unit three-sphere

$$d\Omega_{(3)}^2 = d\chi^2 + \sin^2\chi(d\theta^2 + \sin^2\theta d\phi^2) . \qquad (2.21)$$

The closed geometry (three-sphere), instead of the open or flat universes, arises because we make the physical assumption that the star is at rest ($R = R_0$) at the initial time $\tau = 0$. Flat or open three-geometries would have meant that the star collapsed from infinite radius.

Let us now briefly recall how to determine the function $R(\tau)$. The conservation laws

$$\nabla^\mu T_{\mu\nu} = 0 \qquad (2.22)$$

give the equation

$$\frac{d}{d\tau}\rho R^3 = 0 . \qquad (2.23)$$

Therefore

$$\rho = \frac{A}{R^3} , \qquad (2.24)$$

---

[6]See, for instance, [Misner et al. (1973); Stephani (1982)].

where A is a constant that, as we will see, is related to the initial radius R_0. The only relevant Einstein equation to be determined is the $\tau\tau$ component, namely $R_{\tau\tau} = 8\pi T_{\tau\tau}$ giving

$$\frac{1}{R^2}\left(\frac{dR}{d\tau}\right)^2 = -\frac{1}{R^2} + \frac{8}{3}\pi\rho = -\frac{1}{R^2} + \frac{8\pi A}{3R^3} . \tag{2.25}$$

We can then write

$$\left(\frac{dR}{d\tau}\right)^2 = -1 + \frac{R_0}{R} , \tag{2.26}$$

where we have defined

$$R_0 = \frac{8}{3}\pi A . \tag{2.27}$$

R_0 can be interpreted to be the value of $R(\tau)$ at the initial time $\tau = 0$, when the star starts to contract ($R(\tau) \leq R_0$). Actually, this is the maximum radius the ball can have, after which ($\tau > 0$) the collapse proceeds until $R = 0$ in a *finite time* $\Delta\tau$. This can be seen by integrating Eq. (2.26) close to $R \to 0$, leading to

$$(\tau - \Delta\tau) \sim \frac{R^{3/2}}{\sqrt{R_0}} . \tag{2.28}$$

The exact form of $R(\tau)$ can be given in parametric form by introducing a cycloidal coordinate η such that

$$\tau = \frac{R_0}{2}(\eta + \sin\eta) . \tag{2.29}$$

In terms of η the solution for R is

$$R = \frac{R_0}{2}(1 + \cos\eta) . \tag{2.30}$$

It is easy to see that for $\eta = \pi$ we reach $R = 0$. Therefore, we can determine $\Delta\tau$ immediately

$$\Delta\tau = \frac{\pi R_0}{2} . \tag{2.31}$$

At the point $R = 0$ the star has reached *infinite density* and moreover geodesics followed by the dust particles *cannot be extended beyond* $\tau = \Delta\tau$. For this reason they are called *incomplete* and this signals the presence of spacetime singularities [Geroch (1968); Wald (1984)]. In fact, all these features conspire to make the spacetime to develop there a *curvature singularity*,[7] as it can be inspected directly by calculating the curvature tensor.

Now that we have determined the spacetime metric associated to the region interior to the spherical star, let us have a look at what happens outside. There we have $T_{\mu\nu} = 0$ and therefore we must select a vacuum solution to the field equations. This is not difficult, since Birkhoff's theorem [Hawking and Ellis (1973)] ensures that the only spherically symmetric vacuum solution to the equations of general relativity is given by the Schwarzschild metric

$$ds^2 = -(1 - \frac{2M}{r})dt^2 + \frac{dr^2}{(1 - \frac{2M}{r})} + r^2 d\Omega^2 \ . \tag{2.32}$$

Here M is the mass of the star and $d\Omega^2 = d\theta^2 + \sin^2\theta d\phi^2$ is the metric on the unit two-sphere. We still have to relate, and this is very important, the interior and the exterior metrics along the star surface. For the interior metric the surface corresponds to a fixed coordinate $\chi = \chi_0$, for all τ, since fixed values of χ, θ and ϕ are timelike geodesics. In contrast, for the exterior metric the surface of the star is described by a (timelike geodesic) curve $r = r(t)$. Continuity of the metric implies that the induced metric must be the same on both sides of the star surface. This means that

$$-d\tau^2 + R^2(\tau)\sin^2\chi_0 d\Omega^2 = -\frac{(1-\frac{2M}{r})^2 + (\frac{dr}{dt})^2}{(1-\frac{2M}{r})}dt^2 + r(t)^2 d\Omega^2 \ . \tag{2.33}$$

In particular, by comparing the radius of the two-spheres we have that

$$r(t) = R(\tau)\sin\chi_0 \ , \tag{2.34}$$

which implies, evaluated at the initial point, that

$$r_0 = R_0 \sin\chi_0 \ , \tag{2.35}$$

where r_0 is the initial radius of the star. The relation between the mass of the star M and the internal parameters R_0 and χ_0 can be determined easily in the following way. By considering a dust particle on the surface of the star, due to the absence of any pressure then it must follow a geodesic both of the interior cosmological-type metric and of the exterior Schwarzschild spacetime. The interior trajectory has already been worked out in Eqs. (2.29) and (2.30). In the exterior Schwarzschild spacetime we similarly need the equation for a radial (timelike) geodesic starting at the maximum

---

[7]In the interior "cosmological" region this corresponds to the "big crunch" singularity.

radius r_0 at proper time $\tau = 0$ and then falling towards $r = 0$. The conserved quantity for geodesics $\tilde{E} \equiv -u_0 = -g_{0\mu}u^\mu$, together with the relation $u^\mu u_\mu = -1$, imply the following equations

$$\frac{dt}{d\tau} = \frac{\tilde{E}}{1 - 2M/r} , \qquad (2.36)$$

$$\left(\frac{dr}{d\tau}\right)^2 = \tilde{E}^2 - (1 - 2M/r) , \qquad (2.37)$$

where the numerical constant $\tilde{E}$, energy per unit mass, takes the value $\tilde{E}^2 = 1 - 2M/r_0$. We observe that Eq. (2.37) is similar to Eq. (2.26), which corresponds to the interior geometry. Therefore the solution $r = r(\tau)$ can be given in a similar parametric form [Misner et al. (1973)]

$$r = \frac{r_0}{2}(1 + \cos\eta) ,$$

$$\tau = \left(\frac{r_0^3}{8M}\right)^{1/2} (\eta + \sin\eta) . \qquad (2.38)$$

By comparing the two trajectories one finally gets

$$M = \frac{R_0}{2} \sin^3 \chi_0 . \qquad (2.39)$$

It is easy to check, from the above relations, that the full equality (2.33) is guaranteed. Moreover, it can be shown that not only the metric is continuous along the star's surface, but also the extrinsic curvature (which is proportional to the first derivatives of the metric) is continuous too. This means, as it should for physical consistency, that the matching is smooth [Misner et al. (1973)].

We have therefore succeeded in finding the full solution describing the spacetime of a collapsing sphere of dust. From the point of view of an ideal observer located on the star's surface the collapse starts at proper time $\tau = 0$, when the star has radius r_0, and then, as we have already remarked, after a finite amount of proper time $\Delta\tau$ the star and the observer will be destroyed at the spacetime singularity $r = 0$. It is very interesting to inspect what is the description of the collapse from the point of view of an asymptotic observer located very far from the star's surface. Straightforward integration of the geodesic equations gives

$$t = \int_{r_0}^{r} \frac{\sqrt{1 - 2M/r_0}}{\sqrt{2M/r - 2M/r_0}} \frac{dr}{(1 - 2M/r)} . \qquad (2.40)$$

The remarkable fact is that this integral diverges as $r \to 2M$. This means that the description of the process given by the external observer is quite different from that of the comoving one located on the surface of the star, as represented in Fig. 2.2.

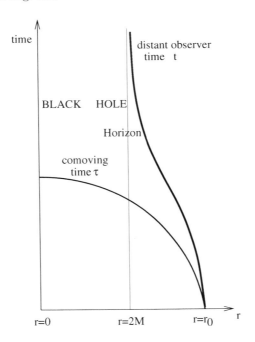

Fig. 2.2 The two alternative descriptions of the gravitational collapse.

For the latter the relation between r and τ is given in Eq. (2.38), and nothing special happens when the star eventually reaches the *Schwarzschild* or *gravitational radius* $r = 2M$. Then the collapse continues until the end-point $r = 0$ in a total finite time $\Delta\tau = \pi R_0/2 \equiv \pi\sqrt{r_0^3/8M}$. In contrast, the external observer will never be able to notice the end-point of the collapse at $r = 0$. And even more surprising, the external observer will never see the star reaching its gravitational radius as this requires, according to Eq. (2.40), an infinitely large amount of time t. Therefore, although in the late stages the collapse proceeds faster and faster in the proper time τ, the outside observer (measuring time t) sees rather the opposite thing: the collapse appears to get slower and slower, with the star eventually "frozen" at its gravitational radius $r = 2M$.

The surface $r = 2M$ is special for the above apparently disturbing features and it is called, for reasons to be explained in the next section, the *event horizon*. It marks the boundary of the black hole region. Any communication between the two observers is impossible once $r(\tau) < 2M$. Anything that may happen to the comoving observer once he has crossed the gravitational radius is lost to the external observer. It is very difficult to see and understand this fact in the two sets of coordinates introduced by Oppenheimer and Snyder (comoving coordinates for the interior and Schwarschild for the exterior). In the absence of a global picture of the dynamics, offered by new coordinate systems regular across $r = 2M$, it is difficult to reconcile the interior and exterior pictures. This will be the focus of Section 2.2. However, before proceeding it will be useful to make a short account on the history of gravitational collapse and the black hole concept. This can be easily done on the basis of the main drawbacks of the Oppenheimer–Snyder model:

- It ignores the internal pressure.

This also implies no possibility of heating up the material and emitting radiation. Therefore there is no way for the star to reduce its mass below the neutron star limit to prevent the collapse. In the late fifties, realistic simulations of the collapse (taking into account: pressure, heat, nuclear reactions, shock waves, radiation and mass ejection) confirmed the physical picture of the Oppenheimer–Snyder model for the formation of a black hole, for sufficiently massive stars. Such results on the gravitational collapse, together with the mathematical understanding of the "Schwarzschild singularity" (see next section), contributed to the general acceptance of the black hole idea at the beginning of the sixties. Finally, the name "black hole" was coined by Wheeler in 1967. In this period neutron stars were discovered, also quasars in 1963, and research to detect black hole signals in binary systems were proposed by Zel'dovich and Novikov in 1966. Since then, and after some initial skepticism, observations have shown the presence of very massive objects as powerful sources of X-rays, due to the accretion of matter around them. Many such objects have been discovered in binary systems, the first being Cygnus X-1, for which the estimates of the mass of the compact object are safely higher than the critical mass of neutron stars. Also, it is believed that a supermassive black hole lies at the center of our galaxy. Nowadays the existence of black holes is generally accepted, and they constitute an important source of dark matter in our Universe. For a fascinating account on the history of black holes we refer to [Israel (1987); Thorne (1994)].

- It assumes perfect spherical symmetry.

A second major limitation of the Oppenheimer–Snyder model is the assumption of perfect spherical symmetry. So a natural question arises: what happens if the collapsed star is not spherical? Should we still expect the formation of a black hole? The answer to this question is at the basis of the so called "no-hair theorem". Moreover, even if a black hole is formed, should we expect the formation of an internal singularity, as suggested by the Oppenheimer–Snyder model? The answer to the latter question appeared in the form of the "singularity theorems", initiated by Penrose. We shall briefly report on these issues in the following subsection.

2.1.2 *Non-spherical collapse*

The first investigations on the possible formation of black holes in non-spherical collapse were initiated in the sixties by [Ginzburg and Ozernoy (1965)] and [Doroshkevich et al. (1965)]. This analysis suggested the conclusion that small perturbations around spherical symmetry do not prevent the formation of a black hole and that the shape of the horizon is, when the system has settled down to a stationary configuration, exactly spherical. The problem for large deviations from spherical symmetry was attacked from a completely different perspective by [Israel (1967)]. Israel was able to produce a remarkable theorem. Without technicalities, it can be stated as: *a static, vacuum black hole with a regular event horizon has to be the Schwarzschild solution*. This result put on a solid basis the above suggestions and was the starting point for a series of theorems [Israel (1968); Carter (1971); Hawking (1972); Robinson (1975)][8] that culminated in the "no-hair theorem":[9] *stationary black holes are characterized only by mass, angular momentum and electric charge*. However, after Israel's theorem subsequent studies and, in particular, the detailed analysis of [Price (1972)] concerning small deformations from spherical symmetry, unraveled the responsible physical mechanism for the no-hair theorem. In a gravitational

---

[8] See also [Heusler (1996)].
[9] This terminology was also invented by J.A. Wheeler. It must be mentioned, however, that in the presence of additional fields some sort of "hair" is allowed (see for instance [Bekenstein (1996); Wilczek (1998)]).

collapse all multipole moments of the asymmetric body are radiated away in the form of gravitational (and electromagnetic) waves. Part of the radiated waves spread off to infinity and part fall downward through the horizon. Only mass, electric charge and angular momentum are protected due to physical conservation laws. This can be stated by the so called Price's theorem: *anything which can be radiated gets radiated away completely.*

Let us now briefly consider the other major problem posed by the Oppenheimer–Snyder model: is the singularity encountered at $r = 0$ a simple artifact of spherical symmetry? Can deviations from spherical symmetry remove the singularity? The standard view, with exceptions, was to reject the physical existence of singularities, mainly because this implies a breakdown of general relativity inside a black hole. This was supported by the work of [Lifshitz and Khalatnikov (1963)], who claimed that the random deformations of a real star will grow up with the implosion and stop it before a singularity is formed. However, Wheeler did not share this view and claimed, as it is now the common opinion, that the existence of singularities is a signal of the failure of classical general relativity, which requires being replaced by a quantum theory of gravity. The situation was largely clarified by [Penrose (1965)]. Penrose's first singularity theorem states that a spacetime singularity is unavoidable at a certain stage of the gravitational collapse, once a *trapped surface* is formed. As we shall explain later in this chapter, a trapped surface is a surface from which light cannot escape outward, and a black hole always contains a trapped surface. As this result was proven rigorously for a generic gravitational collapse, the problem of existence of singularities was solved. Nevertheless the nature of the singularity was not clarified by the theorem.[10] Moreover, a major problem still unsolved is whether a singularity can be visible to an external observer. [Penrose (1969)] formulated the *cosmic censorship conjecture*, stating that "in physically realistic situations singularities are always hidden inside event horizons". The precise formulation of this conjecture and its validity is still under debate.

---

[10]Later, Belinsky, Lifshitz and Khalatnikov [Belinsky *et al.* (1970)] modified their earlier conclusion and encountered the type of (curvature) singularity that is likely to be present inside a black hole formed from randomly deformed stars.

2.2 The Schwarzschild Black Hole

Let us now focus on the physical features of the Schwarzschild solution. As we have already stressed, it naturally describes the vacuum geometry outside a spherical star. The existence and nature of the surface $r = 2M$, which is of physical relevance in models of gravitational collapse is cause of much historical controversy. It is easily seen that the metric components in (t, r) coordinates are singular there and it took much time[11] to understand and, later, to accept its existence. Of particular help was the construction of a coordinate system which is regular on the "critical" surface $r = 2M$. This showed that the "Schwarzschild singularity" is not a real singularity, but a mere consequence of working with a coordinate system that is indeed ill-defined for $r = 2M$. In fact, the evaluation of the curvature invariant

$$R_{\mu\nu\rho\sigma}R^{\mu\nu\rho\sigma} = \frac{48M^2}{r^6} \qquad (2.41)$$

indicates that nothing special happens at $r = 2M$, in contrast with the point $r = 0$ (which is a real spacetime point in the Oppenheimer–Snyder model) where we encounter a curvature singularity. Despite this result, which suggests by itself that no real singularity is given at $r = 2M$, the situation is so subtle that it was not answered satisfactorily until 1960.

2.2.1 Eddington–Finkelstein coordinates

One can approach to the solution of this problem by constructing the light cones, at each point, of the Schwarzschild geometry. Generically, they allow to define the region of the spacetime where signals can be propagated. Since we are interested in knowing whether the points $r = 2M$ belong to the physical spacetime, a natural way is to analyse the propagation of radial light rays. Radial null geodesics verify the equation

$$dt^2 = \frac{dr^2}{(1 - \frac{2M}{r})^2} . \qquad (2.42)$$

Defining the *Regge–Wheeler* or *tortoise* coordinate

$$r^* \equiv r + 2M \ln \frac{|r - 2M|}{2M} , \qquad (2.43)$$

---

[11] For a historical account see [Thorne (1994)].

one finds a simple relation for the ingoing and outgoing radial null geodesics

$$d(t \pm r^*) = 0 \,. \tag{2.44}$$

Note that, as r ranges from $2M$ to $+\infty$, r^* ranges from $-\infty$ to $+\infty$. One can naturally parametrize an ingoing radial null geodesic by fixing the value of the coordinate

$$v \equiv t + r^* \,. \tag{2.45}$$

Therefore, one can parametrize the spacetime points by giving values for either the pair (v, r) or (v, t), in addition to the angular coordinates (θ, ϕ). In the first case, the Schwarzschild metric takes the form

$$ds^2 = -\left(1 - \frac{2M}{r}\right) dv^2 + 2dr\,dv + r^2 d\Omega^2 \,, \tag{2.46}$$

while in the second case we have

$$ds^2 = \left(1 - \frac{2M}{r}\right)(dv^2 - 2dv\,dt) + r^2 d\Omega^2 \,. \tag{2.47}$$

It is clear that only in the first situation we can analytically continue the metric to all possible values of the radial coordinate $r > 0$. In the second case we still have a singularity at $r = 2M$. The coordinates (v, r, θ, ϕ) are called the *ingoing (or advanced) Eddington–Finkelstein* coordinates [Finkelstein (1958)] and because of the cross-term $dr\,dv$ the metric is not singular at $r = 2M$ (see Fig. 2.3).[12]

In the interior region $r < 2M$, and for null or time-like curves ($ds^2 \leq 0$), we have the relation

$$2dr\,dv = -\left[-ds^2 + \left(\frac{2M}{r} - 1\right) dv^2 + r^2 d\Omega^2\right] \leq 0 \,. \tag{2.48}$$

Since we have $dv > 0$ for future-directed timelike and null geodesics we find immediately that $dr \leq 0$. Note that every outgoing null geodesic at $r = 2M$ will remain at $r = 2M$ (i.e., $dr = 0$). Therefore a particle entering the region $r < 2M$ will never escape to the exterior region $r > 2M$ and will unavoidably approach, in a finite proper time, the singular point $r = 0$. Due to this property we call the region $r \leq 2M$ a *black hole*. The boundary of this region $r = 2M$, which separates those events which are visible and invisible for an external observer, is called the *future event horizon*.

---

[12]We must note, however, that the first form of the metric without coordinate singularities was given by [Painlevé (1921); Gullstrand (1922)].

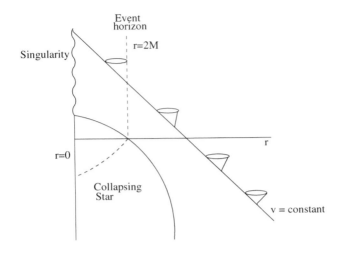

Fig. 2.3 A spherical gravitational collapse viewed in the ingoing Eddington–Finkelstein coordinates.

In a parallel way to the definition of the ingoing radial null coordinate (2.45) we can also define the outgoing radial null coordinate u by

$$u \equiv t - r^* \,. \tag{2.49}$$

The Schwarzschild metric, in the so-called *outgoing (or retarded) Eddington–Finkelstein coordinates* (u, r, θ, ϕ) then takes the form

$$ds^2 = -\left(1 - \frac{2M}{r}\right) du^2 - 2drdu + r^2 d\Omega^2 \,. \tag{2.50}$$

We can see that also the metric (2.50) is well behaved for $r = 2M$ and it can be analytically continued for all the values $r > 0$. At this point we should remark that the points with $r = 2M$ and fixed u correspond to $t = -\infty$. In the process of gravitational collapse the initial condition for $t = -\infty$ is a regular configuration describing a star with radius greater than $r = 2M$. This means that the region $r \leq 2M$ covered by Eq. (2.50) is not physically relevant (note this is not true for the metric given in Eq. (2.46) since there $r = 2M$ corresponds instead to $t = +\infty$). Although this region is non-physical, we will nevertheless study its interesting features. The interior region $r < 2M$ is now very different to the interior of the future event horizon described above. Here we have, for null or timelike signals

$ds^2 \leq 0$, that

$$2drdu = \left[-ds^2 + \left(\frac{2M}{r} - 1\right)du^2 + r^2 d\Omega^2\right] \geq 0 . \qquad (2.51)$$

Therefore, since for future directed worldlines $du > 0$ this implies $dr > 0$. We recover again equality for radial ingoing null geodesics at the special hypersurface $r = 2M$. What we now have is the timereversal of the previous situation. A particle at $r < 2M$ will always escape to $r > 2M$, crossing the null surface $r = 2M$. Because of these properties, the surface $r = 2M$ cannot be the same as the future event horizon. In fact it acts as a one-way membrane, but in the opposite direction. It is called the *past event horizon* and the interior region, with boundary $r = 2M$, is called the *white hole*, as opposite to the black hole. We have to remark again that this region does not exist in a gravitational collapse geometry. Only the future event horizon and the black hole region form in this case.

2.2.2 Kruskal coordinates

Until now we have seen three coordinate systems. The Schwarzschild coordinates (t, r) only cover the exterior region $r > 2M$ and cease to be valid at the horizons $r = 2M$. The advanced Eddington–Finkelstein coordinates (v, r) allow to cover also the black hole interior but not the white hole region. On the other hand, the retarded Eddington–Finkelstein coordinates (u, r) can parametrize the white hole interior, but not the black hole region. One could think that working simultaneously with the (u, v) coordinates one would reach all the possible spacetime points. However, this is not true and one gets the following metric

$$ds^2 = -\left(1 - \frac{2M}{r}\right) du dv + r^2 d\Omega^2 . \qquad (2.52)$$

This double null form of the metric has the same drawback of the standard Schwarzschild form. We cannot reach both future and past event horizons since they are located, respectively, at the limiting points $u = +\infty$ and $v = -\infty$. However, in terms of new coordinates (U, V) defined by

$$V = 4M e^{v/4M}$$
$$U = -4M e^{-u/4M}, \qquad (2.53)$$

called the *Kruskal coordinates* [Kruskal (1960); Szekeres (1960)], points in the future horizon are simply parametrized by the coordinates ($U =$

$0, V, \theta, \phi)$ and those in the past horizon by $(U, V = 0, \theta, \phi)$. The fact that U and V have finite values on the horizons suggests that the corresponding form of the metric is regular there. In fact we have

$$ds^2 = \frac{2Me^{-r/2M}}{r}(-dT^2 + dX^2) + r^2 d\Omega^2 , \qquad (2.54)$$

where we have introduced the Kruskal time and space coordinates $T = (U + V)/2$ and $X = (V - U)/2$, and $r(T, X)$ is given implicitly by

$$16M^2 \left(\frac{r}{2M} - 1\right) e^{r/2M} = -UV = X^2 - T^2 . \qquad (2.55)$$

We observe immediately that there is no trace of any singularity in the metric at $r = 2M$. The only singularity which persists in this frame is the unavoidable singularity at $r = 0$. Being a true curvature singularity it cannot be eliminated by any choice of frame. The relation between the Schwarzschild coordinates (t, r) and the Kruskal coordinates can be completed with the following relation

$$\frac{t}{2M} = \ln\left(-\frac{V}{U}\right) . \qquad (2.56)$$

Joining all the above formulae we can get a full and global picture of the so called *maximal analytical extension* of the Schwarzschild spacetime. From a physical point of view this assumes that no matter is covering up the regular spacetime region. The black hole has not been produced as the result of a collapse. It is assumed that it has always existed and for this reason it is called an *eternal black hole*. The range of the coordinates U, V is without restriction, up to the condition $UV = 16M^2$ which corresponds to the irremovable curvature singularity $r = 0$. We can see the resulting global picture in Fig. 2.4.

Due to the double null form of the metric the radial null geodesics take the form $U = constant$ and $V = constant$, and are represented by straight lines of unit slope. In particular the lines $U = 0$ and $V = 0$ represent the future and past event horizons. The extended Schwarzschild spacetime has four different regions, depending on the signs of U and V, which are bounded by the horizons. Region I, with $U < 0, V > 0$ corresponds to the usual region $r > 2M$ of the Schwarzschild spacetime. This region is clearly relevant for describing the spacetime of a spherical body undergoing gravitational collapse. From Eq. (2.56) we see that the hypersurfaces of constant t are straight lines crossing the origin $V = 0 = U$ and the hypersurfaces of constant r are hyperbolas (in the liming case $r = 2M$ they transform

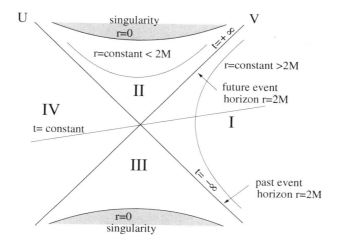

Fig. 2.4 The Kruskal diagram of the Schwarzschild spacetime.

into the horizon null lines). A radially free falling observer will cross the future horizon and enter region II ($U > 0, V > 0$). In this region the curves of constant r are spacelike while the curves of constant t are timelike. An observer worldline is such that the coordinate r is always decreasing and therefore it cannot leave region II: this is the black hole region. This region is produced at the late stages of the gravitational collapse. Region III ($U < 0, V < 0$) is the time reversal of region II. An observer there will always leave this region (i.e., the white hole region) and will enter into regions I or IV ($U > 0, V < 0$). The latter represents, surprisingly, another region similar to the original region I. However, as we can see easily from Fig. 2.4, the two regions are causally disconnected. No signal can be transmitted (or received) from region I to region IV. This region is also non-physical in a collapse process. No more regions can be found by analytical continuation. We have generated in this way the maximal analytical extension of the initial Schwarzschild solution. It is maximal in the sense that any geodesic can be either extended indefinitely with respect to its affine parameter or terminate at the curvature singularity $r = 0$.

The mathematical existence of the four regions is a consequence of the time reversibility of Einstein's equations. The exotic regions III and IV are time reversal of regions I and II. In the physical scenario of a spherical body undergoing gravitational collapse we have a time-asymmetric situation where the regions III and IV do not exist, and they are replaced by a regular metric describing the interior of the collapsing star.

2.3 Causal Structure and Penrose Diagrams

This section is devoted to introduce the reader to a common diagrammatic technique used to understand the causal structure of the different spacetime geometries. This is particularly important in the case of black holes, where there exists a region, the black hole region, where no signal can escape and reach infinity. This feature is crucial and it is important to recognize it immediately in the spacetime diagrams. The main idea is to project the whole spacetime into a finite diagram, called *Penrose* or *Carter–Penrose* diagram, which conserves the causal properties of the original geometry. This is possible by means of a Weyl (or conformal) transformation

$$ds^2 \to d\bar{s}^2 = \Omega^2(x^\mu)ds^2 , \tag{2.57}$$

where the factor $\Omega^2(x^\mu)$ depends on the space-time point and is in general not vanishing and positive. It is clear that this operation will alter the distance between the points. Timelike, spacelike and null trajectories do not change their character under this operation. However, the geodesic character is not always preserved: timelike geodesics are not mapped to timelike geodesics, but null geodesics are. It is not difficult to see that under the Weyl transformation (2.57) [Wald (1984)] the geodesic equation

$$\frac{dx^\mu}{d\lambda}\nabla_\mu\frac{dx^\nu}{d\lambda} = 0 , \tag{2.58}$$

where λ is the affine parameter, is transformed to ($\bar{\nabla}_\mu$ is the covariant derivative associated with $\bar{g}_{\mu\nu}$)

$$\frac{dx^\mu}{d\lambda}\bar{\nabla}_\mu\frac{dx^\nu}{d\lambda} = 2\frac{dx^\nu}{d\lambda}\frac{dx^\alpha}{d\lambda}\nabla_\alpha \ln\Omega - \left(g_{\alpha\beta}\frac{dx^\alpha}{d\lambda}\frac{dx^\beta}{d\lambda}\right)g^{\nu\mu}\nabla_\mu \ln\Omega . \tag{2.59}$$

Only in the null case, where $g_{\alpha\beta}\frac{dx^\alpha}{d\lambda}\frac{dx^\beta}{d\lambda} = 0$, the above equation turns out to be a geodesic equation in the transformed spacetime in terms of a non-affine parameter λ. The new affine parameter $\bar{\lambda}$ is such that

$$\frac{d\bar{\lambda}}{d\lambda} = c\Omega^2 , \tag{2.60}$$

where c is a constant. In particular, the way to represent points at infinity within a finite diagram is to use Ω^2 such that

$$\Omega^2 \to 0 \qquad (2.61)$$

asymptotically. This is so because as it can be seen from Eq. (2.60) an infinite affine distance in the original metric is transformed into a finite affine distance in the new metric. Therefore points at infinity, which strictly speaking do not belong to the original spacetime, are now represented in this *conformal compactification*. We shall now illustrate this technique, that we will use extensively in the course of this book, for the simple case of Minkowski spacetime.

2.3.1 Minkowski space

The starting point is to write the Minkowski metric in terms of the null radial coordinates $v = t + r$, $u = t - r$

$$ds^2 = -du\,dv + \frac{(v-u)^2}{4} d\Omega^2 \ . \qquad (2.62)$$

The range of these coordinates is from $-\infty$ to $+\infty$. Only half ($r > 0$) of the two-dimensional ($t - r$) plane is physically relevant. It is important to identify the different asymptotic regions before they are mapped into the conformal diagram:

- past timelike infinity i^-: this is defined by $t \to -\infty$ at fixed r or, equivalently, $v \to -\infty$, $u \to -\infty$;

- future timelike infinity i^+: $t \to +\infty$ at fixed r, or $v \to +\infty$, $u \to +\infty$;

- spacelike infinity i^0: $r \to +\infty$ at fixed t, or $v \to +\infty$ and $u \to -\infty$;

- past null infinity I^-: $t \to -\infty$, $r \to +\infty$ with fixed $t + r$, or $u \to -\infty$ and v fixed;

- future null infinity I^+: $t \to +\infty$, $r \to +\infty$ with fixed $t - r$, or $v \to +\infty$ and u fixed.

This reflects the different way that massive particles (following timelike geodesics), null light rays and also spacelike geodesics reach infinity. See

Fig. 2.5.

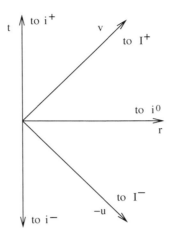

Fig. 2.5 The various ways the different geodesics reach infinity in Minkowski space.

To motivate the choice of the function Ω we shall first perform the following change of coordinates

$$v = \tan \bar{v},$$
$$u = \tan \bar{u}. \tag{2.63}$$

The range $-\infty < u, v < +\infty$ is now mapped to $-\frac{\pi}{2} < \bar{u}, \bar{v} < +\frac{\pi}{2}$. In terms of the coordinates $(\bar{u}, \bar{v})$ the metric now reads

$$ds^2 = (2\cos\bar{u}\cos\bar{v})^{-2} \left[-4d\bar{u}d\bar{v} + \sin^2(\bar{v} - \bar{u})d\Omega^2 \right]. \tag{2.64}$$

Now we perform the Weyl transformation (2.57) with conformal factor

$$\Omega^2 = (2\cos\bar{u}\cos\bar{v})^2. \tag{2.65}$$

The transformed spacetime has metric

$$d\bar{s}^2 = -4d\bar{u}d\bar{v} + \sin^2(\bar{u} - \bar{v})d\Omega^2. \tag{2.66}$$

The conformal compactification of Minkowski spacetime is then obtained by adding to all finite points all the above regions at infinity, which we now identify:

- i^-: $\bar{u} = -\frac{\pi}{2}$, $\bar{v} = -\frac{\pi}{2}$,

- i^+: $\bar{u} = \frac{\pi}{2}, \bar{v} = \frac{\pi}{2}$,
- i^0: $\bar{u} = -\frac{\pi}{2}, \bar{v} = \frac{\pi}{2}$,
- I^-: $\bar{u} = -\frac{\pi}{2}, \bar{v} \neq \pm\frac{\pi}{2}$,
- I^+: $u \neq \frac{\pi}{2}, \bar{v} = \frac{\pi}{2}$.

They are all represented in the triangle of Fig. 2.6, where generically

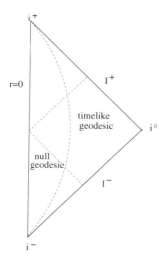

Fig. 2.6   Penrose diagram of Minkowski spacetime.

each point is a two-sphere of radius $\sin(\bar{v} - \bar{u})$. The origin $r = 0$ is the vertical line $\bar{u} = \bar{v}$ and the restriction $r \geq 0$ is transformed to $\bar{v} \geq \bar{u}$. The qualitative behavior of geodesics in Minkowski spacetime can be easily understood in the corresponding Penrose diagram. Timelike geodesics (except those that are asymptotically null) start at i^- in the past and end at i^+ in the future. Null geodesics, starting from I^-, reach the origin, where they get "reflected" and end on I^+. Finally, i^0 is the asymptotic point of all spacelike geodesics.

2.3.2   Schwarzschild spacetime

Let us now turn to the Schwarzschild spacetime, where this technique proves to be important to understand all the subtleties concerning the causal structure. Starting from the maximal extension of the metric provided by the

Kruskal coordinates (U, V)

$$ds^2 = -\frac{2Me^{-r/2M}}{r}dUdV + r^2d\Omega^2 \ , \tag{2.67}$$

one performs on the null Kruskal coordinates the same type of change of coordinates considered in the previous example

$$V = 4M\tan\bar{V} \ ,$$
$$U = 4M\tan\bar{U} \ . \tag{2.68}$$

This leads to

$$ds^2 = -\frac{32M^3 e^{-r/2M}d\bar{U}d\bar{V}}{r(\cos\bar{U}\cos\bar{V})^2} + r^2(\bar{U},\bar{V})^2 d\Omega^2 \ . \tag{2.69}$$

Choosing conformal factor

$$\Omega^2 = \frac{re^{r/2M}(\cos\bar{U}\cos\bar{V})^2}{8M^3} \tag{2.70}$$

the transformed metric becomes

$$d\bar{s}^2 = -4d\bar{U}d\bar{V} + \frac{r^3 e^{r/2M}(\cos\bar{U}\cos\bar{V})^2}{8M^3}d\Omega^2 \ . \tag{2.71}$$

The crucial relation that we need to construct the Penrose diagram is between the null coordinates $\bar{U}$, $\bar{V}$ and r:

$$\left(\frac{r}{2M}-1\right)e^{r/2M} = -\tan\bar{U}\tan\bar{V} \ . \tag{2.72}$$

Now the line $r = 0$ is no more a vertical line as in Minkowski, but is made of two disconnected horizontal curves, namely

$$\tan\bar{U}\tan\bar{V} = 1 \ , \tag{2.73}$$

that is

$$\bar{U} + \bar{V} = \pm\frac{\pi}{2} \ . \tag{2.74}$$

The important feature of these curves is that they are spacelike, as opposed to timelike in the previous case. The restriction $r \geq 0$ is now translated to the condition

$$-\frac{\pi}{2} \leq \bar{U} + \bar{V} \leq +\frac{\pi}{2} \ . \tag{2.75}$$

Therefore the full domain of the coordinates $\bar{U}, \bar{V}$ is

$$-\frac{\pi}{2} \leq \bar{U} + \bar{V} \leq +\frac{\pi}{2} , \quad -\frac{\pi}{2} < \bar{U} < \frac{\pi}{2} , \quad -\frac{\pi}{2} < \bar{V} < \frac{\pi}{2} . \qquad (2.76)$$

The Penrose diagram of the eternal Schwarzschild black hole is given in Fig. 2.7.

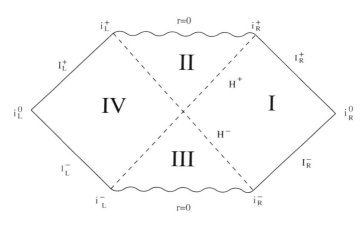

Fig. 2.7   Penrose diagram of the Schwarzschild spacetime.

The horizons, being formed by null geodesics, are unaffected by the Weyl transformation, and are located along the null lines $\bar{U} = 0$, H^+ (future horizon), and $\bar{V} = 0$, H^- (past horizon). They divide the diagram into four different regions, which are called I, II, III and IV. Region I describes the exterior region. The fact that asymptotically it is Minkowski means that it has the same conformal structure at infinity. II is the black hole region, containing the black hole singularity at $\bar{U} + \bar{V} = \frac{\pi}{2}$. III is the white hole region, containing the white hole singularity $\bar{U} + \bar{V} = -\frac{\pi}{2}$. Finally, region IV is another asymptotically flat region which is causally disconnected from I. It has the same conformal structure at infinity and in order to recognise the two asymptotic regions the one in I is denoted with the suffix R (right) and the other with L (left). The spacelike character of the singularity is simply understood by the fact that $r = 0$ is represented by horizontal lines. The behavior of geodesics is of course richer as compared to Minkowski. Timelike geodesics starting from i^- can either end on i^+ or cross the future event horizon and reach, in a finite proper time, the singularity at $r = 0$.

As already mentioned, not all these regions are significant in the process of gravitational collapse. In fact, the real Penrose diagram representing this process is given in Fig. 2.8. We clearly see that only regions I and II are

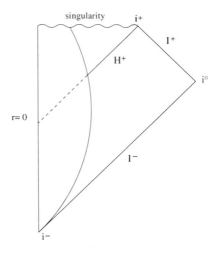

Fig. 2.8   Penrose diagram of a Schwarzschild black hole formed by gravitational collapse.

present, whereas III and IV are replaced by some regular geometry covered up by the matter.

2.4   Kruskal and Locally Inertial Coordinates at the Horizons

Although we have introduced the Kruskal coordinates for global purposes, they are also interesting from a local point of view. Indeed, in addition to being regular on the horizons we will now show that they are closely related to the *locally inertial coordinates* there. Before entering into details let us briefly remind the reader, following the notation of [Weinberg (1972)], how to construct a locally inertial frame about the generic point X in a general spacetime (this is the mathematical realization of Einstein's equivalence principle). In the coordinate system ξ_X^α the metric and the connection behave, just for the point X, as those of Minkowski space in (global) inertial coordinates. This means that

$$ds^2|_X = \eta_{\alpha\beta} d\xi_X^\alpha d\xi_X^\beta|_X ,$$
$$\Gamma^\lambda_{\mu\nu}(\xi_X^\alpha)|_X = 0 . \tag{2.77}$$

This last condition is equivalent to require the vanishing of the first derivatives of the metric

$$\frac{\partial}{\partial \xi_X^\alpha} g_{\mu\nu}(\xi_X^\beta)|_X = 0 \,. \tag{2.78}$$

Therefore, if we expand the metric at the point X in the coordinates ξ_X^α we get

$$g_{\mu\nu}(\xi_X) = \eta_{\mu\nu} + O((\xi_X - \xi_X(X))^2) \,. \tag{2.79}$$

The first term is the Minkowski metric at the point X and the leading (linear) correction vanishes. The first non-zero corrections are of second order and they can never be eliminated completely by appropriate choice of the locally inertial coordinates, otherwise this will make the Riemann curvature tensor vanish at X.

The above pair of conditions (2.77) can determine, up to second order and up to Poincaré transformations, the power expansion of the locally inertial coordinates ξ_X^α in terms of a generic coordinate system x^μ, regular at the point X:

$$\begin{aligned}\xi_X^\alpha &= a^\alpha + b_\mu^\alpha (x^\mu - x^\mu(X)) \\ &+ \frac{1}{2} b_\lambda^\alpha \Gamma_{\mu\nu}^\lambda (x(X))(x^\mu - x^\mu(X))(x^\nu - x^\nu(X)) + \cdots \,,\end{aligned} \tag{2.80}$$

where a^α are four arbitrary constants and the coefficients b_μ^α are such that

$$b_\mu^\alpha b_\nu^\beta \eta_{\alpha\beta} = g_{\mu\nu}(X) \,. \tag{2.81}$$

This condition defines them up to Lorentz transformations $b_\mu^\alpha \to \Lambda_\beta^\alpha b_\mu^\beta$. Irrespective of the higher order terms, the coordinates ξ_X^α in Eq. (2.80) are locally inertial. Every two coordinate systems ξ_X^α and $\xi_X'^\alpha$ which differ for terms which are third order or higher are equally locally inertial.

Now we are going to apply these formulae at the horizons of the Schwarzschild spacetime. To be as clear as possible we shall start our discussion with the ingoing Eddington–Finkelstein form of the metric (2.46). The point X is chosen to be at the future horizon and is parametrized by $(v = v_0, r = 2M, \theta = \theta_0, \phi = \phi_0)$. Moreover we shall concentrate on the $(t-r)$ part of the metric, that we will call $ds_{(2)}^2$, since the angular variables do not play any fundamental role. For this sector it is also convenient to work with locally inertial coordinates in double null form, such that

$$ds_{(2)}^2|_X = -d\xi_X^+ d\xi_X^-|_X \,. \tag{2.82}$$

The non-zero components of the metric and the connection in (v, r) coordinates on the future event horizon H^+ are:

$$g_{vr} = 1 ,$$
$$\Gamma^v_{vv} = \frac{1}{4M} , \quad \Gamma^r_{rv} = -\frac{1}{4M} . \tag{2.83}$$

According to this, and since $\eta_{+-} = -1/2$, $\eta_{++} = 0 = \eta_{--}$, the conditions for the coefficients b^α_μ are

$$b^+_r b^-_r = 0 , \quad b^+_r b^-_v + b^+_v b^-_r = -2 , \quad b^+_v b^-_v = 0 . \tag{2.84}$$

Therefore, either $b^+_r = 0 = b^-_v$ or $b^+_v = 0 = b^-_r$. We can make the first choice without loosing generality. We then obtain, up to additive constants,

$$\xi^+_{H^+} = b^+_v \left[(v - v_0) + \frac{1}{8M}(v - v_0)^2 + \cdots \right]$$
$$\xi^-_{H^+} = b^-_r \left[(r - 2M) - \frac{1}{4M}(v - v_0)(r - 2M) + \cdots \right] \tag{2.85}$$

with the condition $b^+_v b^-_r = -2$. A simple look at the above expansion suggests the introduction of the exponential of the v coordinate, since it plays the role of a null locally inertial coordinate. So defining

$$V = 4M e^{v/4M} , \tag{2.86}$$

we can then rewrite the first relation of (2.85) as

$$\xi^+_{H^+} = b^+_v e^{-v_0/4M} [(V - V_0) + O((V - V_0)^3)] . \tag{2.87}$$

This allows us to identify the Kruskal coordinate V as a locally inertial coordinate at the generic point $v = v_0$ of the future event horizon. The choice $b^+_v = e^{v_0/4M}$ makes the identification immediate. Maintaining the parameter b^+_v free means that we have the freedom of making an arbitrary Lorentz transformation, in the radial direction, to construct locally inertial coordinates. We should stress that the locally inertial coordinates are defined up to second order, so any departure of the V coordinate to the third order can also be considered as a valid locally inertial coordinate.

The analysis for the other locally inertial coordinate $\xi^-_{H^+}$ allows us to construct a regular null coordinate on the future horizon linked to the Eddington–Finkelstein null coordinate u, which, we remember, is defined by

$$u = t - r - 2M \ln \frac{|r - 2M|}{2M} \tag{2.88}$$

Classical Black Holes

and goes to $+\infty$ on the future horizon. There, at an arbitrary fixed point $v = v_0$ we have that $t \to -2M \ln |r - 2M| \to +\infty$ and therefore

$$u \to -4M \ln |r - 2M| . \tag{2.89}$$

A regular null coordinate on the future horizon can be constructed from the leading term of $\xi^-_{H^+}$ and we denote it, by convenience, by

$$U \sim (r - 2M) . \tag{2.90}$$

Therefore this implies that

$$u \sim -4M \ln U . \tag{2.91}$$

We have constructed in this way the other regular Kruskal coordinate.

For the past event horizon H^- the situation is similar, under the replacement $v \leftrightarrow u$. Then one can also identify a null locally inertial coordinate $\xi^-_{H^-}$ as

$$\xi^-_{H^-} = b^-_u e^{u_0/4M}[(U - U_0) + O((U - U_0)^3)] , \tag{2.92}$$

where

$$U = -4M e^{-u/4M} . \tag{2.93}$$

In the same way as in (2.91) we can now construct the other regular Kruskal coordinate V on the past event horizon satisfying

$$v \sim +4M \ln V . \tag{2.94}$$

Note that the pair of coordinates (U, V) cannot be simultaneously locally inertial, for the radial sector of the metric, at any horizon. V is locally inertial at the future event horizon but not at the past event horizon. In a similar way U is locally inertial at the past horizon but not at the future event horizon. We can see this in a clear way by explicitly constructing the pair of locally inertial null coordinates starting from the metric in the Kruskal frame (2.54).

At a generic point on the future horizon $H^+ \equiv (U = 0, V = V_0)$ the locally inertial coordinates $\xi^{\pm}_{H^+}$ are defined by the expansion (with an appropriate choice of constants b^{α}_{μ})

$$\begin{aligned} \xi^+_{H^+} &= b^+_V [(V - V_0) + O((V - V_0)^3)] , \\ \xi^-_{H^+} &= \frac{1}{eb^+_V} \left[U + \frac{V_0}{16M^2 e} U^2 + O(U^3) \right] . \end{aligned} \tag{2.95}$$

At a generic point on the past horizon $H^- \equiv (U = U_0, V = 0)$ we have the following relations

$$\xi_{H^-}^+ = \frac{1}{eb_U^-}\left[V + \frac{U_0}{16M^2 e}V^2 + O(V^3)\right]$$
$$\xi_{H^-}^- = b_U^-\left[(U - U_0) + O((U - U_0)^3)\right] . \qquad (2.96)$$

It is immediate to see, as we have already stressed, that the pair of Kruskal coordinates (U, V) are not, strictly speaking, locally inertial. At the future horizon only one of them, V, can be considered locally inertial, in the sense that together with $\xi_{H^+}^-$ it forms a set of locally inertial coordinates for the radial part of the metric. The other Kruskal coordinate, U, is instead locally inertial at the past horizon. Only at the *bifurcation two-sphere* $(U = 0, V = 0)$ both Kruskal coordinates are proportional to locally inertial coordinates.

We should also stress the fact that the identification of V as a locally inertial coordinate $\xi_{H^+}^+$ on the future horizon holds irrespective of the particular point V_0 selected. This is not trivial since, in general, the locally inertial coordinates change point by point. In particular this implies that the connection component Γ^V_{VV} vanishes along the entire future event horizon. Therefore a null outgoing radial geodesic moving along the future event horizon will verify the equation

$$\frac{d^2 \xi_{H^+}^+}{dV^2} = 0 . \qquad (2.97)$$

This means that the Kruskal coordinate V is an *affine parameter* $\lambda \equiv V$ for these null geodesics that form the future horizon. The same thing can be repeated for the past event horizon and it turns out that it is now the Kruskal coordinate U which is identified with the affine parameter for the radial ingoing null geodesics along the past event horizon.

2.4.1 Redshift factors and surface gravity

In relation with the above discussion we shall now introduce two important functions that will appear later on. These two functions are naturally introduced when one compares the locally inertial coordinates along the future horizon with the null Eddington–Finkelstein coordinates (u, v), which are instead locally inertial at infinity. The comparison between the pair of "inertial" coordinates $\xi_{H^+}^-$ and u given by

$$\frac{d\xi_{H^+}^-}{du} = \frac{1}{eb_V^+}e^{-u/4M} + O((e^{-u/4M})^2) \qquad (2.98)$$

is related to the *redshift factor* for *outgoing radiation*. Notice that it is, remarkably, *exponentially decreasing*. This means that a light ray emitted in the vicinity of the horizon with frequency w will reach infinity at late times $u \to +\infty$ with a highly redshifted frequency $w' \propto we^{-u/4M}$. This fact will be very important for the discussion of the Hawking effect that we will study in the next chapter.

In addition we can also consider the relation between $\xi_{H^+}^+$ and v

$$\frac{d\xi_{H^+}^+}{dv} = b_V^+\frac{dV}{dv} = b_V^+ e^{v/4M} \; . \qquad (2.99)$$

This quantity is the *redshift* for *ingoing radiation*, but in contrast to the previous situation, it is always *finite*. The arbitrary parameter b_V^+ is interpreted as a pure kinematical Doppler effect, due to the freedom of making Lorentz transformations between locally inertial coordinates. The physical relevance of this formula emerges as a way of measuring the failure of the coordinate v, inertial at infinity, to be inertial also at the future horizon. This is measured by the finite parameter $\frac{1}{4M}$, which is related to the strength of the gravitational field at the horizon. This quantity, which turns out to be independent of the point V_0 at the horizon, being V locally inertial everywhere along H^+, is called the *surface gravity* $\kappa = 1/4M$.

For clarity we should compare, at this point, the above redshift factors (between locally inertial coordinates at the future horizon and at infinity) with the more conventional *redshift factor for static observers*, who are associated to trajectories with constant r, θ, ϕ. These trajectories $x^\mu = x^\mu(t)$ have tangent vector

$$\chi^\mu = \frac{dx^\mu(t)}{dt} \; , \qquad (2.100)$$

$\chi^\mu = (1, 0, 0, 0)$. The quantity $(-\chi^\mu\chi_\mu)^{1/2}$ is related to the redshift factor for static observers

$$(-\chi^2)^{1/2} = (-g_{tt})^{1/2} = \left(1 - \frac{2M}{r}\right)^{1/2} \; . \qquad (2.101)$$

A light ray of frequency w_1 emitted at r_1 is detected at r_2 with frequency $w_2 = w_1(1-\frac{2M}{r_1})^{1/2}(1-\frac{2M}{r_2})^{-1/2}$.[13]

The vector field χ^μ has an important application concerning the definition of the surface gravity, which can be understood intuitively following the example given in [Wald (1984)]. The four-velocity vector of a particle at rest with $r = constant$ is given by $u^\mu = (-\chi^\nu\chi_\nu)^{-1/2}\chi^\mu$. The energy E_∞ of the particle per unit mass as measured by an inertial observer at infinity is $E_\infty = -\chi_\mu u^\mu$. Suppose the particle is held stationary by a string with the other end of the string held by the inertial asymptotic observer. The force, per unit mass, to be exerted is $F_\infty^\mu = -\nabla^\mu E_\infty$. A simple calculation gives $F_\infty^\mu = -\nabla^\mu(-\chi^\nu\chi_\nu)^{1/2}$. The value of this force $(F_\infty^\mu F_{\infty\mu})^{1/2}$ when the particle is close to the horizon $r \to 2M$, coincides with the surface gravity κ. More concretely[14]

$$\kappa = \lim_{r \to 2M}(F_\infty^\mu F_{\infty\mu})^{1/2}$$
$$= \lim_{r \to 2M} \frac{1}{2}\frac{d}{dr}(-g_{tt}) = \frac{1}{4M} \;. \qquad (2.102)$$

The vector field $\chi^\mu = (\frac{\partial}{\partial t})^\mu$ can be used to define the surface gravity from a more geometrical point of view. It is called a *Killing vector* for the stationary Schwarzschild metric since it verifies

$$\delta_{\chi^\sigma}g_{\mu\nu} = 2\nabla_{(\mu}\chi_{\nu)} = 0 \;, \qquad (2.103)$$

which means that the metric is left invariant under translations of the t coordinate. It is normalized such that it has $\chi^2 = -1$ at infinity ($r \to +\infty$), but it vanishes at the event horizon $\chi^2 = 0$, and this is why the event horizon is also called a *Killing horizon*. The vector field χ^μ is normal to the event horizon and since $\chi^\nu\chi_\nu = 0$ there this implies that also the vector $\nabla^\mu(\chi^\nu\chi_\nu)$ is normal. Therefore they must be proportional. The function relating them defines the *surface gravity* κ of the horizon through the relation

$$\nabla^\mu(\chi^\nu\chi_\nu) = -2\kappa\chi^\mu \;. \qquad (2.104)$$

It is easy to see that the above defining relation for κ coincides with the definitions given previously. Since the field χ^μ is a Killing vector it verifies

---

[13] Note that for $r_2 \to +\infty$ and $r_1 \to 2M$ we have an infinite redshift. Instead, for $r_2 \to 2M$ and $r_1 \to +\infty$ we have an infinite blueshift, just the inverse effect.

[14] This provides a generic expression for the surface gravity of a static metric in the form $ds^2 = g_{tt}dt^2 - g_{tt}^{-1}dr^2 + r^2 d\Omega^2$. We have $\kappa = -\frac{1}{2}g'_{tt}(r)|_{r=r_H}$, evaluated at the horizon r_H ($g_{tt}(r_H) = 0$).

the relation $\nabla_\mu \chi_\nu + \nabla_\nu \chi_\mu = 0$. Using it we can reexpress Eq. (2.104) as

$$\chi^\nu \nabla_\nu \chi_\mu = \kappa \chi_\mu \ . \tag{2.105}$$

This is just a geodesic equation in a non-affine parametrization. On the horizon $\chi^\mu = (\partial/\partial v)^\mu$ (i.e., v is the Killing parameter). In terms of the "Kruskal" coordinate $V = \kappa^{-1} e^{\kappa v}$ the above equation takes the geodesic form with an affine parametrization. So we encounter again the meaning of κ as measuring the failure of the Killing parameter v to agree with the affine parameter V on the future horizon. For completeness we shall give another equivalent definition of the surface gravity [Wald (1984)]

$$\kappa^2 = -\frac{1}{2} \nabla^\mu \chi^\nu \nabla_\mu \chi_\nu \ , \tag{2.106}$$

where it is understood that the right hand side is evaluated at the horizon. A simple calculation shows that this expression is equivalent to the first line of Eq. (2.102).

2.5 Charged Black Holes

After having reviewed the main aspects of the Schwarzschild geometry that we will need most for the rest of the book, it is natural to extend our analysis to other stationary black hole spacetimes. According to the no-hair theorem the most general stationary black hole solution of the Einstein–Maxwell equations

$$R_{\mu\nu} - \frac{1}{2} g_{\mu\nu} R = 2 \left(F_{\mu\sigma} F_\nu^\sigma - \frac{1}{4} g_{\mu\nu} F_{\rho\sigma} F^{\rho\sigma} \right)$$
$$\nabla_\mu F^{\mu\nu} = 0 \ , \tag{2.107}$$

where $F_{\mu\nu}$ is the electromagnetic tensor field, is uniquely characterized by three parameters (the so called Kerr–Newman family): the mass M, the angular momentum J and the charge Q. The solution with general (M, J) and $Q = 0$ was first constructed by [Kerr (1963)], and subsequently generalized by [Newman et al.(1965)] to include the charge (electric and/or magnetic). In the non-rotating (static) case $J = 0$ we are left with the Reissner–Nordström solution [Reissner (1916); Nordström (1918)]

$$ds^2 = -(1 - 2M/r + Q^2/r^2) dt^2 + \frac{dr^2}{(1 - 2M/r + Q^2/r^2)} + r^2 d\Omega^2 \ , \tag{2.108}$$

where in the electrically charged case the radial component of the electric field is

$$E_r = F_{rt} = \frac{Q_e}{r^2} \qquad (2.109)$$

and in the magnetically charged case

$$B_r = F_{\theta\phi} = \frac{Q_m}{r^2} \; . \qquad (2.110)$$

This looks like a "simple" generalization of the Schwarzschild solution,[15] with the addition of a charge (in general $Q^2 = Q_e^2 + Q_m^2$ where Q_e is the electric and Q_m the magnetic charge). Nonetheless, it presents important differences, especially in the causal structure, which we are now going to study. First of all in these coordinates the solution takes the static form. At infinity it is asymptotically Minkowskian. At $r = 0$ the metric has a curvature singularity, as for Schwarzschild. The novelty appears at the horizon, which in this case is defined by the equation

$$(1 - 2M/r + Q^2/r^2) = 0 \; . \qquad (2.111)$$

This equation has two solutions and this means that we have indeed two horizons located at

$$r_\pm = M \pm \sqrt{M^2 - Q^2} \; . \qquad (2.112)$$

Note that in the limit $Q \to 0$ we recover the Schwarzschild metric with only one horizon. Another interesting limit is $|Q| = M$, in geometrized units, which corresponds to the so called *extremal Reissner–Nordström black hole* that we will study in detail later. The extremal black hole can be regarded as a special frontier separating black holes ($M > |Q|$) from more exotic solutions possessing naked singularities: this refers to the fact that when $M < |Q|$ the singularity at $r = 0$ is not hidden behind any horizon.[16]

Let us now focus on the *non-extremal* black hole solutions $M > |Q|$. We shall proceed by analogy with the Schwarzschild metric and extend the geometry across the coordinate singularities at $r = r_+$ and $r = r_-$ by introducing new coordinate systems well behaved there. The radial ingoing

---

[15] Birkhoff's theorem still holds: the only spherically symmetric solution of the Einstein–Maxwell system is the static Reissner–Nordström solution. The no-hair theorem avoids instead the assumption of spherical symmetry.

[16] According to the cosmic censorship hypothesis this case should not occur in gravitational collapse.

null geodesics are determined by

$$dt + \frac{dr}{(1 - 2M/r + Q^2/r^2)} = 0 \ . \tag{2.113}$$

This allows to construct a tortoise coordinate r^* as

$$\begin{aligned}r^* &= \int \frac{dr}{(1 - 2M/r + Q^2/r^2)} \\ &= r + \frac{1}{2\kappa_+} \ln \frac{|r - r_+|}{r_+} + \frac{1}{2\kappa_-} \ln \frac{|r - r_-|}{r_-} \ ,\end{aligned} \tag{2.114}$$

where $\kappa_\pm$ are given by

$$\kappa_\pm = \frac{(r_\pm - r_\mp)}{2r_\pm^2} \ . \tag{2.115}$$

Defining the advanced null coordinate v

$$v = t + r^* \ , \tag{2.116}$$

we can then reexpress the metric in the ingoing Eddington–Finkelstein coordinates (v, r, θ, ϕ)

$$ds^2 = -\frac{(r - r_+)(r - r_-)}{r^2} dv^2 + 2 dv dr + r^2 d\Omega^2 \ . \tag{2.117}$$

It is easy to see that this metric is regular everywhere except at the physical singularity $r = 0$. The hypersurfaces $r = r_\pm$ are null and the exterior one $r = r_+$ is a future event horizon and inside there is the black hole region. The quantity κ_+ is just the surface gravity, defined in the same way as for Schwarzschild. In terms of the physical parameters (M, Q) it is given by

$$\kappa_+ = \frac{\sqrt{M^2 - Q^2}}{(M + \sqrt{M^2 - Q^2})^2} \ . \tag{2.118}$$

Notice that in the limit $Q \to 0$ we recover the Schwarzschild surface gravity $\kappa_+ \to 1/4M$. Moreover it is worth remarking that in the extremal limit $Q = M$ the surface gravity vanishes. This very peculiar feature is caused by the presence of the charge, and has no analogue in the Schwarzschild case.

We can also construct the time reversal extension by defining the retarded null coordinate $u = t - r^*$ leading to the metric

$$ds^2 = -\frac{(r - r_+)(r - r_-)}{r^2} du^2 - 2 du dr + r^2 d\Omega^2 \ . \tag{2.119}$$

The hypersurface $r = r_+$ is the past event horizon and inside we have the white hole region. In the double null coordinate system (u, v) we find the metric

$$ds^2 = -\frac{(r - r_+)(r - r_-)}{r^2} du dv + r^2 d\Omega^2 \ . \tag{2.120}$$

As already remarked for Schwarzschild, this coordinate system looses validity at both future and past event horizons. In order to construct a coordinate system regular on the event horizon one has to introduce the Kruskal coordinates so defined[17]

$$U^+ = -\frac{1}{\kappa_+} e^{-\kappa_+ u} \ , \quad V^+ = \frac{1}{\kappa_+} e^{\kappa_+ v} \ . \tag{2.121}$$

In terms of these coordinates the metric becomes

$$ds^2 = -r_+ r_- \frac{e^{-2\kappa_+ r}}{r^2} \left(\frac{r - r_-}{r_-}\right)^{1-\frac{\kappa_+}{\kappa_-}} dU^+ dV^+ + r(U^+, V^+)^2 d\Omega^2 \ , \tag{2.122}$$

where r is defined implicitly by

$$U^+ V^+ = -e^{2\kappa_+ r} \left(\frac{r - r_+}{r_+}\right) \left(\frac{r - r_-}{r_-}\right)^{\frac{\kappa_+}{\kappa_-}} \frac{1}{\kappa_+^2}. \tag{2.123}$$

It is now easily seen that the metric (2.122) is regular at $r = r_+$, but unlike in Schwarzschild these coordinates do not cover the whole spacetime manifold. In the Kruskal spacetime diagram drawn from the metric (2.122) in Fig. 2.9 we identify 4 different regions.

Regions I and IV are the asymptotically flat exterior regions, $U^+ = 0$ and $V^+ = 0$ correspond to the (future and past) event horizons, and regions II and III describe the interior $r < r_+$ up to $r = r_-$, which cannot be reached as it is defined by

$$U^+ V^+ = +\infty \ . \tag{2.124}$$

Therefore this coordinate system describes the solution in the region $r > r_-$. Indeed, it can be easily seen from Eq. (2.122) that we still have a coordinate singularity at $r = r_-$ which can be eliminated by defining new Kruskal coordinates around r_- (note that $\kappa_- < 0$).

$$U^- = -\frac{1}{\kappa_-} e^{-\kappa_- u} \ , \quad V^- = \frac{1}{\kappa_-} e^{\kappa_- v} \ . \tag{2.125}$$

---
[17]See for instance [Townsend (1997)].

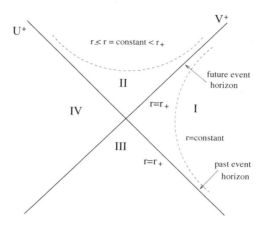

Fig. 2.9 Kruskal diagram of a Reissner–Nordström black hole based on the event horizon.

This leads to

$$ds^2 = -r_+ r_- \frac{e^{-2\kappa_- r}}{r^2} \left(\frac{r_+ - r}{r_+}\right)^{1-\frac{\kappa_-}{\kappa_+}} dU^- dV^- + r(U^-, V^-)^2 d\Omega^2 \ , \quad (2.126)$$

where

$$U^- V^- = -e^{2\kappa_- r} \left(\frac{r_- - r}{r_-}\right) \left(\frac{r_+ - r}{r_+}\right)^{\frac{\kappa_-}{\kappa_+}} \frac{1}{\kappa_-^2} \ , \quad (2.127)$$

which is regular at r_-, but now singular at r_+. Therefore this coordinate system covers the region $r < r_+$. This means that we need at least two patches to describe the full spacetime. The surface $r = r_-$ is called the *inner horizon*. The region $r < r_-$ is a new region that did not exist in Schwarzschild. Indeed, inside the event horizon of the Schwarzschild black hole t does not act as a time coordinate (this is because the metric component $g_{tt} = 1 - 2M/r$ changes sign) and the same thing happens in the Reissner–Nordström spacetime in the region $r_- < r < r_+$. However, when $r < r_-$ $g_{tt} = 1 - 2M/r + Q^2/r^2$ is positive and so t is again a time coordinate. The consequence of this is that the singularity at $r = 0$ is not spacelike, but timelike. In the Kruskal spacetime diagram (see Fig. 2.10) based on the metric (2.126) the singularity is represented, see Eq. (2.127), by the two timelike branches of the hyperbolae

$$U^- V^- = -1/\kappa_-^2 \ , \quad (2.128)$$

44   Modeling Black Hole Evaporation

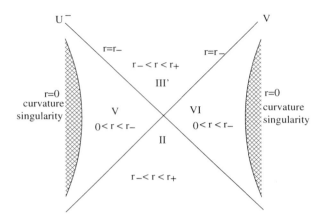

Fig. 2.10   Kruskal diagram of a Reissner–Nordström black hole based on the inner horizon.

and the (future and past) inner horizon by the null lines $U^- = 0$, $V^- = 0$.

The fact of having timelike singularities implies that the metric (2.126) can be extended to the future, namely to region III' encountering a coordinate singularity at r_+ and so on. In this way one can generate the full spacetime diagram by patching together infinitely many times metrics of the types in Eqs. (2.122) and (2.126). Turning to Penrose diagrams, the Kruskal diagram of Fig. 2.9 is mapped to the *building block* of Fig. 2.11 and

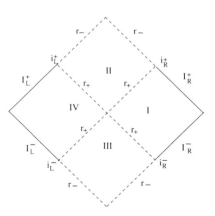

Fig. 2.11   Penrose diagram of Fig. 2.9.

the diagram of Fig. 2.10 becomes the second fundamental building block

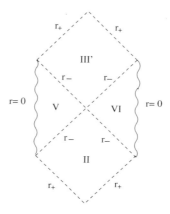

Fig. 2.12   Penrose diagram of Fig. 2.10.

given in Fig. 2.12.

These two building blocks have a common region, namely $r_- < r < r_+$, which allows to make the full analytic extension. The Penrose diagram of the maximal extension is then obtained by gluing these two building blocks infinitely many times in the future and in the past as represented in Fig. 2.13 (see, for instance, [D'Eath (1996)]). It should be remembered, however, that this beautiful mathematical construction has to be taken with much care from the physical point of view. In fact, the surface $r = r_-$ is also called the Cauchy horizon because the behavior of matter fields beyond it cannot be determined from initial data. To do this one needs to impose boundary conditions at the timelike singularities, but we cannot deal adequately with singularites within the classical theory. Moreover, the Cauchy horizon is an infinite blueshift surface in the sense that

$$\frac{dv}{dV^-} = e^{-\kappa_- v} \to +\infty \qquad (2.129)$$

as $v \to +\infty$ ($r \to r_-$), meaning that if we send a wave of finite frequency inside the horizon a free falling observer close to r_- will detect it as infinitely blueshifted. This fact makes the Cauchy horizon unstable against small perturbations in the external region, as it happens for instance in the process of gravitational collapse. This produces a curvature singularity there called mass inflation [Poisson and Israel (1990)].

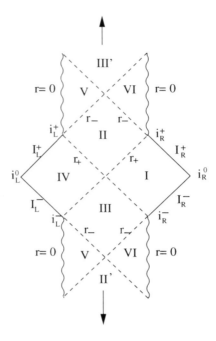

Fig. 2.13 Penrose diagram of the maximal analytical extension of the Reissner–Nordström black hole.

2.6 The Extremal Reissner–Nordström Black Hole

Extremal black holes are rather peculiar objects and deserve a separate study. They are defined by $|Q| = M$, and in this case we do not have any more two horizons, but only one at $r = M$. The metric takes the form

$$ds^2 = -\left(1 - \frac{M}{r}\right)^2 dt^2 + \frac{dr^2}{\left(1 - \frac{M}{r}\right)^2} + r^2 d\Omega^2 \,. \tag{2.130}$$

Naively, at this level one could be tempted to think that extremal black holes can be obtained by taking the limit of non-extremal black holes when $M \to |Q|$. However this could be misleading as we will see in the following. We start our analysis of the geometry by constructing the corresponding ingoing Eddington–Finkelstein coordinates $(v = t + r^*, r, \theta, \phi)$, where the tortoise coordinate is given by

$$r^* = r + 2M \ln \frac{|r - M|}{M} - \frac{M^2}{r - M} \,. \tag{2.131}$$

The metric takes the form

$$ds^2 = -\left(1 - \frac{M}{r}\right)^2 dv^2 + 2dvdr + r^2 d\Omega^2 , \qquad (2.132)$$

which shows, as in previous solutions analysed, that no real singularity exists on the hypersurface $r = M$ (the future event horizon). Analogously, one can define the outgoing (or retarded) Eddington–Finkelstein coordinate $u = t - r^*$ and the corresponding metric

$$ds^2 = -\left(1 - \frac{M}{r}\right)^2 du^2 - 2dudr + r^2 d\Omega^2 , \qquad (2.133)$$

where $r = M$ is now the past event horizon. Using the double null coordinates (u, v) the metric becomes

$$ds^2 = -\left(1 - \frac{M}{r}\right)^2 dudv + r^2 d\Omega^2 . \qquad (2.134)$$

The following step would be to construct the Kruskal-type coordinates. Naively one would consider the standard expressions $V = (\kappa_+)^{-1} e^{\kappa_+ v}$, $U = -(\kappa_+)^{-1} e^{-\kappa_+ u}$, and then take the limit $r_+ \to r_-$. However, in this limit

$$\kappa_+ \to 0 \qquad (2.135)$$

and the above change of coordinates becomes ill-defined. The fact that the surface gravity κ_+ vanishes for the extremal black hole is its main characteristic and will have important consequences when we discuss, in the next chapters, quantum effects on this particular background.

In analogy with the analysis in Section 2.4 we shall now construct the locally inertial coordinates at the future event horizon. Starting from the metric (2.132) a simple calculation leads to

$$\begin{aligned} \xi_X^+ &= b_v^+ [(v - v_0) + \quad \text{3rd order}] \\ \xi_X^- &= b_r^- [(r - M) + \quad \text{3rd order}] . \end{aligned} \qquad (2.136)$$

The interpretation is clear. The ingoing Eddington–Finkelstein coordinate v is, unlike in Schwarzschild, locally inertial. It is easy to understand the reason for this result. Indeed, in Schwarzschild the surface gravity measures the deviation of v from the locally inertial coordinate on the horizon. For the extremal black hole the surface gravity vanishes, so v is locally inertial. Moreover the second relation serves to identify the other locally inertial coordinate ξ_X^- in terms of u. Close to the future event horizon, where v is finite, the coordinate u behaves as

$$u \sim -4M \ln \frac{|r-M|}{M} + \frac{2M^2}{(r-M)} . \tag{2.137}$$

The locally inertial coordinate there ξ_X^-, which we denote by U, reads

$$U \sim -(r-M) , \tag{2.138}$$

where we have chosen $b_r^- = -1$. From Eq. (2.84) this implies $b_v^+ = 2$. We can then identify the following relation

$$u \sim -4M \ln \frac{|U|}{M} - \frac{2M^2}{U} . \tag{2.139}$$

This provides an implicit definition for the outgoing Kruskal-type coordinate U.[18] We can repeat the same arguments on the past event horizon based on the metric (2.133) to discover that u is a locally inertial coordinate and that the other (Kruskal-type) locally inertial coordinate V is given by

$$v \sim 4M \ln \frac{|V|}{M} - \frac{2M^2}{V} . \tag{2.140}$$

From the above results a nice picture emerges to describe the locally inertial coordinates at the horizons. For the future event horizon a null locally inertial pair of coordinates is given by

$$(\xi_{H+}^+, \xi_{H+}^-) \equiv (2v, U) , \tag{2.141}$$

while for the past event horizon we have

$$(\xi_{H-}^+, \xi_{H-}^-) \equiv (V, 2u) . \tag{2.142}$$

We can immediately appreciate the differences between the extremal and Schwarzschild black holes. In the latter case we have found (Section 2.4) the relations (for simplicity we consider now $b_V^+ = 1 = b_U^-$)

$$(\xi_{H+}^+, \xi_{H+}^-) \equiv (V, e^{-1}(U + \kappa^2 e^{-1} V_0 U^2)) , \tag{2.143}$$

and

$$(\xi_{H-}^+, \xi_{H-}^-) \equiv (e^{-1}(V + \kappa^2 e^{-1} U_0 V^2), U) . \tag{2.144}$$

---
[18] In the literature a relation of this type has been found in a different way in [Liberati et al. (2000)].

To understand the causal structure of the extremal black hole spacetime the Kruskal type extension is however unnecessary, since the advanced and retarded Eddington–Finkelstein coordinate systems can be fitted together to give the maximal extension of the geometry [Carter (1966)]. The basic building blocks constructed from the metrics (2.132) and (2.133) are given in Fig. 2.14. By gluing infinitely many times these two basic building blocks

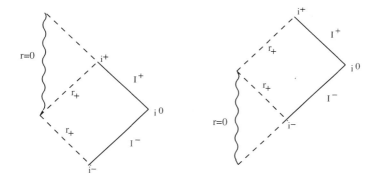

Fig. 2.14 Penrose diagrams of the building blocks of the extremal Reissner–Nordström black hole.

one obtains the full Penrose diagram given in Fig. 2.15.

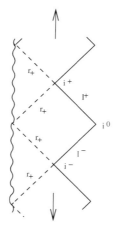

Fig. 2.15 Penrose diagram of the maximal analytical extension of the extremal Reissner–Nordström black hole.

To finish, we shall mention one peculiarity of the extremal horizon. The spacelike distance between any exterior point and the horizon $r = M$ is infinite. This can be understood by considering a line of constant t, θ and ϕ and integrating the line element (2.130) between r_0 and M:

$$s = \int_{r_0}^{M} \frac{dr}{(1 - M/r)} \sim \ln(r - M) \to \infty \ . \qquad (2.145)$$

The spatial geometry takes the form of a *semi-infinite throat*, as given in Fig. 2.16. However, timelike curves and light rays reach the horizon, crossing the throat in a finite affine time.

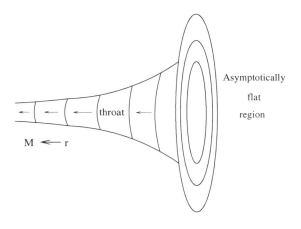

Fig. 2.16 Semi-infinite throat corresponding to the spatial geometry of an extremal Reissner–Nordström black hole.

2.7 Rotating Black Holes

In this section we shall briefly analyse a different set of stationary black hole solutions describing rotating generalizations of the Schwarzschild black hole. Due to the no-hair theorem, they are the unique black hole geometries expected as final states from an uncharged gravitational collapse. For this reason the Kerr solution [Kerr (1963)] is of special relevance. The metric in Boyer–Lindquist coordinates [Boyer and Lindquist (1967)] (t, r, θ, ϕ) is given by

$$ds^2 = -\left(\frac{\Delta - a^2 \sin^2\theta}{\rho^2}\right) dt^2 - 2a\frac{2Mr\sin^2\theta}{\rho^2} dtd\phi$$
$$+ \left[\frac{(r^2+a^2)^2 - \Delta a^2 \sin^2\theta}{\rho^2}\right] \sin^2\theta d\phi^2 + \frac{\rho^2}{\Delta} dr^2 + \rho^2 d\theta^2 \;, (2.146)$$

where

$$\rho^2 \equiv r^2 + a^2 \cos^2\theta \;, \quad \Delta \equiv r^2 - 2Mr + a^2 \;. \tag{2.147}$$

We mention that in the charged case the only difference in the metric is that $\Delta = r^2 - 2Mr + a^2 + Q^2$. The parameters M and a have a clear physical significance. In the asymptotic region $r \to +\infty$, the g_{tt} term of the metric behaves as in the Schwarzschild metric, so M is the mass of the solution, as measured at spatial infinity. Moreover the asymptotic behavior of the off-diagonal term (which makes the metric non-static) is

$$g_{t\phi} \sim -aM\frac{2r\sin^2\theta}{r^2} \;, \tag{2.148}$$

and the proportionality parameter aM gives the angular momentum J of the solution. Therefore, a is just the angular momentum per unit mass $a = J/M$. It is easily seen that in the non-rotating case $a = 0$ the metric describes the static Schwarzschild solution. The Kerr metric is stationary (but non-static, since it is not invariant under $t \to -t$) and axisymmetric. The latter property means that the metric is invariant under rotations around the axis of symmetry. In addition to the timelike Killing vector $\xi^\mu = \left(\frac{\partial}{\partial t}\right)^\mu$, out of the three Killing vectors of the Schwarzschild solution, related to rotational invariance, we maintain the one related to rotational invariance around the axis of symmetry $\psi^\mu = \left(\frac{\partial}{\partial \phi}\right)^\mu$.

A simple inspection reveals that the above metric is singular along the surfaces where

$$\Delta = 0 \tag{2.149}$$

and

$$\rho = 0 \;. \tag{2.150}$$

Moreover, curvature invariants blow up when $\rho = 0$, meaning that a curvature singularity develops along this surface. This singularity is, however, rather peculiar. In fact, it is a *ring singularity* [Carter (1973);

Hawking and Ellis (1973); Chandrasekhar (1983)] on the locus

$$r = 0, \quad \theta = \frac{\pi}{2} . \qquad (2.151)$$

As we will see in the following, however, this curvature singularity develops beyond a Cauchy horizon, as in the Reissner–Nordström case, and therefore one cannot expect that it will be maintained, as such, in a realistic collapse. So we shall not enter into the mathematical details of the Kerr singularity. We shall focus instead on the "coordinate singularities" at $r = r_\pm$, where $r_\pm = M \pm \sqrt{M^2 - a^2}$, in the case $M > |a|$. To see that these singularities can be removed we can introduce, in analogy with the Eddington–Finkelstein coordinates, the coordinates v and χ by

$$dv = dt + \frac{r^2 + a^2}{\Delta} dr,$$
$$d\chi = d\phi + \frac{a}{\Delta} dr . \qquad (2.152)$$

We obtain the metric

$$ds^2 = -\frac{(\Delta - a^2 \sin^2 \theta)}{\rho^2} dv^2 + 2 dv dr$$
$$- \frac{4Mar \sin^2 \theta}{\rho^2} dv d\chi - 2a \sin^2 \theta d\chi dr$$
$$+ \frac{((r^2+a^2)^2 - \Delta a^2 \sin^2 \theta)}{\rho^2} \sin^2 \theta d\chi^2 + \rho^2 d\theta^2 . \qquad (2.153)$$

This metric is clearly regular at $\Delta = (r - r_+)(r - r_-) = 0$. The surfaces with $r = constant$ are timelike for $r > r_+$, spacelike for $r_- < r < r_+$ and timelike again for $r < r_-$. The above metric describes a black hole with an event horizon at $r = r_+$ [Hawking and Ellis (1973)]. To realize this fact is not as simple as it was in the static cases analysed previously. Moreover, as in the Reissner–Nordström case, we also find a Cauchy horizon at $r = r_-$, which due to the infinite blueshift is likely to be unstable under small perturbations.

The most interesting novel feature of the Kerr black hole is the emergence of a region, external to the event horizon, where the stationary Killing vector ξ^μ is spacelike (see Fig. 2.17). This region is called the *ergosphere* and it is characterized by

$$r_+ < r < M + \sqrt{M^2 - a^2 \cos^2 \theta} . \qquad (2.154)$$

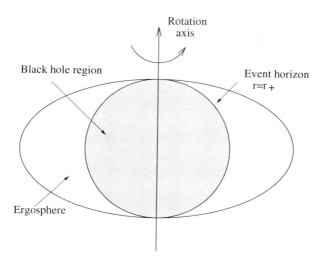

Fig. 2.17 The ergosphere and the event horizon of a Kerr black hole.

All timelike curves in this region must have

$$\frac{d\phi}{dt} > 0 \;, \tag{2.155}$$

which means that any observer will be necessarily dragged by the black hole. For instance, for any particle or light ray at fixed r and θ the rotation has a lower bound given by [Misner et al. (1973); Frolov and Novikov (1998)]

$$\frac{d\phi}{dt} \geq \frac{a\sin\theta - \sqrt{\Delta}}{(r^2 + a^2)\sin\theta - \sqrt{\Delta}\, a \sin^2\theta} \;. \tag{2.156}$$

In particular, for a photon on the horizon the angular velocity is constant and fixed by

$$\Omega_H = \frac{a}{r_+^2 + a^2} \;, \tag{2.157}$$

which naturally defines the angular velocity of the horizon. From a more mathematical point of view it is just the combination of Killing vectors

$$\chi^\mu = \xi^\mu + \Omega_H \psi^\mu \tag{2.158}$$

which becomes null on the horizon. This Killing vector allows to define the surface gravity at the event horizon as in Subsection 2.4.1 and the result is

$$\kappa = \frac{r_+ - r_-}{2(r_+^2 + a^2)} . \tag{2.159}$$

We finally mention that, as in the Reissner–Nordström case, there is an extremal black hole solution $M = |a|$ characterized by a single horizon $r_+ = r_-$ and vanishing surface gravity $\kappa = 0$. This horizon is also at an infinite spacelike distance from any point in the exterior region. The remaining possibility $M < |a|$ exhibits a naked singularity.

2.7.1 Energy extraction from rotating black holes: the Penrose process

The ergosphere plays an important role, as it was pointed out in [Penrose (1969)], because it leads to the possibility of extracting energy from the black hole. The main feature of this region is that the Killing vector ξ^μ is spacelike, but unlike in Schwarzschild this happens in a region outside the event horizon.

We can sketch Penrose's argument as follows. Particles with four-momenta p^μ, following a geodesic trajectory, have the following constant of motion

$$E = -p_\mu \xi^\mu . \tag{2.160}$$

At infinity, this constant is just $E = p^0$, and coincides with the energy of the particle. Let us now suppose that the initial particle, when inside the ergosphere, decays into two particles, A and B, of which one, A, falls through the horizon and the other, B, escapes to the exterior region. This is shown in Fig. 2.18. Local conservation of energy and momentum implies that $p^\mu = p_A^\mu + p_B^\mu$, and therefore

$$E = E_A + E_B . \tag{2.161}$$

However, although E and E_B are positive this is not necessarily true for E_A. In fact, as we have already remarked, in the ergosphere ξ^μ is spacelike and the scalar product with the timelike vector p_A^μ is not negative definite. If we take E_A to be negative (this means that the "gravitational" binding energy of the particle is bigger than its rest mass) then the energy of the outgoing particle E_B should exceed the energy of the initial particle

$$E_B > E . \tag{2.162}$$

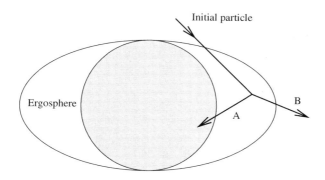

Fig. 2.18 Penrose process to extract energy from a rotating black hole.

As a consequence the mass of the black hole will decrease to $M + \delta M$ where $\delta M = E_A$. This provides a mechanism for extracting energy from a rotating black hole.

Let us study in detail this process. Any particle outside the event horizon must satisfy the inequality

$$-p_\mu \chi^\mu > 0 \ . \qquad (2.163)$$

This implies that, from Eq. (2.158),

$$-E_A + \Omega_H L_A < 0 \ , \qquad (2.164)$$

where L_A is the angular momentum of the particle A. By rewriting it as $L_A < E_A/\Omega_H$ one sees that L_A is negative as well, and this in turn implies that the black hole decreases also its angular momentum to $J + \delta J$ where $\delta J = L_A$. This inequality can be rewritten in the following elegant form

$$\delta M_{irr} > 0 \qquad (2.165)$$

where the *irreducible mass* M_{irr} is defined by [Christodoulou (1970); Christodoulou and Ruffini (1971)]

$$M_{irr}^2 = \frac{1}{2}\left[M^2 + \sqrt{M^4 - J^2}\right] \ . \qquad (2.166)$$

The fact that M_{irr} cannot decrease gives an upper limit on the energy that can be extracted from a rotating black hole. For instance, for an extremal rotating black hole $J = M^2$ the maximum energy that can be extracted is $M(1 - 1/\sqrt{2})$ which is approximately 29% of its initial mass.

The wave analog of the Penrose process is called superradiance [Zel'dovich (1972); Starobinsky (1973)]. One can check, by solving the wave equation in the Kerr geometry, that an initial wave of definite amplitude A_{in} with frequency $w < m\Omega_H$, where m is its angular momentum, which is scattered by the geometry will produce a final wave with a bigger amplitude $A_{out} > A_{in}$ and frequency w accompanied by a negative energy flux through the horizon. There is also a quantum version of superradiance [Unruh (1974)] that implies a spontaneous creation of particles from rotating black holes. We postpone a more detailed description of these effects to Chapter 3.

2.8 Trapped Surfaces, Apparent and Event Horizons

In this section we shall analyse in more detail the concept of "horizon". The intuitive idea which is at the basis of the analysis presented so far, is that the horizon is the border of the region from which it is impossible to escape to infinity. What we have just given concerns the definition of the future event horizon. In a more precise and technical language, *the future event horizon can be seen as the boundary of the past light-cone of future null infinity I^+*. An important global feature of the future event horizon, proven by Penrose [Penrose (1968)], is that once it is formed *its generators are null geodesics that never intersect each other and never leave the horizon*. However, a delicate point involving the definition of the event horizon is its *global character*. It requires to know the entire spacetime geometry for all times before knowing its exact location. This is clearly possible for the stationary solutions studied until now, but it is much more difficult for a general time-dependent spacetime. In the classical theory of gravitational collapse we have the additional input that the end point configurations, when the geometry has settled down and almost all multipole moments have been radiated away, are given by the known stationary solutions. This allows to identify the (global) event horizon. However, at the quantum level things are more complicated than that and, in general, it is an open question to know the spacetime geometry at all times. This clearly makes it very hard to construct the actual event horizon. For these reasons, which play a central role in this book, it is convenient to also use a local definition of horizon, called the *apparent horizon*, introduced in the sixties. Before defining what an apparent horizon is we need to introduce another basic concept, namely the *trapped surface*. A trapped surface is a closed spacelike surface such that both the ingoing and the outgoing null geodesics orthogonal to it are converging. The physical idea is clear: the pull of gravity is so strong that it forces the dragging back of light. We can better explain this concept by first visualizing what happens in Minkowski spacetime. From any two-sphere of fixed r there are two families of null geodesics emanating from it, those that go out (outgoing) and those that go in (ingoing). The former diverge (i.e., they propagate towards bigger values of r), while the latter converge (they evolve to smaller values of r). In the Schwarzschild spacetime the same situation happens if one takes any two-sphere for $r > 2M$, but in the region $r < 2M$ this is no more true, as all null geodesics, even those in the outward direction, converge. The apparent horizon is then defined as the outer boundary of the spacetime region containing trapped surfaces. The notion of trapped surface is especially important for the singularity theorems [Penrose (1965)], since it has been shown that under very general assumptions, in particular when the stress-energy tensor of matter satisfies the weak energy condition $T_{\mu\nu}v^\mu v^\nu \geq 0$ for any timelike vector v^μ, the formation of a trapped surface unavoidably implies the formation of a (black hole) singularity as well.

Let us see how we can determine in a simple situation, and from the above definition, the location of the apparent horizon. Let us consider a time dependent spherically symmetric metric of the form

$$ds^2 = -e^{2\rho}dx^+ dx^- + r^2 d\Omega^2 , \qquad (2.167)$$

where ρ and r are regarded now as functions of the null coordinates $x^\pm$. Ingoing and outgoing null rays are just given by $x^+ = const.$ and $x^- = const.$ The collection of two-spheres of area $4\pi r^2$ that are trapped surfaces is determined by the condition

$$\partial_+ r^2 \leq 0 . \qquad (2.168)$$

This means that the outward wavefront of a flash of light emitted from a given sphere decreases in area. Moreover, since we always have $\partial_- r^2 > 0$ (we exclude white hole regions), these qualitative arguments lead to the following formula for the location of the apparent horizon

$$|\nabla r|^2 = 0 . \qquad (2.169)$$

Applying this condition to the static Schwarzschild and Reissner–Nordström solutions in $(t - r)$ coordinates we immediately see that this

gives
$$g_{rr}^{-1} = 0 , \qquad (2.170)$$
i.e., the apparent horizon coincides with the event horizon.

However, this is no more true in the general dynamical case, as we will now see.[19] Figure 2.19 shows a Schwarzschild black hole of mass M_1

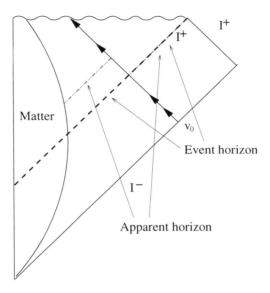

Fig. 2.19 Formation of a black hole followed by the collapse of a null shell.

(created from collapse) increasing its mass by absorbing matter in the form of a spherical thin null shell of mass M_2. The final configuration is a black hole of mass $M = M_1 + M_2$. The null shell is located at $v = v_0$ and the full spacetime geometry is constructed by matching the "initial" geometry

$$ds^2 = -\left(1 - \frac{2M_1}{r_{in}}\right) du_{in} dv + r_{in}^2 d\Omega^2 , \qquad (2.171)$$

valid for $v < v_0$ and outside the collapsing body, with the final geometry

$$ds^2 = -\left(1 - \frac{2M}{r_{out}}\right) du_{out} dv + r_{out}^2 d\Omega^2 \qquad (2.172)$$

---
[19]This example is important also because it introduces some of the technicalities that will be used several times in this book.

for $v > v_0$. The matching of the induced metrics at both sides of $v = v_0$ requires

$$r_{in}(u_{in}, v_0) = r_{out}(u_{out}, v_0) \,, \qquad (2.173)$$

where

$$\frac{v_0 - u_{in}}{2} = r^*_{in} = r_{in} + 2M_1 \ln \frac{|r_{in} - 2M_1|}{2M_1}$$

$$\frac{v_0 - u_{out}}{2} = r^*_{out} = r_{out} + 2M \ln \frac{|r_{out} - 2M|}{2M} \,. \qquad (2.174)$$

In the region $v > v_0$ the future event horizon is simply given by $r_{out} = 2M$ or $u^H_{out} = +\infty$, but in the region $v < v_0$ using the above relations we find that it corresponds to

$$u^H_{in} = v_0 - 4M - 4M_1 \ln \frac{2|M - M_1|}{2M_1} \,. \qquad (2.175)$$

Therefore, along u^H_{in} it is

$$r^*_{in} = \frac{v - u^H_{in}}{2} = r^*_{in}(2M) + \frac{v - v_0}{2} \,. \qquad (2.176)$$

The radius of the horizon r_H is obtained by inverting the relation $r^*_{in}(r_{in})$ given in the first line of Eq. (2.174). This shows that in the initial region r^*_{in} increases, along the event horizon, with v and is always bigger than $r^*_{in}(r_{ah} = 2M_1) = -\infty$, which identifies instead the apparent horizon. This example clearly shows the difference between apparent and event horizons.

In general, in the classical theory apparent horizons can be either null or spacelike, located inside or at most coincident with the event horizon, but in the quantum theory (where the weak energy condition can be violated), they can be timelike and lie outside the event horizon.

2.8.1 The area law theorem

A crucial property of the future event horizon is given by the so called *area law theorem* [Hawking (1972)]. It states, under general conditions like the absence of naked singularities and the validity of the weak energy condition, that the cross-sectional area A of the future event horizon never decreases with time

$$\delta A \geq 0 \,. \qquad (2.177)$$

The proof is essentially based on the "focusing theorem" and the properties of the generators of the horizon mentioned at the beginning of the previous section. To give an intuitive idea we shall follow [Misner et al. (1973)]. Let us consider the family of null geodesics propagating along the future event horizon and its two-dimensional cross section with area A. The focusing theorem tells us that (assuming the weak energy condition)

$$\frac{d^2 A^{1/2}}{d\lambda^2} \leq 0 , \qquad (2.178)$$

where λ is an affine parameter along a fiducial null geodesic. If the first derivative $dA^{1/2}/d\lambda$ were negative at some λ_0 this would imply that, due to Eq. (2.178), there must exist some $\lambda_1 > \lambda_0$ for which $A(\lambda_1) = 0$. At this point the null geodesics considered would cross each other, in clear contradiction with the properties of the generators of the horizon. Assuming that no singularity is hit by the future event horizon, to prevent the crossing of the geodesics before λ_1, the conclusion is that $dA^{1/2}/d\lambda \geq 0$, or, in other words, that the area of the future event horizon never decreases.

To clarify the meaning of this theorem let us go back to the example given in Fig. 2.19. The area of the future event horizon

$$A = 4\pi r_H^2 \qquad (2.179)$$

is given by

$$A = 16\pi M^2 , \qquad v > v_0 ,$$
$$A = 16\pi r_H^2 \leq 16\pi M^2 , \quad v < v_0 . \qquad (2.180)$$

A simple calculation, using the relations derived in the previous section, gives

$$\frac{dA}{dv} = 32\pi r_H \frac{dr_H}{dv}$$
$$= 16\pi r_H \left(1 - \frac{2M_1}{r}\right) > 0 \qquad (2.181)$$

because, as already stressed, the event horizon lies outside the apparent horizon $r_{ah} = 2M_1$.

Another illustrative example concerns stationary rotating black holes. A straightforward calculation of the area of the event horizon gives

$$A = \int_{r=r_+} d\theta d\phi \sqrt{g_{\theta\theta}g_{\phi\phi}} = 4\pi(r_+^2 + a^2) . \qquad (2.182)$$

This quantity turns out to be proportional to the square of the irreducible mass (see Eq. (2.166))

$$A = 16\pi M_{irr}^2 \,. \tag{2.183}$$

Therefore, for infinitesimal transformations from one initial (nearly) stationary black hole to a final (nearly) stationary black hole the area law theorem turns out to be equivalent to the statement that $\delta M_{irr}^2 \geq 0$, as found when analysing the Penrose process.

Finally, a more extreme application involves the collision of two black holes (see Fig. 2.20). To simplify the discussion we shall assume that the

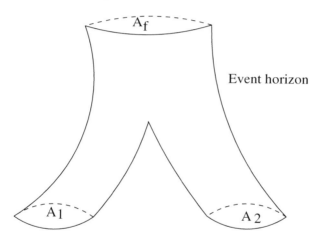

Fig. 2.20 Collision of two black holes to form a new one.

initial black holes are nonrotating and have masses M_1 and M_2. The final product of the collision is a black hole of mass M together with the emission of gravitational radiation of energy E. We can make a crude estimate of the upper bound of $E = M_1 + M_2 - M$ using the area law theorem and assuming that the areas can be approximated by

$$A_f = 16\pi M^2 \geq A_i = 16\pi(M_1^2 + M_2^2) \,. \tag{2.184}$$

The energy of the radiation therefore satisfies the following inequality

$$E \leq M_1 + M_2 - \sqrt{M_1^2 + M_2^2} \,. \tag{2.185}$$

2.9 The Laws of Black Hole Mechanics

At the beginning of the seventies, just before the discovery of black hole radiation, a series of theorems were proven suggesting that the black hole event horizons satisfy relations which exhibit a surprising analogy with the laws of thermodynamics. We have already encountered a signal of this in the area law theorem. This theorem suggests a formal analogy with the second law of thermodynamics. The area A of the future event horizon behaves as the entropy of a closed thermodynamical system, in the sense that both quantities never decrease with time. This analogy can be further extended to the other laws of thermodynamics. The zeroth law states the existence of a quantity, the *temperature*, that remains constant in the thermodynamic equilibrium. Is it possible to define an analogous quantity for the black hole event horizon? The answer is positive for a *stationary* black hole. All such black holes admit the existence of a Killing field χ^μ, with respect to which the black hole horizon is also a Killing horizon. As we have already explained in Subsection 2.4.1, the surface gravity κ of the horizon is given by the relation

$$\kappa^2 = -\frac{1}{2}\nabla^\mu\chi^\nu\nabla_\mu\chi_\nu . \qquad (2.186)$$

Taking now the derivative with respect to χ^μ one finds that κ is constant along each orbit of the Killing field. In principle, the function κ could change value from orbit to orbit. However, in the case of a stationary black hole in vacuum, or in the presence of electromagnetic fields, a direct calculation based on the Kerr–Newman solution shows that the value of κ

$$\kappa = \frac{(M^2 - a^2 - Q^2)^{1/2}}{2M[M + (M^2 - a^2 - Q^2)^{1/2}] - Q^2} \qquad (2.187)$$

is constant on the event horizon. In the presence of matter one can prove this using Einstein's equations together with the so called dominant energy condition for the stress tensor [Bardeen et al. (1973)]. Therefore we can state the analog of the *zeroth law* of thermodynamics in the form:

"For a stationary black hole the surface gravity κ is constant on the horizon."

With the above formal relation between surface gravity and temperature, and between area and entropy, we can naturally ask whether the analog of the *first law* of thermodynamics holds. It was proven in [Bardeen *et al.* (1973)] that the parameters of two nearby stationary black holes must be related according to the theorem:[20]

$$c^2 \delta M = \frac{c^2}{8\pi G}\kappa \delta A + \Omega_H \delta J + \Phi_H \delta Q , \qquad (2.188)$$

where Φ_H denotes the electrostatic potential, which turns out to be constant too on the event horizon, just as Ω_H. We have reestablished the physical constants because they will be important for the discussion we will make at the end of this section. Note the similarity between the above expression and the corresponding first law of thermodynamics

$$\delta E = T\delta S - P\delta V + \mu \delta N , \qquad (2.189)$$

where μ is the chemical potential. One can try to further extend the analogy to the *third law* of thermodynamics. Now the situation is more subtle. The analog of "Nernst version" of the third law, which states that *absolute zero temperature* cannot be reached by a finite number of steps, was proven in [Israel (1986)]. [Wald (1974)] showed that the closer the state of the black hole is to the state of an extremal ($\kappa = 0$) black hole the harder it is to get even closer. However, the stronger "Planck version", stating that the entropy goes to zero (or to a universal constant) when the temperature approaches zero, is not satisfied in the case of black holes: the area of the event horizon of Kerr–Newmann black holes does not vanish when $\kappa \to 0$.

Comparing the zero, first and second laws of thermodynamics with the corresponding laws (theorems) of black hole mechanics we find the correspondence[21]

$$E \leftrightarrow Mc^2 \qquad T \leftrightarrow \alpha \frac{c^2 \kappa}{G} \qquad S \leftrightarrow \frac{1}{8\pi \alpha} A , \qquad (2.190)$$

where α is an undetermined constant. Can we consider that this analogy is something more than a mere formal coincidence? The identification of E

---

[20] In a vacuum this can be easily checked by straightforward computation, taking into account (2.187) and that $A = 4\pi(r_+^2 + a^2)$ and $\Omega_H = a/(r_+^2 + a^2)$.

[21] Notice that a few extensive thermodynamic parameters E, V, etc., characterize univocally an equilibrium state. In the same way, due to no-hair theorem, the stationary black holes are characterized by few parameters: M, J and Q. Also, for equilibrium states one can define the intensive parameters T, P, etc. For black holes, and only for stationary configurations, one has a well defined notion of κ, Ω_H and Φ_H.

with M is clearly not purely formal. Both represent indeed the total energy of the black hole system. In sharp contrast, the identification $T \leftrightarrow \alpha c^2 \kappa /G$ is completely non-physical (in the context of the classical theory). The physical temperature of a black hole in classical general relativity is absolute zero. The existence of the event horizon prevents the black hole to emit anything. It can be regarded as a perfect absorber, with absolute zero temperature. According to this, the identification $S \leftrightarrow \frac{1}{8\pi\alpha}A$ cannot be more than a mathematical curiosity. This was the general view of most physicists before the discovery of the Hawking effect. The most notable exception was Bekenstein, who put forward the idea of a real physical connection between entropy and area of the event horizon [Bekenstein (1973)], even before the work of [Bardeen et al. (1973)]. He suggested a *generalized second law*: "the sum of the black holes entropy and the entropy of matter outside black holes would never decrease"

$$\delta \left(S + \frac{1}{8\pi\alpha}A \right) \geq 0 . \qquad (2.191)$$

The existence of black holes is not compatible with the ordinary second law of thermodynamics. If matter can fall into a black hole and disappear the entropy of matter for the external observer decreases. However, the area of the event horizon increases. Bekenstein suggested that the generalized entropy $S' \equiv S + A/8\pi\alpha$ does not decrease. A more detailed inspection of this suggestion shows that it is not consistent. We can consider a black hole immersed in a thermal bath at a temperature lower that $\alpha c^2 \kappa / G$. Since the black hole will absorb part of the radiation without emitting anything we have a heat flow from a cold thermal radiation to a "hotter" black hole. This would disagree with the generalized second law because the loss of entropy from the thermal radiation would be greater than the increase in black hole entropy.

An additional physical input is required to pass from a formal to a physical analogy. Some insights can be gained analyzing the dimension of the constant α (see, for instance, [Kiefer (1999)]). A simple look unravels that, since S has the dimension of Boltzmann's constant k_B, αk_B must have dimensions of length squared. With the physical constants that we have in classical general relativity (i.e., Newton's constant G and the velocity of light c) we cannot form a constant (to be identified with αk_B) with dimensions of length squared. We need the Planck's constant $\hbar$ for that.

From G, c and $\hbar$ we can form the Planck length

$$l_P = \sqrt{\frac{G\hbar}{c^3}}. \qquad (2.192)$$

With this fundamental length available ($k_B \alpha \propto l_P^2$) one can go further in the analogy and write

$$T \propto \frac{\hbar \kappa}{k_B c} \qquad S \propto \frac{k_B c^3 A}{G\hbar}. \qquad (2.193)$$

The lesson of this brief discussion is that the input required to properly establish a physical analogy between black holes and thermodynamics involves considering quantum effects. This will be the topic of the next chapter. Therefore, we postpone a thorough discussion of the relation between black holes and thermodynamics to the end of Chapter 3, when we will have more familiarity with the Hawking effect.

Before ending this section we would like to point out an important feature shared by both the laws of thermodynamics and the laws of black hole mechanics, which is their *universality*. In thermodynamics this is clear, and it is the reason why we call them "laws": they can be applied to any macroscopic physical system. For black holes the origin of these "laws" is different, they are in fact "theorems" within the theory of general relativity. However, these theorems are likely to apply generically to any black hole. Even more, as it has been shown in [Wald (1993); Iyer and Wald (1994)], the first law of black hole mechanics holds in an arbitrary, and metric based, generally covariant theory of gravity. Also the second law is likely to hold in a wide class of theories [Wald (1998)]. One consequence of this "universal" behavior of black holes, which goes beyond Einstein's theory, is that any hint towards a further understanding of the thermodynamic nature of black hole physics in a particular scenario will be of great value. These comments can serve as a motivation to justify next section, where we shall describe briefly the properties of a family of black holes that appear in the low-energy regime of string theory. They have interesting features concerning both geometric and thermodynamic behavior. Moreover, as we will see later in Chapter 6, they allow a more comprehensible study of black holes from the point of view of the quantum theory [Callan *et al.* (1992)].

2.10 Stringy Black Holes

A wider perspective of the properties and physics of black holes can be acquired by studying other types of black hole solutions that appear in theories that aim to generalize Einstein's theory of gravity. Of particular interest is considering those black hole configurations that emerge as classical solutions of the *low-energy limit of superstring theory*. They were discovered when the field content of Einstein–Maxwell theory was enlarged to include a *dilaton* scalar field ϕ which couples in a non trivial way to the metric and the gauge field [Gibbons and Maeda (1988); Garfinkle et al. (1991)]. When the electromagnetic field vanishes the only static, spherically symmetric black hole solutions are the Schwarzschild black holes with a constant dilaton. However, since the dilaton has an exponential coupling to F^2, it cannot remain constant for charged black holes. This causes the charged stringy black holes to differ significantly from the Reissner–Nordström solution of pure Einstein–Maxwell theory. Nevertheless, these black hole solutions satisfy the laws of black hole mechanics, as described in the previous section. They have however some thermodynamic novelties, like the fact that entropy, as measured formally by the area of the horizon, vanishes for the extremal solution with zero surface gravity. We shall consider in detail the quantum implications of the stringy black holes in Chapter 6. In this section we shall focus on the classical aspects of these black hole solutions.

In the low-energy limit of superstring theory [Polchinski (1998)], the interaction of the basic massless modes of the closed string (graviton $g_{\mu\nu}$, dilaton ϕ and an abelian gauge field A_μ) is described by the action[22]

$$S = \frac{1}{16\pi} \int d^4x \sqrt{-g}\, e^{-2\phi} \left(R + 4(\nabla\phi)^2 - \frac{1}{2}F^2 \right) . \tag{2.194}$$

In the pure "stringy" form the Einstein term in the action does not have its canonical form, since it is multiplied by the exponential of the dilaton. Moreover, the kinetic term of the dilaton field has an unconventional positive sign, relative to the Einstein term. This is potentially dangerous to maintain the theorems of black hole mechanics, since then the weak energy condition is not satisfied. One can reestablish the canonical form of the

---

[22] The effective action can also contain additional terms associated to other massless modes, like antisymmetric tensor fields, massless fermions as well as other gauge fields. Moreover the dilaton can also acquire a mass if supersymmetry is broken. These issues are not important for our purpose of discussing black holes outside the Einstein–Maxwell theory (here, in this section) and their quantum behavior (later, in Chapter 6).

Einstein action by performing a Weyl (conformal) transformation

$$g^S_{\mu\nu} \to g^E_{\mu\nu} = e^{-2\phi} g^S_{\mu\nu} , \tag{2.195}$$

where $g^S_{\mu\nu}$ is the metric in the so called *string frame* and it is the one that appears is the action (2.194). The metric $g^E_{\mu\nu}$ is referred to as the metric in the *Einstein frame* and this is so because the action takes canonical form

$$S = \frac{1}{16\pi} \int d^4x \sqrt{-g} \left(R - 2(\nabla\phi)^2 - \frac{1}{2} e^{-2\phi} F^2 \right) . \tag{2.196}$$

We maintain the same notation for the metric, despite the fact that (2.196) refers to the Einstein frame while (2.194) refers to the string frame. We mention that to pass from one frame to the other we have used the fact that, under the above Weyl transformation (2.195), we have

$$\sqrt{-g^S} = e^{4\phi} \sqrt{-g^E}$$
$$R^S = e^{-2\phi}(R^E - 6(\nabla^E)^2 \phi - 6(\nabla^E \phi)^2) , \tag{2.197}$$

where the indices S and E refer to the corresponding geometric quantities in the string and Einstein frames, respectively.

The equations of motion derived from the low-energy string action (2.196), in the Einstein frame, are as follows. The Maxwell equations are replaced by

$$\nabla_\mu (e^{-2\phi} F^{\mu\nu}) = 0 , \tag{2.198}$$

where we still maintain the usual relation $F_{\mu\nu} = \partial_\mu A_\nu - \partial_\nu A_\mu$. Einstein's equations take the form

$$G_{\mu\nu} = e^{-2\phi}(F_{\mu\sigma} F_\nu{}^\sigma - \frac{1}{4} g_{\mu\nu} F^2) + 2[\nabla_\mu \phi \nabla_\nu \phi - \frac{1}{2} g_{\mu\nu} (\nabla\phi)^2] , \tag{2.199}$$

and the dilaton equation is

$$\nabla^2 \phi = -\frac{1}{4} e^{-2\phi} F^2 . \tag{2.200}$$

When $F_{\mu\nu} = 0$, these equations reduce to Einstein's equations coupled to a massless scalar field. The no-hair theorem implies that the only static black hole solution is Schwarzschild with a constant dilaton field [Bekenstein (1972)]. In the presence of an *electric charge* the dilaton cannot be constant and the static, spherically symmetric solutions are given by

$$ds^2 = -\left(1 - \frac{2M}{r}\right) dt^2 + \frac{dr^2}{(1 - \frac{2M}{r})} + r\left(r - \frac{Q^2}{2M} e^{-2\phi_0}\right) d\Omega^2 ,$$

$$e^{2\phi} = e^{-2\phi_0}\left(1 - \frac{Q^2}{2Mr} e^{-2\phi_0}\right) ,$$

$$F_{rt} = \frac{Q}{r^2} , \qquad (2.201)$$

where ϕ_0 is the asymptotic constant value of the dilaton. One can generate *magnetically charged* solutions by applying the duality transformations $F_{\mu\nu} \to \frac{1}{2} e^{-2\phi} \epsilon_{\mu\nu}{}^{\lambda\rho} F_{\lambda\rho}$ and $\phi \to -\phi$. This transformation does not modify the right hand side of Einstein's equations (2.199) and hence the Einstein's metric in (2.201). But the solution for the dilaton and the electromagnetic field are now given by

$$e^{-2\phi} = e^{-2\phi_0}\left(1 - \frac{Q^2}{2Mr} e^{-2\phi_0}\right) ,$$

$$F_{\theta\phi} = Q \sin\theta , \qquad (2.202)$$

where in the above equations Q refers now to the magnetic charge. From now on, and for simplicity, we shall take $\phi_0 = 0$. Since the Einstein metric is the same for both electrically and magnetically charged black holes, we can discuss the spacetime geometry simultaneously. Comparing this solution to Reissner–Nordström we see important differences immediately. First, there is only one horizon at $r = 2M$ and not two as for the pure Einstein–Maxwell theory. In fact, the $(t - r)$ part of the metric is identical to Schwarzschild. This implies that also the surface gravity coincides with Schwarzschild

$$\kappa = \frac{1}{4M} . \qquad (2.203)$$

Important differences arise in the angular part. There is a curvature (spacelike) singularity, hidden inside the horizon, when the radius of the two-sphere vanishes at $r = \frac{Q^2}{2M}$. Moreover, the area of the event horizon is given by

$$A = 8\pi M \left(2M - \frac{Q^2}{2M}\right) . \qquad (2.204)$$

Another difference with respect to the Reissner–Nordström solution concerns the extremal configuration. Here it is given by

$$M = \frac{1}{2}|Q|, \tag{2.205}$$

instead of the condition $M = |Q|$ for the Reissner–Nordström black holes. In the extremal limit the Einstein metric becomes

$$ds^2 = -\left(1 - \frac{2M}{r}\right)dt^2 + \frac{dr^2}{(1 - \frac{2M}{r})} + r^2\left(1 - \frac{2M}{r}\right)d\Omega^2. \tag{2.206}$$

The event horizon shrinks down to zero area and turns out to be singular. The Penrose diagram of the extremal black hole is identical to the exterior region $r > 2M$ of Schwarzschild, with both future and past event horizons at $r = 2M$ replaced by null singularities.

The above discussion concerns the solution for the Einstein metric. If we go back to the string frame we find important differences. In fact, a different behavior between electrically and magnetically charged solutions emerges. This is so because the dilaton changes sign under duality and this affects the corresponding string metric. The string metric for electrically charged black holes is given by

$$ds^2 = -\left(1 - \frac{2M}{r}\right)\left(1 - \frac{Q^2}{2Mr}\right)dt^2 + \frac{(1 - \frac{Q^2}{2Mr})}{(1 - \frac{2M}{r})}dr^2$$
$$+ r^2\left(1 - \frac{Q^2}{2Mr}\right)^2 d\Omega^2. \tag{2.207}$$

In the extremal limit the above expression gives

$$ds^2 = -(1 - \frac{2M}{r})^2 dt^2 + dr^2 + (r - 2M)^2 d\Omega^2. \tag{2.208}$$

We observe that the g_{tt} component is similar to the one of the extremal Reissner–Nordström solution. However, the spatial metric is flat. As the corresponding metric in the Einstein's frame, there is a singularity at $r = 2M$.

The string metric for the magnetically charged black holes is more interesting and reads

$$ds^2 = -\frac{(1 - \frac{2M}{r})}{(1 - \frac{Q^2}{2Mr})}dt^2 + \frac{dr^2}{(1 - \frac{Q^2}{2Mr})(1 - \frac{2M}{r})} + r^2 d\Omega^2. \tag{2.209}$$

Of particular interest is the extremal solution

$$ds^2 = -dt^2 + \frac{dr^2}{(1 - \frac{2M}{r})^2} + r^2 d\Omega^2 \ . \tag{2.210}$$

The spatial part of the metric is similar to the extremal Reissner–Nordström black hole. Therefore, this implies that the horizon $r = 2M$ is infinitely far away along spacelike geodesics. However, since now the g_{tt} component is trivial, the "horizon" has also moved off to infinity along timelike and null directions. This can be easily seen by introducing the familiar coordinate r^*, defined by $dr^* = (1 - \frac{2M}{r})^{-1} dr$. As one approaches the "horizon" $r \to 2M$ we have $r^* \to 2M \ln(r - 2M)$, and therefore $r^* \to -\infty$. The string metric there behaves

$$ds^2 = -dt^2 + dr^{*2} + 4M^2 d\Omega^2 \ . \tag{2.211}$$

We then see that the "horizon" $r^* = -\infty$ is indeed at an infinite distance away in all directions. Moreover, as we approach it the geometry behaves as a two-dimensional Minkowski space times a two-sphere of fixed radius. In addition, when $r \to +\infty$ the geometry approaches that of flat four-dimensional space. The corresponding Penrose diagram for this solution is the same as that of two-dimensional Minkowski space (see Fig. 2.21). For

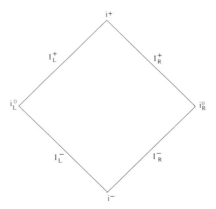

Fig. 2.21 Penrose diagram of the extremal magnetically charged dilatonic black hole.

near-extremal solutions

$$M = \frac{1}{2}|Q| + \Delta m \ , \tag{2.212}$$

where $\Delta m \ll |Q|$, the surface gravity reads

$$\kappa \approx \frac{1}{2|Q|}(1 - \frac{2\Delta m}{|Q|}) \,. \qquad (2.213)$$

Although the surface gravity formally approaches the constant value $\kappa \to 1/2|Q|$ in the extremal limit, the extremal configuration has no surface gravity. The black hole is indeed replaced by a narrow "bottomless hole" in space, as represented in Fig. 2.16. In the throat region, $r \to 2M$, the dilaton field is linear in the coordinate r^*: $\phi \to -r^*/2Q$. For this reason the solution in this region is called the *linear dilaton vacuum*.

Finally, let us remark on the difference in the behavior of the surface gravity κ (2.213) with respect to the Reissner–Nordström solution. For near-extremal configurations $M = |Q| + \Delta m$ we have

$$\kappa \approx \sqrt{\frac{2\Delta m}{Q^3}} \,. \qquad (2.214)$$

In this case κ vanishes in the limit $\Delta m \to 0$. The classical and quantum properties of both the near-extremal stringy and Reissner–Nordström black holes will be studied extensively in Chapter 6.

Chapter 3

The Hawking Effect

3.1 Canonical Quantization in Minkowski Space

The aim of this section is to present the main ingredients of the canonical quantization of a field in Minkowski spacetime. We shall not devote much time to this, but we shall stress those points which are relevant to prepare the reader to the transition to curved spacetime quantization. To this end it is enough to study the simplest case, namely a real massless scalar field f with action given by

$$S = -\frac{1}{2}\int d^4x \partial_\mu f \partial^\mu f \ . \tag{3.1}$$

The equation of motion is the massless Klein–Gordon equation

$$\partial_\mu \partial^\mu f = 0 \ . \tag{3.2}$$

To quantize the field f one splits a generic solution to the Klein–Gordon equation into positive and negative frequency solutions

$$f(t,\vec{x}) = \sum_i [a_i u_i(t,\vec{x}) + a_i^\dagger u_i^*(t,\vec{x})] \ , \tag{3.3}$$

with respect to a *global inertial time* "t" of Minkowski space. This means that

$$\frac{\partial}{\partial t} u_j(t,\vec{x}) = -iw_j u_j(t,\vec{x}) \ , \quad w_j > 0 \ . \tag{3.4}$$

Moreover, the elementary solutions $u_i(t,\vec{x})$ are chosen to form a complete orthonormal basis with respect to the Klein–Gordon product

$$(f_1, f_2) = -i \int d^3\vec{x}(f_1 \partial_t f_2^* - f_2^* \partial_t f_1) \ . \tag{3.5}$$

On the space of positive frequency solutions this scalar product turns out to be positive definite allowing to construct the so called *one-particle Hilbert space*. From it one builds the *many-particle Fock space*. It can be described from the commutators between the operators a_i and $a_j^\dagger$:

$$[a_i, a_j^\dagger] = (u_i, u_j)\hbar = \delta_{ij}\hbar , \qquad (3.6)$$

$$[a_i, a_j] = 0 = [a_i^\dagger, a_j^\dagger] , \qquad (3.7)$$

obtained when the classical field f is promoted to a quantum field operator obeying the equal time commutation relations

$$[f(t,\vec{x}), \pi(t,\vec{x}')] = i\hbar \delta^3(\vec{x} - \vec{x}') , \qquad (3.8)$$

$$[f(t,\vec{x}), f(t,\vec{x}')] = 0 = [\pi(t,\vec{x}), \pi(t,\vec{x}')] , \qquad (3.9)$$

where $\pi = \partial_t f$ is the canonical conjugate variable to f. The Fock space is constructed from a vacuum state $|0\rangle$ annihilated by the a_i operators

$$a_i |0\rangle = 0 . \qquad (3.10)$$

The states obtained by acting with the creation operators $a_j^\dagger$ on the vacuum

$$|1_i\rangle = \hbar^{-1/2} a_i^\dagger |0\rangle \qquad (3.11)$$

span the one-particle Hilbert space. The many-particle states are then constructed in a similar way

$$|n_{i_1}^{(1)}, n_{i_2}^{(2)}, ..., n_{i_k}^{(k)}\rangle = (n^{(1)}! n^{(2)}! ... n^{(k)}!)^{-\frac{1}{2}} (\hbar^{-1/2} a_{i_1}^\dagger)^{n^{(1)}} (\hbar^{-1/2} a_{i_2}^\dagger)^{n^{(2)}} ...$$
$$(\hbar^{-1/2} a_{i_k}^\dagger)^{n^{(k)}} |0\rangle , \qquad (3.12)$$

where all the indices $i_1, i_2, ..., i_k$, labeling one-particle states, are different.

The usual choice for the orthonormal basis is given by the plane wave modes

$$u_{\vec{k}} \equiv \frac{1}{\sqrt{16\pi^3 w}} e^{-iwt + i\vec{k}\vec{x}} , \qquad (3.13)$$

where the generic index i is given now by the vector $\vec{k}$ and the frequency w is given by $w = |\vec{k}|$. The commutation relations for the ladder operators are then

$$[a_{\vec{k}}, a_{\vec{k}'}^\dagger] = \hbar \delta^3(\vec{k} - \vec{k}') , \qquad (3.14)$$

$$[a_{\vec{k}}, a_{\vec{k}'}] = 0 = [a_{\vec{k}}^\dagger, a_{\vec{k}'}^\dagger] \ . \tag{3.15}$$

A crucial property of the quantization is that it is independent of the particular inertial time "t" chosen. Any other choice of time, related via *Poincaré transformations*, leads to the same characterizacion of positive frequency modes. Therefore the splitting between positive and negative frequency modes assumed in the expansion (3.3) is invariant. As a consequence the vacuum state is also invariant and, therefore, the full Fock space.[1]

3.2 Quantization in Curved Spacetimes

When the spacetime is not Minkowski and Poincaré symmetry is lost one has to generalize the steps involved in the construction of a quantum field theory given in the previous section. Most of the steps can be extended to a curved spacetime without problems.

- The wave equation in flat spacetime (3.2) can be naturally generalized by means of the generally covariant d'Alembertian operator $\Box \equiv \nabla_\mu \nabla^\mu$

$$\Box f = 0 \ .\text{[2]} \tag{3.16}$$

- The Klein–Gordon product in Minkowski space (3.5) can also be extended to curved spacetime

$$(f_1, f_2) = -i \int_\Sigma d\Sigma^\mu (f_1 \partial_\mu f_2^* - f_2^* \partial_\mu f_1) \ , \tag{3.17}$$

where Σ is an "initial data" Cauchy hypersurface and $d\Sigma^\mu = d\Sigma n^\mu$, with $d\Sigma$ being the volume element and n^μ a future directed unit normal vector to Σ. Using Gauss's theorem one can show that the inner product is indeed independent of the choice of hypersurface.

---

[1] A simple way to realize this is evaluating the action of the quantum conserved charges, associated with the Poincaré symmetry, on the vacuum.
[2] We still have the freedom to add a coupling of f with the scalar curvature R. This leads to the wave equation:$\Box f - \xi R f = 0$ where ξ is a new coupling constant. The simplest choice for ξ is the minimal coupling ($\xi = 0$). Another commonly used choice is the so called conformal coupling ($\xi = 1/6$). In any case, the above wave equation turns out to be equivalent to (3.2) in Minkowski spacetime.

The difference with respect to Minkowski spacetime is that there is not a natural splitting of modes between positive and negative frequency solutions. Different choices of positive frequency solutions lead, in general, to different definitions of the vacuum state and therefore of the corresponding Fock space. The *absence* of a *unique notion of vacuum state* is the main difference between the quantum theory in Minkowski space and the quantization in a curved spacetime. However, there is a situation for which one can naturally select a space of positive frequency solutions. This arises when the curved spacetime is *stationary*. As we have explained in the previous chapter, this means that we have a timelike vector field ξ^μ leaving invariant the spacetime metric $\delta_{\xi^\alpha} g_{\mu\nu} = 0$, where δ_{ξ^α} is the infinitesimal transformation generated by the vector ξ^α. This allows to define a natural notion of positive frequency modes u_i

$$\xi^\mu \nabla_\mu u_j = -iw_j u_j \ , \qquad w_j > 0 \ . \tag{3.18}$$

Introducing the *Killing time t* obeying $\xi^\mu \nabla_\mu t = 1$, it is easy to see that this definition is equivalent and generalizes the condition (3.4) for Minkowski spacetime. The standard particle interpretation of states in the Fock space in Minkowski spacetime can be extended to stationary spacetimes.

When the spacetime is *not stationary*, as it happens for instance in the process of gravitational collapse, one looses the natural criterium to define positive frequency modes. Therefore, the unambiguous concept of particle states of Minkowski space disappears. Nevertheless one can still manage to find a particle interpretation for those spacetimes that possess *asymptotic stationary regions* in the past and in the future. These regions will be called "in" and "out" and for them we have a natural particle interpretation.

One can construct an orthonormal set of modes, exact solutions of the wave equation in the whole spacetime, such that they have positive frequencies with respect to the inertial time in the past. We can call these solutions u_i^{in}. In addition, one can also find a new set of orthonormal modes u_i^{out} with positive frequencies in the future. Assuming that our quantum field is the "in" vacuum state $|in\rangle$ (i.e., a state with no particle content in the past), one can investigate the effect of the dynamical (non-stationary) gravitational field at late times, when the geometry has settled down again to a stationary configuration. We will see later that, surprisingly, the state $|in\rangle$ is not, in general, perceived as a vacuum state in the Fock space of the "out" region. This is summarized in Fig. 3.1. To quantify this result one needs then to introduce the formalism of the so called *Bogolubov transformations*.

The Hawking Effect

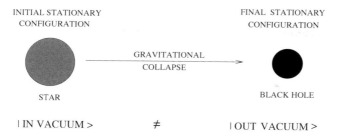

Fig. 3.1 In the process of black hole formation the vacuum state is not invariant.

3.2.1 *Bogolubov transformations and particle production*

Let us expand the field f in terms of the positive frequency solutions u_i^{in} in the initial stationary region

$$f = \sum_i [a_i^{in} u_i^{in} + a_i^{in\dagger} u_i^{in*}] \, . \qquad (3.19)$$

Choosing instead the positive frequency solutions in the final stationary region, labeled by u_i^{out}, we can perform the alternative expansion

$$f = \sum_i [a_i^{out} u_i^{out} + a_i^{out\dagger} u_i^{out*}] \, . \qquad (3.20)$$

The modes u_i^{in} satisfy the orthonormal relations

$$(u_i^{in}, u_j^{in}) = \delta_{ij} \, , \quad (u_i^{in*}, u_j^{in*}) = -\delta_{ij} \, , \quad (u_i^{in}, u_j^{in*}) = 0 \, , \qquad (3.21)$$

and the ladder operators verify the commutation relations

$$[a_i^{in}, a_j^{in\dagger}] = \hbar \delta_{ij} \, , \qquad (3.22)$$

$$[a_i^{in}, a_j^{in}] = 0 = [a_i^{in\dagger}, a_j^{in\dagger}] \, . \qquad (3.23)$$

Similar relations hold for the u_i^{out} modes and the corresponding creation and annihilation operators. Since both sets of modes are complete one can also expand one set of modes, for instance u_i^{out} and u_i^{out*}, in terms of the other. Moreover, it is not guaranteed that positive frequency solutions u_i^{out} can be expanded in terms of pure positive frequency solutions u_i^{in}. So, in general we have

$$u_j^{out} = \sum_i (\alpha_{ji} u_i^{in} + \beta_{ji} u_i^{in*}) \, . \qquad (3.24)$$

These relations are the Bogolubov transformations and the matrices α_{ij}, β_{ij} are called the Bogolubov coefficients. Taking into account the relations (3.21), the scalar product between both sets of modes gives directly the Bogolubov coefficients

$$\alpha_{ij} = (u_i^{out}, u_j^{in}) \ , \quad \beta_{ij} = -(u_i^{out}, u_j^{in*}) \ . \tag{3.25}$$

The orthonormality relations (3.21) also lead to the conditions

$$\sum_k (\alpha_{ik}\alpha_{jk}^* - \beta_{ik}\beta_{jk}^*) = \delta_{ij} \ , \tag{3.26}$$

and

$$\sum_k (\alpha_{ik}\beta_{jk} - \beta_{ik}\alpha_{jk}) = 0 \ . \tag{3.27}$$

One can also invert the relations (3.24) and obtain

$$u_i^{in} = \sum_j (\alpha_{ji}^* u_j^{out} - \beta_{ji} u_j^{out*}) \ , \tag{3.28}$$

and using the fact that $a_i^{in} = (f, u_i^{in})$ and $a_i^{out} = (f, u_i^{out})$ we can expand one of the two sets of creation and annihilation operators in terms of the other

$$a_i^{in} = \sum_j (\alpha_{ji} a_j^{out} + \beta_{ji}^* a_j^{out\dagger}) \ , \tag{3.29}$$

or

$$a_i^{out} = \sum_j (\alpha_{ij}^* a_j^{in} - \beta_{ij}^* a_j^{in\dagger}) \ . \tag{3.30}$$

If the coefficients β_{ij} do not vanish the vacuum states $|in\rangle$ and $|out\rangle$, defined as

$$a_i^{in}|in\rangle = 0 \ , \tag{3.31}$$

$$a_i^{out}|out\rangle = 0 \ , \tag{3.32}$$

are different. One can realize this immediately by computing the expectation value of the "out" particle number operator for the ith mode

$$N_i^{out} \equiv \hbar^{-1} a_i^{out\dagger} a_i^{out} \ , \tag{3.33}$$

in the state $|in\rangle$

$$\begin{aligned}\langle in|N_i^{out}|in\rangle &= \hbar^{-1}\langle in|a_i^{out\dagger}a_i^{out}|in\rangle \\ &= \hbar^{-1}\langle in|\sum_j(-\beta_{ij}a_j^{in})\sum_k(-\beta_{ik}^*a_k^{in\dagger})|in\rangle \\ &= \sum_j|\beta_{ij}|^2 \; .\end{aligned} \quad (3.34)$$

If any of the β_{ij} coefficients are different from zero the particle content of the $|in\rangle$ vacuum state, with respect to the "out" Fock space, is non-trivial. In contrast, if all the coefficients β_{ij} vanish, the relation (3.26) turns out to be

$$\sum_k \alpha_{ik}\alpha_{jk}^* = \delta_{ij} \; . \quad (3.35)$$

This means that the positive frequency mode basis u_i^{in} and u_i^{out} are related by a unitary transformation, with matrix α_{ij}, and therefore the definition of the vacuum remains unchanged: $|in\rangle = |out\rangle$.

To unravel all the details concerning the particle content of the $|in\rangle$ state in the final stationary region we need to give the explicit expression of $|in\rangle$ in terms of the "out" Fock space. This can be achieved by acting with the relation (3.29) on the $|in\rangle$ state

$$\sum_j(\alpha_{ji}a_j^{out}+\beta_{ji}^*a_j^{out\dagger})|in\rangle = 0 \; , \quad (3.36)$$

and then[3]

$$(a_k^{out}+\sum_{ij}\beta_{ji}^*\alpha_{ik}^{-1}a_j^{out\dagger})|in\rangle = 0 \; . \quad (3.37)$$

Defining the symmetric matrix V_{jk}

$$V_{jk} \equiv -\sum_i \beta_{ji}^*\alpha_{ik}^{-1} \; , \quad (3.38)$$

one can find recursively the expression of $|in\rangle$ in terms of V_{jk}. The final result is

$$|in\rangle = \langle out|in\rangle \exp\left(\frac{1}{2\hbar}\sum_{ij}V_{ij}a_i^{out\dagger}a_j^{out\dagger}\right)|out\rangle \; . \quad (3.39)$$

---

[3] The existence of the inverse matrix α_{kj}^{-1} is guaranteed by the properties (3.26), (3.27) of the Bogolubov coefficients [Wald (1975)].

An immediate consequence of this result is that the amplitude for producing an odd number of particles vanishes

$$\langle 1_{j_1}, 1_{j_2}, ..., 1_{j_n} | in \rangle = 0 \quad n \text{ odd} . \tag{3.40}$$

Particles are produced in pairs, and only for n even the corresponding amplitudes are non-trivial. In particular, for the vacuum to two-particle amplitudes we get just the matrix V_{ij}, up to the normalization constant $\langle out | in \rangle$

$$\langle 1_{j_1}, 1_{j_2} | in \rangle = \langle out | in \rangle V_{j_1 j_2} . \tag{3.41}$$

Note that the symmetry of the matrix V_{ij} is necessary to recover compatibility with the spin-statistic theorem, which requires a symmetrized tensor product for the two-particle (and n-particle, in general) sector of the Fock space. Had we started with a spinor field, the corresponding V_{ij} would have been antisymmetric, in agreement with the spin-statistic connection.

Moreover, to have a meaningful result the right hand side of Eq. (3.39) should be a normalizable state within the "out" Fock space. It is not difficult to see [Wald (1975)] that this is equivalent to the condition

$$\sum_{ij} |\beta_{ij}|^2 < +\infty , \tag{3.42}$$

which has a clear physical meaning: the total number of particles produced by the gravitational field, during the transition between the two asymptotic stationary configurations "in" and "out", should be finite.

In the next sections we shall apply this framework, which was first applied in cosmology by [Parker (1969)], to analyze the particle production in the non-stationary background describing the formation of a black hole due to gravitational collapse. We shall mainly follow the original paper [Hawking (1975)] and the complementary ones [Wald (1975); Parker (1975)]. Since this is the main aim of this chapter we shall do it step by step starting from a very simple model. The reader can complement the analysis presented in this chapter with that given in the books [Birrell and Davies (1982); Frolov and Novikov (1998)]. Special emphasis on conceptual and mathematical problems can be found in [Wald (1994)], and the membrane viewpoint in [Thorne et al. (1986)]. Finally a list of selected reviews on this topic is: [DeWitt (1975); Parker (1976); Isham (1977); Gibbons (1979); Ford (1997); Wipf (1998); Kiefer (1999); Jacobson (2003)].

3.3 Hawking Radiation in Vaidya Spacetime

The aim of this section is to introduce the Hawking effect in the simplest possible scenario. We want to present the basic ingredients to understand it leaving aside, momentarily, all the additional complications that are present in a real gravitational collapse. Our purpose is first to give to the reader the main idea lying behind the Hawking effect. Once this is clear then it is easier to analyse it in more realistic scenarios, where we will see that the bulk of the result obtained in the simplified background is unchanged. This will be considered later.

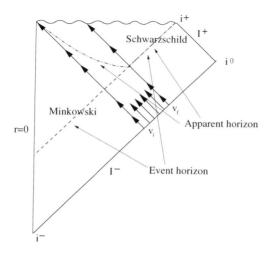

Fig. 3.2 A Schwarzschild black hole produced by collapse of ingoing radiation.

Let us now consider the simplest solution to Einstein's equations describing the formation of a black hole. This is given by the *Vaidya spacetime* with line element

$$ds^2 = -(1 - \frac{2M(v)}{r})dv^2 + 2dvdr + r^2 d\Omega^2 \ . \tag{3.43}$$

When $M(v) = M = constant$ this is nothing but the Schwarzschild solution in the advanced Eddington–Finkelstein gauge. The important fact is that if the mass depends only on the null coordinate v then the above turns out to be an exact solution to Einstein's field equations with a stress-energy tensor of the form

$$T_{vv} = \frac{L(v)}{4\pi r^2}, \tag{3.44}$$

where

$$\frac{dM(v)}{dv} = L(v). \tag{3.45}$$

The physical interpretation is that this solution describes a general purely ingoing radial flux of radiation described by the function $L(v)$. This is certainly a simplification of the physical gravitational collapse process where both ingoing and outgoing fluxes of radiation, as well as non-spherical deformations, are present before the geometry settles down to the final black hole configuration. Nevertheless this simplification is the one that allows the most straightforward approach to the problem keeping all the main ingredients. If we imagine the influx of radiation to be turned on at some finite advanced time v_i and turned off at v_f then the spacetime geometry can be naturally divided into three regions (see Fig. 3.2.):

- A Minkowski vacuum region for $v < v_i$;
- An intermediate collapse region $v_i < v < v_f$;
- The final Schwarzschild black hole configuration $v > v_f$.

Since the important regions in order to derive the Hawking effect are the first and the third ones we can freely narrow the intermediate region down to, ideally, a single null surface. Therefore we shall consider an ingoing shock wave located at some $v = v_0$ of the form

$$L(v) = M\delta(v - v_0), \tag{3.46}$$

leading to

$$M(v) = M\theta(v - v_0). \tag{3.47}$$

The resulting spacetime is then obtained by patching portions of Minkowski and Schwarzschild spacetimes along $v = v_0$. This is the simplest non-stationary spacetime which allows particle production. We shall call the Minkowski portion the "in" region and the Schwarzschild portion the "out" region. The Penrose diagram of this process is depicted in Fig. 3.3.

Following this reasoning we shall also consider the simplest form of matter field to be quantized in this background. Starting from the massless

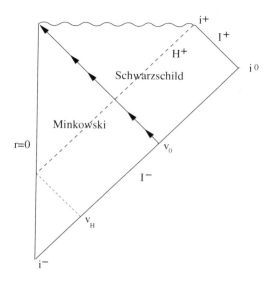

Fig. 3.3 A Schwarzschild black hole produced by a shock wave at $v = v_0$

Klein–Gordon equation

$$\Box f = 0 \,, \tag{3.48}$$

since the background is spherically symmetric we can always expand the field f in the following way

$$f(x^\mu) = \sum_{l,m} \frac{f_l(t,r)}{r} Y_{lm}(\theta, \varphi) \,, \tag{3.49}$$

where Y_{lm} are the spherical harmonics. The Klein–Gordon equation for $f(x^\mu)$ is then converted into a two-dimensional wave equation for $f_l(t,r)$, for each angular momenta l, with a non vanishing potential. In the Minkowski "in" region $v < v_0$ we have

$$\left(-\frac{\partial^2}{\partial t^2} + \frac{\partial^2}{\partial r^2} - \frac{l(l+1)}{r^2}\right) f_l(t,r) = 0 \,, \tag{3.50}$$

while in the Schwarzschild "out" region $v > v_0$ it is

$$\left(-\frac{\partial^2}{\partial t^2} + \frac{\partial^2}{\partial r^{*2}} - V_l(r)\right) f_l(t,r) = 0 \,, \tag{3.51}$$

where the potential is given by

$$V_l(r) = \left(1 - \frac{2M}{r}\right)\left[\frac{l(l+1)}{r^2} + \frac{2M}{r^3}\right] . \tag{3.52}$$

We notice that the potential vanishes both at $r = +\infty$ ($r^* = +\infty$) and at the horizon $r = 2M$ ($r^* = -\infty$). We shall now take what could seem to be a very crude approximation. Motivated by the fact that the important physics happens at the horizon, where the potential vanishes, we shall neglect the potential everywhere. This is better justified for the "s-wave component" ($l = 0$), since it is the one less affected by the potential. This approximation will find a justification in Section 3.6, in the context of the *geometric optics approximation*, and we shall apply it in the next chapter when considering the *near-horizon approximation*.

This means that we deal with the field f satisfying the free field equation in both the Minkowski and Schwarzschild regions. In Minkowski we have the equation

$$\left(-\frac{\partial^2}{\partial t^2} + \frac{\partial^2}{\partial r^2}\right) f(t,r) = 0 , \tag{3.53}$$

with the additional (regularity) condition that the field vanishes at $r = 0$

$$f(t, r=0) = 0 . \tag{3.54}$$

In the Schwarzschild region we have the free field equation

$$\left(-\frac{\partial^2}{\partial t^2} + \frac{\partial^2}{\partial r^{*2}}\right) f(t,r) = 0 . \tag{3.55}$$

Since we want to look for positive frequency solutions we can further assume a harmonic time dependence

$$f(t,r) = e^{-iwt} f(r) . \tag{3.56}$$

The above wave equations transform into

$$\frac{d^2 f(r)}{dr^2} + w^2 f(r) = 0 , \tag{3.57}$$

and

$$\frac{d^2 f(r)}{dr^{*2}} + w^2 f(r) = 0 . \tag{3.58}$$

The Hawking Effect

The solutions to these equations are better expressed in terms of null coordinates. The initial Minkowski metric can be written as

$$ds^2 = -du_{in}dv + r_{in}^2 d\Omega^2 , \tag{3.59}$$

where $u_{in} = t_{in} - r_{in}$, $v = t_{in} + r_{in}$. For the Schwarzschild metric we have

$$ds^2 = -(1 - \frac{2M}{r_{out}})du_{out}dv + r_{out}^2 d\Omega^2 , \tag{3.60}$$

where $u_{out} = t_{out} - r_{out}^*$, $v = t_{out} + r_{out}^*$. The relation between u_{out} and u_{in} will be derived in the next subsection. The solutions of the above wave equations will be generically *ingoing waves*, e^{-iwv}, and *outgoing waves*, $e^{-iwu_{out}}$ in the Schwarzschild region or $e^{-iwu_{in}}$ in the Minwoski region.

3.3.1 Ingoing and outgoing modes

We now apply the general scheme introduced in Section 3.2 to work out the particle production in the above background geometry describing the formation of a Schwarzschild black hole. We shall define the "in" Fock space associated with the natural time parameter at I^-. The positive frequency modes are

$$u_w^{in} = \frac{1}{4\pi\sqrt{w}} \frac{e^{-iwv}}{r} , \tag{3.61}$$

obeying the normalization condition

$$(u_w^{in}, u_{w'}^{in}) = -i \int_{I^-} dv r^2 d\Omega (u_w^{in} \partial_v u_{w'}^{in*} - u_{w'}^{in*} \partial_v u_w^{in}) = \delta(w - w') . \tag{3.62}$$

We can also define the Fock space at I^+ associated to the natural time parameter at I^+. The natural set of positive frequency modes is then

$$u_w^{out} = \frac{1}{4\pi\sqrt{w}} \frac{e^{-iwu_{out}}}{r} , \tag{3.63}$$

where we use the notation "out" to refer to the modes and coordinates at I^+. These modes obey an analogous normalization condition

$$(u_w^{out}, u_{w'}^{out}) = -i \int_{I^+} du_{out} r^2 d\Omega (u_w^{out} \partial_{u_{out}} u_{w'}^{out*} - u_{w'}^{out*} \partial_{u_{out}} u_w^{out}) = \delta(w - w') . \tag{3.64}$$

Note that we are apparently choosing I^+ as an alternative Cauchy surface. However this is not strictly correct since in the "out" region I^+ is not a proper Cauchy surface. We must add the future horizon H^+ to have a complete Cauchy surface: $I^+ \bigcup H^+$. In other words, the modes u_w^{out} are not complete and we have to add, to construct the proper "out" Fock space, those modes that cross the future horizon H^+. We shall denote them by u_w^{int}. To give an explicit expression for them is problematic since there is no natural choice of time in this region. However we shall not need any expression of u_w^{int} to evaluate the particle production at I^+. This is because the result is insensitive to the particular choice of the modes u_w^{int}.

Our next step is to evaluate the Bogolubov coefficients relating the "in" and "out" basis solution. For our purposes the most important coefficients to work out are

$$\beta_{ww'} = -(u_w^{out}, u_{w'}^{in*}) = i \int_{I^-} dv r^2 d\Omega (u_w^{out} \partial_v u_{w'}^{in} - u_{w'}^{in} \partial_v u_w^{out}) \,, \qquad (3.65)$$

where we have chosen I^- as the Cauchy surface for the scalar product. We remark that we have the freedom to choose any "initial data hypersurface", since the scalar product is insensitive to this choice. Therefore, we pick up I^- for mathematical and physical convenience. To evaluate this integral one needs to know the behavior of the modes u_w^{out} at I^-. In the simplified model that we are using this can be easily done due to the free propagation implied by the wave equation of Eqs. (3.53) and (3.55). The form of the modes u_w^{out} (3.63) at I^+ remains unaltered in the Schwarzschild portion of the spacetime ($v > v_0$). In the Minkowski region ($v < v_0$) we can determine the form of the modes by imposing two conditions:

- Matching condition along the line $v = v_0$. This means that just before the shock wave the form of the modes in the u_{in} coordinate is

$$u_w^{out} = \frac{1}{4\pi\sqrt{w}} \frac{e^{-iwu_{out}(u_{in})}}{r} \,, \qquad (3.66)$$

where the relation $u_{out} = u_{out}(u_{in})$ is determined by the matching conditions for the metric. This is similar to the analysis made in Section 2.8 for the matching of two Schwarzschild metrics. Requiring the metric on the shock wave to be the same on both sides implies

$$r(v_0, u_{in}) = r(v_0, u_{out}) \,, \qquad (3.67)$$

where

$$r(v_0, u_{in}) = \frac{v_0 - u_{in}}{2} \,, \qquad (3.68)$$

and
$$r(v_0, u_{out}) + 2M \ln(\frac{r(v_0, u_{out})}{2M} - 1) = \frac{v_0 - u_{out}}{2}. \qquad (3.69)$$

This gives
$$u_{out} = u_{in} - 4M \ln \frac{|v_0 - 4M - u_{in}|}{4M}. \qquad (3.70)$$

- The regularity condition at $r = 0$ (3.54) forces the following form of the modes in the Minkowski part of the spacetime

$$u_w^{out} = \frac{1}{4\pi\sqrt{w}} \left(\frac{e^{-iwu_{out}(u_{in})}}{r} - \frac{e^{-iwu_{out}(v)}}{r} \theta(v_H - v) \right), \qquad (3.71)$$

where
$$u_{out}(v) = u_{out}(u_{in} \leftrightarrow v), \qquad (3.72)$$

and $v_H = v_0 - 4M$ is the location of the null ray that will form the event horizon at $u_{out} = +\infty$ (see Fig. 3.3).

It is instructive to analyse the behavior of u_w^{out} in two limiting regions. At early times $v \to -\infty$ we have $u_{out}(v) \approx v$. This is so because for $u_{in} \to -\infty$ (3.70) implies that $u_{out}(u_{in}) \approx u_{in}$. Therefore at I^-

$$u_w^{out} \approx -\frac{1}{4\pi\sqrt{w}} \frac{e^{-iwv}}{r}, \qquad (3.73)$$

and so the mode u_w^{out}, for $u_{out} \to -\infty$, is still a pure positive frequency mode with respect to the inertial time at I^-. The related $\beta_{ww'}$ coefficients vanish and this means that asymptotically and at early times $u_{out} \to -\infty$ there is no particle emission reaching I^+. The corresponding $\alpha_{ww'}$ coefficients are those required by the completeness relations (3.26) $\alpha_{ww'} = \delta(w - w')$.

As u_{out} increases the situation changes drastically. At late times $u_{out} \to +\infty$ ($v \to v_H$) we find the relation

$$u_{out}(u_{in}) = v_H - 4M \ln \frac{v_H - u_{in}}{4M}, \qquad (3.74)$$

and then using (3.72)

$$u_{out}(v) = v_H - 4M \ln \frac{v_H - v}{4M}. \qquad (3.75)$$

Therefore, close to v_H and at I^-, the modes u_w^{out} behave as

$$u_w^{out} \approx -\frac{1}{4\pi\sqrt{w}} \frac{e^{-iw(v_H - 4M \ln \frac{v_H - v}{4M})}}{r} \theta(v_H - v) \ . \quad (3.76)$$

The infinite blueshift in the exponent of (3.76) as well as the existence of the critical ingoing time v_H, typical of the process of black hole formation, will largely modify the early time properties (see Fig. 3.4). In fact (3.76) turns

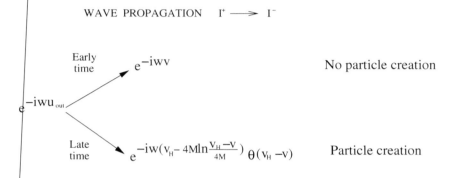

Fig. 3.4 Wave propagation and physical consequences.

out to be a superposition of positive and negative frequency modes with respect to the inertial time at I^-. This is at the basis of the astonishing result of Hawking [Hawking (1974); Hawking (1975)] that we will analyze now in detail.

Before entering into the technical details of Hawking's derivation we can provide some sort of intuitive argument to justify the existence of Hawking radiation, i.e., the spontaneous emission of particles that accompanies the formation of a black hole. Let us imagine an asymptotic observer in the initial region sending a radial light ray with very large frequency, just before the advanced time v_H. This ray is described by a wave of amplitude A and frequency w'. It will propagate through the Minkowski region without modifying its frequency. Once it enters into the Schwarzschild region, crossing the dynamical line $v = v_0$, both amplitude and effective frequency will be modified by the geometry. The frequency suffers a very large redshift. When it finally approaches I^+ the effective frequency w, measured by an inertial observer at I^+, is largely diminished according to

$$w \sim w' e^{-u_{out}/4M} \ . \quad (3.77)$$

The amplitude of the wave has also decreased by a certain amount. This is just to compensate the partial reflection suffered by the wave due to the potential (3.52). However one can leave aside this effect since it is smaller than the redshift effect. In fact, for frequencies much larger that the maximum of the potential (3.52), the wave reflection is negligible. As a consequence of the gravitational redshift the energy of the outgoing wave is much smaller than that of the incoming wave, without having increased the black hole internal energy. To have the possibility of reestablishing energy conservation, the process described should be accompanied by an additional emission of particles, carrying the necessary energy. Within the language of quantum physics, this should correspond to a process of stimulated emission, which in turn suggests the existence of a parallel mechanism of spontaneous emission. Such a process of spontaneous creation of particles during the formation of a black hole is just the Hawking effect. Notice the importance of the fact that the full background geometry is *not stationary*. The presence of the *event horizon* is also crucial. Without it, the redshift is not divergent and the energy deficit in the process would be very small. To compensate it we do not need a steady flux of radiation, as required by the presence of the horizon, but just a transitory emission of radiation.

3.3.2 Wave packets

The quantity that, *a priori*, we are mainly interested to determine is the expectation value of the number of particles of a given frequency emitted at I^+:

$$\langle in | N_w^{out} | in \rangle = \int_0^\infty dw' |\beta_{ww'}|^2 . \qquad (3.78)$$

It has been assumed that the quantum state of the matter field is $|in\rangle$, i.e., the natural vacuum at I^-. The coefficients $\beta_{ww'}$ are given, *a priori*, by the expression (3.65). However, to get physically sensible results we have to be careful in the way we treat the above expression. Let us clarify the physical meaning of the quantity $\langle in | N_w^{out} | in \rangle$. It gives the mean particle number detected at I^+ with a definite frequency w. The use of this type of states, with a definite frequency, implies absolute uncertainty in time. Therefore $\langle in | N_w^{out} | in \rangle$ provides, indeed, the number of particles with frequency w emitted at anytime u_{out}. But we are mainly interested in evaluating the mean particle number produced at late retarded times $u_{out} \to +\infty$, when, in a realistic situation, the black hole has settled down to a stationary configuration. To properly evaluate the late time particle production one has to replace the plane wave type modes, which are completely delocalized, by wave packets.

One can introduce a complete orthonormal set of wave packet modes at I^+, with discrete quantum numbers, as follows:[4]

$$u_{jn}^{out} = \frac{1}{\sqrt{\epsilon}} \int_{j\epsilon}^{(j+1)\epsilon} dw\, e^{2\pi i w n/\epsilon} u_w^{out}, \qquad (3.79)$$

with integers $j \geq 0$, n. These wave packets are peaked about $u_{out} = 2\pi n/\epsilon$ with width $2\pi/\epsilon$. Taking ϵ small ensures that the modes are narrowly centered around $w \simeq w_j = j\epsilon$ (see for instance Fig. 3.5). Therefore, the

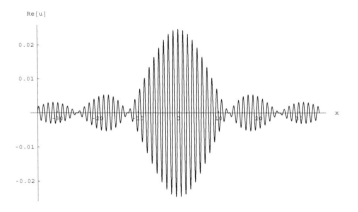

Fig. 3.5 Real part of a wave packet mode u^{out}: $u \equiv r u_{jm}^{out}(x \equiv u_{out})$, for $n = 0$, $j = 10$ and $\epsilon = 0.5$.

computation of $\langle in|N_{jn}^{out}|in\rangle$, associated with the wave packet, has a clear physical interpretation. It gives the counts of a particle detector sensitive only to frequencies within ϵ of w_j which is turned on for a time interval $2\pi/\epsilon$ at time $u_{out} = 2\pi n/\epsilon$. The use of the basis u_{jn}^{out} is not only a mathematical refinement to get rigorous results, it also has a direct interpretation in terms of physical measurements. For instance, the late time modes, and the corresponding measurements, are characterized by having a very large quantum number n. These modes are precisely those that we have to

---

[4]Note that this definition differs in the sign of the exponent from that given in [Hawking (1975); Wald (1975)]. This is due to the different definition of positive frequency e^{iwt}, instead of e^{-iwt}, used in these references. Obviously, the convention does not modify the results.

propagate backwards into the past to see their decomposition in positive and negative frequencies at I^-.

To determine the particle production we need to evaluate the coefficients $\beta_{jn,w'}$

$$\beta_{jn,w'} = -(u_{jn}^{out}, u_{w'}^{in*}) = i \int_{I^-} dv r^2 d\Omega (u_{jn}^{out} \partial_v u_{w'}^{in} - u_{w'}^{in} \partial_v u_{jn}^{out}). \quad (3.80)$$

We can perform a partial integration and discard the boundary terms since the modes u_{jn}^{out}, when traced backwards in time, vanish at $v = -\infty$ and $v = +\infty$

$$\beta_{jn,w'} = 2i \int_{I^-} dv r^2 d\Omega u_{jn}^{out} \partial_v u_{w'}^{in}. \quad (3.81)$$

Since we are interested in the late time particle production we need to compute the above coefficients when $n \to +\infty$. The wave packets at I^+ with large n will be concentrated, when they are propagated backwards in time, in an infinitesimal interval around v_H at I^- (see Fig. 3.6).

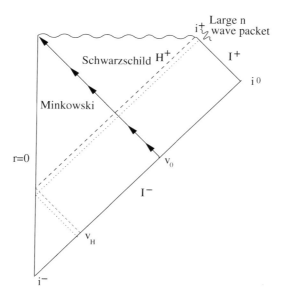

Fig. 3.6 The backward propagation of a late time wave packet mode.

The advantage of this is that in the evaluation of $\beta_{jn,w'}$ for $n \to +\infty$

we can use the relation (3.76). So we can write

$$\beta_{jn,w'} = \frac{-1}{2\pi\sqrt{\epsilon}} \int_{-\infty}^{v_H} dv \int_{j\epsilon}^{(j+1)\epsilon} dw e^{2\pi i w n/\epsilon} \sqrt{\frac{w'}{w}} e^{-iw(v_H - 4M \ln \frac{v_H - v}{4M}) - iw'v} \ . \tag{3.82}$$

It is easy to also find a similar expression for the $\alpha_{jn,w'}$ coefficients

$$\alpha_{jn,w'} = (u_{jn}^{out}, u_{w'}^{in}) = -2i \int_{I^-} dv r^2 d\Omega u_{jn}^{out} \partial_v u_{w'}^{in*}$$

$$= \frac{-1}{2\pi\sqrt{\epsilon}} \int_{-\infty}^{v_H} dv \int_{j\epsilon}^{(j+1)\epsilon} dw e^{2\pi i w n/\epsilon}$$

$$\times \sqrt{\frac{w'}{w}} e^{-iw(v_H - 4M \ln \frac{v_H - v}{4M}) + iw'v} \ . \tag{3.83}$$

For convenience let us introduce the variable x defined by $x = v_H - v$. We then get

$$\beta_{jn,w'} = \frac{-e^{-iw'v_H}}{2\pi\sqrt{\epsilon}} \int_0^{+\infty} dx \int_{j\epsilon}^{(j+1)\epsilon} dw e^{2\pi i w n/\epsilon}$$

$$\times \sqrt{\frac{w'}{w}} e^{-iw(v_H - 4M \ln \frac{x}{4M}) + iw'x} \ . \tag{3.84}$$

The integral over frequencies can be performed immediately taking into account that w varies in a small interval

$$\beta_{jn,w'} = \frac{-e^{-i(w_j + w')v_H}}{\pi\sqrt{\epsilon}} \sqrt{\frac{w'}{w_j}} \int_0^{+\infty} dx e^{iw'x} \frac{\sin \epsilon L/2}{L} e^{iLw_j} \ , \tag{3.85}$$

where $w_j = j\epsilon \approx (j+1/2)\epsilon$ and

$$L = \frac{2\pi n}{\epsilon} + 4M \ln \frac{x}{4M} \ . \tag{3.86}$$

For the $\alpha_{jn,w'}$ coefficients the result is

$$\alpha_{jn,w'} = \frac{-e^{-i(w_j - w')v_H}}{\pi\sqrt{\epsilon}} \sqrt{\frac{w'}{w_j}} \int_0^{+\infty} dx e^{-iw'x} \frac{\sin \epsilon L/2}{L} e^{iLw_j} \ . \tag{3.87}$$

Due to the combination of the oscillating and damping terms the integral

$$I(w') = \int_0^{+\infty} dx e^{-iw'x} \frac{\sin \epsilon L/2}{L} e^{iLw_j} \ , \tag{3.88}$$

appearing in the above expressions ($w' < 0$ for $\beta_{jn,w'}$ and $w' > 0$ for $\alpha_{jn,w'}$) is convergent. See Fig. 3.7. Using a Wick rotation one can find

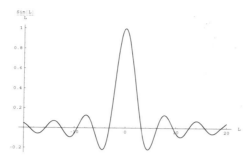

Fig. 3.7 Graphical representation of the function $(\sin \epsilon L/2)/L$, with $\epsilon = 2$.

[Wald (1975)] a crucial relation between $\beta_{jn,w'}$ and $\alpha_{jn,w'}$.

3.3.3 *Wick rotation*

Let us assume that the branch cut of the logarithm function $\ln x$, appearing in the definition of L, lies on the negative real axis (see Fig. 3.8). Therefore the integrand is an analytic function in all complex plane x except in the real negative axis. For $w' > 0$ this permits to rotate the contour of integration to the negative imaginary axis and then set $x = -iy$. Taking into account that $\ln(-iy/4M) = -i\pi/2 + \ln(y/4M)$, we get

$$I(w' > 0) = -i \int_0^{+\infty} dy\, e^{-w'y} \frac{\sin \epsilon L_y/2}{L_y} e^{iL_y w_j} , \qquad (3.89)$$

where

$$\begin{aligned} L_y &= \frac{2\pi n}{\epsilon} + 4M \ln(-\frac{iy}{4M}) \\ &= \frac{2\pi n}{\epsilon} + 4M(-\frac{i\pi}{2} + \ln \frac{y}{4M}) . \end{aligned} \qquad (3.90)$$

Therefore we finally have

$$I(w' > 0) = -i e^{2M\pi w_j} e^{2\pi i n w_j/\epsilon} \int_0^{+\infty} dy\, e^{-w'y} \frac{\sin \epsilon L_y/2}{L_y} e^{i4M \ln(y/4M) w_j} . \qquad (3.91)$$

On the other hand, for $w' < 0$ we can rotate the contour of integration to the positive imaginary axis and define $x = iz$. Now we have $\ln(iz/4M) =$

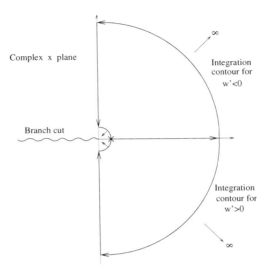

Fig. 3.8 Contour integrals for the evaluation of the Bogolubov coefficients.

$i\pi/2 + \ln(z/4M)$, so we get

$$I(w' < 0) = i \int_0^{+\infty} dz e^{w'z} \frac{\sin \epsilon L_z/2}{L_z} e^{iL_z w_j} , \quad (3.92)$$

where

$$L_z = \frac{2\pi n}{\epsilon} + 4M(\frac{i\pi}{2} + \ln \frac{z}{4M}) . \quad (3.93)$$

The final result is then

$$I(w' < 0) = i e^{-2M\pi w_j} e^{2\pi i n w_j/\epsilon} \int_0^{+\infty} dz e^{w'z} \frac{\sin \epsilon L_z/2}{L_z} e^{i4M \ln(z/4M) w_j} . \quad (3.94)$$

For very narrow wave packets ($\epsilon \ll 1$), centered around w_j, and at late times $n/\epsilon \to +\infty$, we have $\epsilon L_y \approx \epsilon L_z$ and the above two integrals (3.91) and (3.94) are related as follows

$$I(w' > 0) = -e^{4\pi M w_j} I(w' < 0) . \quad (3.95)$$

Therefore

$$\alpha_{jn,w'} = -e^{4\pi M w_j} e^{2iw' v_H} \beta_{jn,w'} . \quad (3.96)$$

This implies the important result

$$|\alpha_{jn,w'}| = e^{4\pi M w_j}|\beta_{jn,w'}|. \tag{3.97}$$

One can also obtain the above result without making explicit use of the wave packets. In terms of pure plane waves with definite frequency one evaluates the coefficients $\beta_{ww'}$ according to a formula analogous with (3.81)

$$\beta_{ww'} = 2i \int_{I^-} dv r^2 d\Omega u_w^{out} \partial_v u_{w'}^{in}, \tag{3.98}$$

and a similar expression for $\alpha_{ww'}$

$$\alpha_{ww'} = -2i \int_{I^-} dv r^2 d\Omega u_w^{out} \partial_v u_{w'}^{in*}. \tag{3.99}$$

Using the late time relation (3.76) one then gets

$$\beta_{ww'} = \frac{-1}{2\pi}\sqrt{\frac{w'}{w}} \int_{-\infty}^{v_H} dv e^{-iw(v_H - 4M \ln \frac{v_H-v}{4M}) - iw'v}, \tag{3.100}$$

and

$$\alpha_{ww'} = \frac{-1}{2\pi}\sqrt{\frac{w'}{w}} \int_{-\infty}^{v_H} dv e^{-iw(v_H - 4M \ln \frac{v_H-v}{4M}) + iw'v}. \tag{3.101}$$

Introducing again the variable $x = v_H - v$ one transforms the above expressions into

$$\beta_{ww'} = \frac{-1}{2\pi}\sqrt{\frac{w'}{w}} (4M)^{-4Mwi} e^{-i(w+w')v_H} \int_0^{+\infty} dx x^{4Mwi} e^{iw'x}, \tag{3.102}$$

and

$$\alpha_{ww'} = \frac{-1}{2\pi}\sqrt{\frac{w'}{w}} (4M)^{-4Mwi} e^{-i(w-w')v_H} \int_0^{+\infty} dx x^{4Mwi} e^{-iw'x}. \tag{3.103}$$

The above integrals do not converge absolutely. It is the additional integration over frequencies in the wave packets that makes the integral convergent. However one can insert into the exponent an infinitesimal negative real part $(-\epsilon)$ to make the integrals to converge. This way one "imitates" the role of the wave packets. Using the identity

$$\int_0^{+\infty} x^a e^{-bx} = b^{-1-a}\Gamma(1+a), \tag{3.104}$$

where $a = 4Mwi$ and $b = \pm w'i + \epsilon$, one finally obtains

$$\beta_{ww'} = \frac{-1}{2\pi}\sqrt{\frac{w'}{w}}(4M)^{-4Mwi}e^{-i(w+w')v_H}(-iw'+\epsilon)^{-1-4Mwi}\Gamma(1+4Mwi) , \qquad (3.105)$$

and

$$\alpha_{ww'} = \frac{-1}{2\pi}\sqrt{\frac{w'}{w}}(4M)^{-4Mwi}e^{-i(w-w')v_H}(iw'+\epsilon)^{-1-4Mwi}\Gamma(1+4Mwi) . \qquad (3.106)$$

Note that $\beta_{ww'} = -i\alpha_{w(-w')}$. Taking into account that

$$\ln(-w' - i\epsilon) = -i\pi + \ln w' , \qquad (3.107)$$

it is now easy to find that

$$\alpha_{ww'} = -e^{4\pi Mw}e^{2iw'v_H}\beta_{ww'} , \qquad (3.108)$$

and

$$|\alpha_{ww'}| = e^{4\pi Mw}|\beta_{ww'}| , \qquad (3.109)$$

in agreement with the previous rigorous derivation (3.97).

3.3.4 Planck spectrum

If one uses the expression (3.78) to evaluate $\langle in|N_w^{out}|in\rangle$ one finds that, due to the factor $(w')^{-1/2}$ in the coefficients $\beta_{ww'}$, it diverges logarithmically. This means that the black hole produces at I^+ an infinite number of particles of frequency w. As we have already noted this type of divergence comes from the use of the continuum-normalization modes u_w^{out}. The quantity $\langle in|N_w^{out}|in\rangle$, which has dimension of time, represents indeed the flux of radiation integrated for all times. In contrast, the number of particles detected at time $u_{out} = 2\pi n/\epsilon \pm \pi/\epsilon$, represented by the dimensionless quantity $\langle in|N_{jn}^{out}|in\rangle$, is *finite* and can be easily calculated from the result (3.97) and the condition (3.26), which in our case reads

$$\int_0^{+\infty} dw'(\alpha_{jn,w'}\alpha^*_{j'n',w'} - \beta_{jn,w'}\beta^*_{j'n',w'}) = \delta_{jj'}\delta_{nn'} . \qquad (3.110)$$

For the particular values $j = j'$, $n = n'$ we have

$$\int_0^{+\infty} dw'(|\alpha_{jn,w'}|^2 - |\beta_{jn,w'}|^2) = 1 . \qquad (3.111)$$

Now taking into account that $|\alpha_{jn,w'}|^2 = e^{8\pi M w_j}|\beta_{jn,w'}|^2$, we can write

$$(e^{8\pi M w_j} - 1)\int_0^{+\infty} dw'|\beta_{jn,w'}|^2 = 1 , \qquad (3.112)$$

and, therefore, at late times $n \to +\infty$

$$\langle in|N_{jn}^{out}|in\rangle = \int_0^{+\infty} dw'|\beta_{jn,w'}|^2 = \frac{1}{e^{8\pi M w_j} - 1} . \qquad (3.113)$$

This expectation value coincides exactly with a *Planck distribution* of thermal radiation for bosons (k_B is Boltzman's constant)

$$\frac{1}{e^{\hbar w_j/k_B T} - 1} \qquad (3.114)$$

with a temperature

$$T_H = \frac{\hbar}{8\pi k_B M} \qquad (3.115)$$

called the *Hawking temperature* of the black hole. For fermions a similar computation can be performed and it also gives a Planck distribution: $\langle in|N_{jn}^{out}|in\rangle = (e^{\hbar w_j/k_B T_H} + 1)^{-1}$, but now with the corresponding Fermi statistic.

This is the famous and very important result obtained in [Hawking (1975)]. We will see later that this result is very robust and, although it has been derived within a very simple model for black hole formation, it remains correct for a generic and complicated process of gravitational collapse ending in a black hole. Putting the corresponding numbers in the expression for the Hawking temperature one finds that $T_H \approx 10^{-7} M_{Sun}/M$ ^0K. For black holes created from gravitational collapse this is indeed a small effect.[5]

It is now worth commenting how this result can also be obtained using the continuum-normalization modes. One has first to evaluate the integral

$$\int_0^{+\infty} dw' \beta_{w_1 w'} \beta_{w_2 w'}^* . \qquad (3.116)$$

Inserting the previous result (3.105) the above integral turns out to be

$$\int_0^{+\infty} dw' \beta_{w_1 w'} \beta_{w_2 w'}^* =$$

---

[5]However, one must not exclude the possibility that in the early universe mini black holes have formed by density fluctuations. For black holes of masses $M \approx 10^{15}$g the Hawking temperature is huge $T_H \approx 10^{11}$ ^0K.

$$\frac{(4M)^{4M(w_2-w_1)i}}{(2\pi)^2\sqrt{w_1 w_2}} e^{-i(w_1-w_2)v_H} i^{-4M(w_1-w_2)i} \Gamma(1+4Mw_1 i)\Gamma(1-4Mw_2 i) \times$$

$$\int_0^{+\infty} \frac{dw'}{w'} e^{-4Mw_1 i \ln(-w'-i\epsilon)} e^{4Mw_2 i \ln(w'-i\epsilon)} \ . \quad (3.118)$$

Using the relation (3.107) and integrating over the variable $y = \ln w'$ the result is

$$\int_0^{+\infty} dw' \beta_{w_1 w'} \beta^*_{w_2 w'} = \frac{1}{8\pi M w_1} |\Gamma(1+4Mw_1 i)|^2 e^{-4\pi M w_1} \delta(w_1 - w_2) \ . \quad (3.119)$$

We immediately get

$$\int_0^{+\infty} dw' \beta_{w_1 w'} \beta^*_{w_2 w'} = \frac{1}{e^{8\pi M w_1} - 1} \delta(w_1 - w_2) \ , \quad (3.120)$$

where the coefficient in front of the delta function $\delta(w_1 - w_2)$ represents the steady thermal flow of radiation of frequency w_1. It is easy to see how we can recover the result (3.113). Introducing wave packets for the Bogolubov coefficients

$$\beta_{jn,w'} = \frac{1}{\sqrt{\epsilon}} \int_{j\epsilon}^{(j+1)\epsilon} dw \, e^{2\pi i w n/\epsilon} \beta_{ww'} \ , \quad (3.121)$$

we have

$$\langle in | N_{jn}^{out} | in \rangle = \int_0^{+\infty} dw' |\beta_{jn,w'}|^2$$

$$= \frac{1}{\epsilon} \int_{j\epsilon}^{(j+1)\epsilon} dw_1 e^{2\pi i w_1 n/\epsilon} \int_{j\epsilon}^{(j+1)\epsilon} dw_2 e^{-2\pi i w_2 n/\epsilon} \int_0^{+\infty} dw' \beta_{w_1 w'} \beta^*_{w_2 w'}$$

$$= \frac{1}{\epsilon} \int_{j\epsilon}^{(j+1)\epsilon} dw_1 \frac{1}{e^{8\pi M w_1} - 1} = \frac{1}{e^{8\pi M w_j} - 1} \ , \quad (3.122)$$

where in the last step we have assumed that the wave packets are sharply peaked around the frequencies w_j.

3.3.5 Uncorrelated thermal radiation

To properly show that the radiation produced by the black hole is *thermal* one has to check that the probabilities of emitting different numbers of particles also agree with those of thermal radiation. For instance, one can compute

$$\langle in|N_{jn}^{out}N_{jn}^{out}|in\rangle = \hbar^{-2}\langle in|a_{jn}^{out\dagger}a_{jn}^{out}a_{jn}^{out\dagger}a_{jn}^{out}|in\rangle \ . \tag{3.123}$$

Using the Bogolubov transformations (3.29) and (3.30) and the commutation relations of the "in" ladder operators, we can rewrite the above expression as

$$\langle in|N_{jn}^{out}N_{jn}^{out}|in\rangle = \int_0^{+\infty} dw'|\beta_{jn,w'}|^2 + 2\left(\int_0^{+\infty} dw'|\beta_{jn,w'}|^2\right)^2$$
$$+ \left|\int_0^{+\infty} dw'\alpha_{jn,w'}\beta_{jn,w'}\right|^2 . \tag{3.124}$$

We already know the first two terms. To evaluate the third one we can make use of (3.121) and also

$$\alpha_{jn,w'} = \frac{1}{\sqrt{\epsilon}} \int_{j\epsilon}^{(j+1)\epsilon} dw e^{2\pi i w n/\epsilon} \alpha_{ww'} \ . \tag{3.125}$$

Then

$$\int_0^{+\infty} dw' \alpha_{jn,w'}\beta_{jn,w'} = \frac{1}{\epsilon} \int_{j\epsilon}^{(j+1)\epsilon} dw_1 e^{2\pi i w_1 n/\epsilon} \int_{j\epsilon}^{(j+1)\epsilon} dw_2 e^{2\pi i w_2 n/\epsilon} \times$$
$$\int_0^{+\infty} dw' \alpha_{w_1 w'}\beta_{w_2 w'} \ . \tag{3.126}$$

The integral

$$\int_0^{+\infty} dw' \alpha_{w_1 w'}\beta_{w_2 w'} \tag{3.127}$$

can be worked out as the previous one (3.116). The result is now proportional to $\delta(w_1 + w_2)$, and since w_1 and w_2 are both positive, this implies that

$$\int_0^{+\infty} dw'\alpha_{jn,w'}\beta_{jn,w'} = 0 \ . \tag{3.128}$$

Therefore, we finally get

$$\langle in|N_{jn}^{out}N_{jn}^{out}|in\rangle = \frac{e^{-8\pi M w_j}(1+e^{-8\pi M w_j})}{(1-e^{-8\pi M w_j})^2} \ , \tag{3.129}$$

which again agrees with a thermal distribution for the expectation value $\langle in|N_{jn}^{out}N_{jn}^{out}|in\rangle$. All higher moments can be calculated in a similar way and coincide with the thermal probability

$$P(N_{jn}) = (1 - e^{-8\pi M w_j})e^{-8\pi N M w_j}, \qquad (3.130)$$

to emit N particles in the mode (jn). For instance,

$$\langle in|N_{jn}^{out} N_{jn}^{out}|in\rangle = \frac{e^{-8\pi M w_j}(1 + e^{-8\pi M w_j})}{(1 - e^{-8\pi M w_j})^2}$$

$$= \sum_{N=0}^{+\infty} N^2 P(N_{jn}) . \qquad (3.131)$$

We can also consider the expectation value $\langle in|N_{jn}^{out} N_{kn}^{out}|in\rangle$ between different modes (jn) and (kn). The result is now

$$\langle in|N_{jn}^{out} N_{kn}^{out}|in\rangle = \left(\int_0^{+\infty} dw'|\beta_{jn,w'}|^2\right)\left(\int_0^{+\infty} dw'|\beta_{kn,w'}|^2\right)$$
$$+ \left|\int_0^{+\infty} dw' \beta_{jn,w'} \beta_{kn,w'}^*\right|^2 + \left|\int_0^{+\infty} dw' \alpha_{jn,w'} \beta_{kn,w'}\right|^2, \qquad (3.132)$$

where, in addition to the Bogolubov transformations and the commutation relations, we have used (3.27). The last term above vanishes due to fact that the integral (3.127) is proportional to $\delta(w_1 + w_2)$. The second one requires to evaluate

$$\int_0^{+\infty} dw' \beta_{jn,w'} \beta_{kn,w'}^* = \frac{1}{\epsilon}\int_{j\epsilon}^{(j+1)\epsilon} dw_1 e^{2\pi i w_1 n/\epsilon} \int_{k\epsilon}^{(k+1)\epsilon} dw_2 e^{-2\pi i w_2 n/\epsilon} \times$$
$$\int_0^{+\infty} dw' \beta_{w_1 w'} \beta_{w_2 w'}^* . \qquad (3.133)$$

Now the integral (3.116) is proportional to $\delta(w_1 - w_2)$ and this together with the fact that $j \neq k$ forces to get a vanishing result. The first term in (3.132) is the only non trivial contribution and corresponds to the product of the expectation values $\langle in|N_{jn}^{out}|in\rangle$ and $\langle in|N_{kn}^{out}|in\rangle$

$$\langle in|N_{jn}^{out} N_{kn}^{out}|in\rangle = \frac{1}{e^{8\pi M w_j} - 1} \frac{1}{e^{8\pi M w_k} - 1}$$
$$= \langle in|N_{jn}^{out}|in\rangle \langle in|N_{kn}^{out}|in\rangle . \qquad (3.134)$$

A similar result is obtained for all higher moments.

This means the complete absence of any *correlation* between different modes, as typically happens in the thermal radiation. Therefore, the quantum state of the late-time radiation at I^+ is exactly described by a *thermal density matrix* with temperature $T_H = \hbar/8\pi k_B M$

$$\rho_{thermal} = \prod_{jn} \sum_{N=0}^{+\infty} P(N_{jn})|N_{jn}\rangle\langle N_{jn}|$$

$$= \prod_{jn}\left(1-e^{-\hbar w_j/k_B T_H}\right)\sum_{N=0}^{+\infty}e^{-N\hbar w_j/k_B T_H}|N_{jn}\rangle\langle N_{jn}|, \quad (3.135)$$

where $|N_{jn}\rangle$ is the state in the Fock space at I^+ with N particles of mode (jn). Any measurement at I^+ is described by $\rho_{thermal}$. For instance $\langle in|N_{jn}^{out}N_{kn}^{out}|in\rangle$ can be reproduced by $\rho_{thermal}$ via the relation

$$\langle in|N_{jn}^{out}N_{kn}^{out}|in\rangle = Tr\{\rho_{thermal}N_{jn}^{out}N_{kn}^{out}\}. \quad (3.136)$$

The fact that the state at I^+ is not a pure state could appear rather surprising, since, in view of (3.39), the pure $|in\rangle$ state is still a pure state when expressed in the "out" Fock space. However, our situation is different since the "out" Fock space is not the Fock space at I^+. We still have an additional sector, corresponding to the modes crossing the event horizon H^+. Since we have ignored this sector we have lost the possible correlations with quanta entering into the black hole, which, when properly taken into account, should restore the purity description of the $|in\rangle$ state. This is a very delicate question that we will consider in the following.

3.3.6 Where are the correlations?

Until now we have ignored the incoming modes at the future horizon H^+. These will enter in the description of the "out" Fock space and cannot be neglected if we want to get a full physical picture of the Hawking effect. In addition to the expansion of the field f in the "in" modes

$$f = \int_0^{+\infty} dw(a_w^{in}u_w^{in} + a_w^{in\dagger}u_w^{in*}), \quad (3.137)$$

we can, alternatively, perform the expansion

$$f = \int_0^{+\infty} dw(a_w^{out}u_w^{out} + a_w^{out\dagger}u_w^{out*} + a_w^{int}u_w^{int} + a_w^{int\dagger}u_w^{int*}), \quad (3.138)$$

where $a_w^{int\dagger}$ and a_w^{int} are creation and annihilation operators of incoming particles at the future horizon. Results of physical measurements at I^+ do not depend on the form of the u_w^{int} modes. However, for our purposes of looking for correlations between the emitted quanta at I^+ and the incoming quanta at H^+ we need to choose a form for u_w^{int}. As we have already stressed there is no natural choice for u_w^{int}, since there is no natural time parameter on H^+. The criterium to follow is simplicity to get the simplest and clearest results showing the presence of physical correlations. Tracing back in time the u_w^{out} modes we found that they have support in the portion of I^- $v < v_H$ ($u_w^{out} \propto \theta(v_H - v)$). This means, obviously, that the emitted quanta at I^+ cannot see the correlations that the $|in\rangle$ state has between points before and after v_H

$$\langle in|f(v < v_H)f(v > v_H)|in\rangle \neq 0 \ . \tag{3.139}$$

To naturally exploit this simple fact one can define [Parker (1975); Wald (1975)] the incoming modes crossing H^+ by reversing the sign of $v_H - v$ in (3.76) and also the sign of w (otherwise the mode is a negative frequency solution, due to the "time reflection" for $v_H - v$)

$$u_w^{int} \equiv -\frac{1}{4\pi\sqrt{w}} \frac{e^{iw(v_H - 4M \ln \frac{v - v_H}{4M})}}{r} \theta(v - v_H). \tag{3.140}$$

A practical advantage of this definition is that the additional Bogolubov coefficients $\gamma_{ww'}$, $\eta_{ww'}$, relating "int" and "in" modes, will have simple relations with the previous ones $\alpha_{ww'}$, $\beta_{ww'}$ (relating the "out" and "in" modes). One has

$$u_w^{out} = \int_0^{+\infty} dw' (\alpha_{ww'} u_{w'}^{in} + \beta_{ww'} u_{w'}^{in*})$$
$$u_w^{int} = \int_0^{+\infty} dw' (\gamma_{ww'} u_{w'}^{in} + \eta_{ww'} u_{w'}^{in*}) \ . \tag{3.141}$$

The additional Bogolubov coefficients are given by

$$\eta_{ww'} = -(u_w^{int}, u_{w'}^{in*}) = 2i \int_{I^-} dv r^2 d\Omega u_w^{int} \partial_v u_{w'}^{in}$$
$$\gamma_{ww'} = (u_w^{int}, u_{w'}^{in}) = -2i \int_{I^-} dv r^2 d\Omega u_w^{int} \partial_v u_{w'}^{in*} \ . \tag{3.142}$$

Using the expression for u_w^{int} we have

$$\eta_{ww'} = \frac{-1}{2\pi} \sqrt{\frac{w'}{w}} \int_{v_H}^{+\infty} dv e^{iw(v_H - 4M \ln \frac{v - v_H}{4M}) - iw'v} \ , \tag{3.143}$$

and

$$\gamma_{ww'} = \frac{-1}{2\pi}\sqrt{\frac{w'}{w}}\int_{v_H}^{+\infty} dv e^{iw(v_H - 4M \ln \frac{v-v_H}{4M}) + iw'v} . \qquad (3.144)$$

Introducing again the variable $x = v - v_H$ one transforms the above expressions into

$$\eta_{ww'} = \frac{-1}{2\pi}\sqrt{\frac{w'}{w}}(4M)^{4Mwi} e^{i(w-w')v_H} \int_0^{+\infty} dx x^{-4Mwi} e^{-iw'x} , \qquad (3.145)$$

and

$$\gamma_{ww'} = \frac{-1}{2\pi}\sqrt{\frac{w'}{w}}(4M)^{4Mwi} e^{i(w+w')v_H} \int_0^{+\infty} dx x^{-4Mwi} e^{iw'x} . \qquad (3.146)$$

By direct comparison of the above expressions with those of Eqs. (3.102) and (3.103) it is easy to find

$$\alpha_{ww'} = e^{2iw'v_H} \gamma_{ww'}^*$$
$$\beta_{ww'} = e^{-2iw'v_H} \eta_{ww'}^* , \qquad (3.147)$$

and taking into account the fundamental relation (see Eq. (3.108)) $\alpha_{ww'} = -e^{4\pi Mw} e^{2iw'v_H} \beta_{ww'}$ we get

$$\beta_{ww'} = -e^{-4\pi Mw} \gamma_{ww'}^* , \qquad (3.148)$$

and

$$\eta_{ww'} = -e^{-4\pi Mw} \alpha_{ww'}^* . \qquad (3.149)$$

The general relation (3.30) takes now the form

$$a_w^{out} = \int_0^{+\infty} dw' (\alpha_{ww'}^* a_{w'}^{in} - \beta_{ww'}^* a_{w'}^{in\dagger})$$
$$a_w^{int} = \int_0^{+\infty} dw' (\gamma_{ww'}^* a_{w'}^{in} - \eta_{ww'}^* a_{w'}^{in\dagger}) , \qquad (3.150)$$

along with the corresponding hermitian relations. The action on the $|in\rangle$ state leads to

$$a_w^{out}|in\rangle = -\int_0^{+\infty} dw' \beta_{ww'}^* a_{w'}^{in\dagger}|in\rangle ,$$
$$a_w^{int}|in\rangle = -\int_0^{+\infty} dw' \eta_{ww'}^* a_{w'}^{in\dagger}|in\rangle , \qquad (3.151)$$

and also
$$a_w^{out\dagger}|in\rangle = \int_0^{+\infty} dw' \alpha_{ww'} a_{w'}^{int\dagger}|in\rangle ,$$
$$a_w^{int\dagger}|in\rangle = \int_0^{+\infty} dw' \gamma_{ww'} a_{w'}^{int\dagger}|in\rangle . \qquad (3.152)$$

Using the relations (3.148) and (3.149) we finally get
$$(a_w^{out} - e^{-4\pi Mw} a_w^{int\dagger})|in\rangle = 0 , \qquad (3.153)$$

$$(a_w^{int} - e^{-4\pi Mw} a_w^{out\dagger})|in\rangle = 0 . \qquad (3.154)$$

We note that these are a particular case of the general equations (3.37) which allow to describe the state $|in\rangle$ in the "out" Fock space. Therefore, the fundamental blocks of the matrix V_{ij} are given, in the continuum-normalization basis we are using, by (up to the factor $\delta(w-w')$)
$$\begin{pmatrix} 0 & e^{-4\pi Mw} \\ e^{-4\pi Mw} & 0 \end{pmatrix} .$$

Using now Eq. (3.39) we get
$$|in\rangle = \langle out|in\rangle \exp\left(\sum_w \hbar^{-1} e^{-4\pi Mw} a_w^{int\dagger} a_w^{out\dagger}\right)|out\rangle$$
$$= \langle out|in\rangle \prod_w \sum_N e^{-4\pi NMw} \frac{1}{N!\hbar^N} (a_w^{int\dagger})^N (a_w^{out\dagger})^N |out\rangle$$
$$= \langle out|in\rangle \prod_w \sum_N e^{-4\pi NMw} |N_w^{out}\rangle \otimes |N_w^{int}\rangle, \qquad (3.155)$$

where $|N_w^{out}\rangle$ and $|N_w^{int}\rangle$ are N-particle states with frequency w at I^+ and H^+, respectively. The amplitude $\langle out|in\rangle$ can be formally obtained by requiring normalization $\langle in|in\rangle = 1$ in the above expression
$$\langle out|in\rangle = \prod_w \sqrt{1 - e^{-8\pi Mw}} . \qquad (3.156)$$

So, finally we obtain
$$|in\rangle = \prod_w \sqrt{1 - e^{-8\pi Mw}} \sum_{N=0}^{+\infty} e^{-4\pi NMw} |N_w^{out}\rangle \otimes |N_w^{int}\rangle . \qquad (3.157)$$

The above expression concerns continuum-normalization states. In terms of finite-normalization modes u_{jn}^{out} and u_{jn}^{int}, where

$$u_{jn}^{int} = \frac{1}{\sqrt{\epsilon}} \int_{j\epsilon}^{(j+1)\epsilon} dw e^{2\pi i w n/\epsilon} u_w^{int} , \qquad (3.158)$$

we would have obtained, at late times in I^+ ($n \to +\infty$)

$$|in\rangle = \prod_{jn} \sqrt{1 - e^{-8\pi M w_j}} \sum_{N=0}^{+\infty} e^{-4\pi N M w_j} |N_{jn}^{out}\rangle \otimes |N_{jn}^{int}\rangle . \qquad (3.159)$$

The physical meaning of this expression is clear. We have an independent emission in frequencies of entangled quantum states representing outgoing and incoming radiation. There are not correlations between particles emitted in different modes. Moreover, late times for the external asymptotic observer ($n \to +\infty$) means "early time" for the horizon states $|N_{jn}^{int}\rangle$, since very large n requires $v - v_H \to 0^+$ in (3.158). Therefore, the correlations ensuring the purity of the "in" vacuum state take place between the (late time) outgoing and (early time) incoming quanta entering the horizon (see Fig. 3.9). The probabilities for occupation of each mode

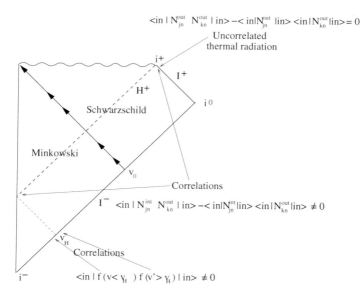

Fig. 3.9 Correlations needed to ensure the purity of the IN vacuum.

are independent and are of the form

$$P(N_{jn}) \equiv |\langle in|(N_{jn}^{out}\rangle \otimes |N_{jn}^{int}\rangle)|^2$$
$$= (1 - e^{-8\pi M w_j})e^{-8\pi N M w_j} , \qquad (3.160)$$

in agreement with the previous thermal result (3.130).

3.3.7 *Thermal density matrix*

Let us see now how we can recover the thermal density matrix (3.135) that describes the physical measurements at I^+. If we are interested in computing expectation values of any operator O at future null infinity $\langle in|O|in\rangle$ we can write

$$\langle in|O|in\rangle = \prod_{j'n'jn} \sqrt{1-e^{-8\pi M w_{j'}}}\sqrt{1-e^{-8\pi M w_j}} \sum_{N'N} e^{-4\pi N' M w_{j'}} e^{-4\pi N M w_j}$$
$$\langle N'^{out}_{j'n'}|O|N^{out}_{jn}\rangle \langle N'^{int}_{j'n'}|N^{int}_{jn}\rangle , \qquad (3.161)$$

and since $\langle N'^{int}_{j'n'}|N^{int}_{jn}\rangle = \delta_{N'N}\delta_{j'j}\delta_{n'n}$ we finally get

$$\langle in|O|in\rangle = \prod_{jn}(1-e^{-8\pi M w_j}) \sum_{N=0}^{+\infty} e^{-8\pi N M w_j} \langle N^{out}_{jn}|O|N^{out}_{jn}\rangle . \qquad (3.162)$$

But this expression can be immediately rewritten as

$$\langle in|O|in\rangle = Tr\{\prod_{jn}(1-e^{-8\pi M w_j}) \sum_{N=0}^{+\infty} e^{-8\pi N M w_j} |N^{out}_{jn}\rangle\langle N^{out}_{jn}|O\}$$
$$= Tr\{\rho_{thermal}O\} . \qquad (3.163)$$

The conclusion is clear. The $|in\rangle$ state can be described as a thermal state, as far as we consider measurements only in the asymptotic region I^+. In other words, by tracing out all quantum degrees of freedom going down the black hole the pure quantum state $|in\rangle$ is described by a thermal density matrix

$$\rho_{thermal} = Tr_{int}|in\rangle\langle in| = \prod_{jn}(1-e^{-8\pi M w_j}) \sum_{N=0}^{+\infty} e^{-8\pi N M w_j}|N^{out}_{jn}\rangle\langle N^{out}_{jn}| . \qquad (3.164)$$

To finish we want to stress here the two crucial calculations leading black holes to emit *exactly* black body radiation:

- The first one is tied to the *divergent redshift* property of the black hole horizon. It is expressed mathematically via the relation $u_{out}(u_{in}) \approx -4M \ln(v_H - u_{in}) + constant$ and leads to the special relation between the Bogolubov coefficients $|\alpha_{jn,w'}| = e^{4\pi M w_j} |\beta_{jn,w'}|$.

- The second important calculation is a generic result of the formalism of the Bogolubov transformations which naturally applies in quantum field theory in curved space. It is the *exponential* appearing in the general expression $|in\rangle = \langle out|in\rangle \exp(\frac{1}{2\hbar} \sum_{ij} V_{ij} a_i^{out\dagger} a_j^{out\dagger})|out\rangle$, together with the above exponential factor $e^{4\pi M w_j}$, which enters into the matrix V_{ij}, responsible for getting just the black body Boltzmann factor $\exp(-N\hbar w_j/k_B T_H)$ for the thermal probability distribution at the Hawking temperature.

We must observe that the first result $|\alpha_{jn,w'}| = e^{4\pi M w_j}|\beta_{jn,w'}|$ is enough to get the Planck spectrum $\langle in|N_{jn}^{out}|in\rangle = (e^{8\pi M w_j}-1)^{-1}$, as was first pointed out in [Hawking (1975)]. Nevertheless, the second result is necessary to show complete agreement with black body thermal emission (this was first pointed out in [Wald (1975); Parker (1975)]).[6]

3.4 Including the Backscattering

Up to now we have assumed that the field f propagates in the Schwarzschild portion of the spacetime as a pure free field, without feeling the potential $V_l(r)$. The aim of this section is to give a complete picture of the radiation process, still in the simplified Vaidya spacetime, including the effects of the *potential barrier* for all angular momenta. By neglecting the effects of the potential one would naively find that each angular momentum component of the field contributes equally to the emitted energy flux (i.e., the *luminosity* L)

$$L_l = \frac{(2l+1)}{2\pi} \int_0^{+\infty} dw \frac{\hbar w}{e^{8\pi M w} - 1} = \frac{(2l+1)\hbar}{768\pi M^2}, \qquad (3.165)$$

---

[6]In principle, there are other probability distributions that agree with $\langle in|N_{jn}^{out}|in\rangle$ but are not thermal.

and therefore by summing over all angular momentum modes one would obtain a divergent result

$$L \equiv \sum_{l=0}^{+\infty} L_l = +\infty \ . \tag{3.166}$$

As we are now going to explain in detail, the effect of the potential (the *backscattering*) is to modify the pure Planckian spectrum by a factor Γ_{wl}

$$\frac{1}{e^{8\pi M w} - 1} \rightarrow \frac{\Gamma_{wl}}{e^{8\pi M w} - 1} \ , \tag{3.167}$$

such that the total luminosity

$$L = \frac{1}{2\pi} \int_0^{+\infty} dw \frac{\hbar w}{e^{8\pi M w} - 1} \sum_{l=0}^{+\infty} (2l+1) \Gamma_{wl} \tag{3.168}$$

is finite. This is so because the potential barrier increases with l ($l = 0$ gives the dominant contribution) and the factor Γ_{wl} decreases with l.

3.4.1 Waves in the Schwarzschild geometry

To start the study of the effects of backscattering on the black hole radiation we need first to discuss the behavior of our massless scalar field in the Schwarzschild geometry. To this end we shall go back to the analysis of the wave equation for the field f started in the previous section. Here, however, for reasons that will be clear later, we shall consider the full static region of the eternal Schwarzschild spacetime ($-\infty < u, v < +\infty$) and not only the physical portion of it ($v > v_0$), outside the collapsing region. Using the expansion of the field in spherical harmonics

$$f(x^\mu) = \sum_{l,m} \frac{f_l(r)}{r} e^{-iwt} Y_{lm}(\theta, \varphi) \ , \tag{3.169}$$

the functions f_l satisfy the ordinary differential equation

$$\left(\frac{d^2}{dr^{*2}} + w^2 - V_l(r) \right) f_l(r) = 0 \ , \tag{3.170}$$

where, we remember, the effective potential has the form $V_l(r) = (1 - 2M/r)[l(l+1)/r^2 + 2M/r^3]$. It describes a single potential barrier, the maximum of which is roughly at $r = 3M$ (see Fig. 3.10).

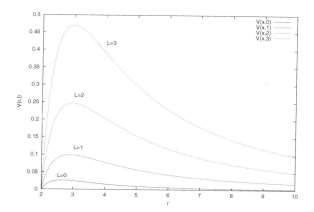

Fig. 3.10 Graphical representation of the potential function $V(r \equiv x, l)$ for the lowest values of l ($M = 1$).

Since the potential vanishes at both spatial infinity ($r^* = +\infty$) and at the event horizon ($r^* = -\infty$), this implies that the two linearly independent solutions to (3.170) behave asymptotically as $e^{\pm i w r^*}$.

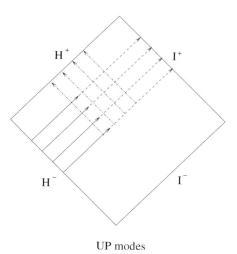

UP modes

Fig. 3.11 Graphical representation of the up modes. Starting from H^-, part of the wave is transmitted to I^+ and part is reflected to H^+.

Let us describe solutions for $f_l(r^*)$ that satisfy these asymptotic conditions and are of special relevance for our purposes. We can define $f_l^{up}(r^*,w)$ satisfying

$$f_l^{up}(r^*,w) \sim e^{iwr^*} + r_l(w)e^{-iwr^*} \quad r^* \to -\infty$$
$$f_l^{up}(r^*,w) \sim t_l(w)e^{iwr^*} \quad r^* \to +\infty , \quad (3.171)$$

where $r_l(w)$ and $t_l(w)$ are reflection and transmission coefficients [DeWitt (1975)] for waves emerging from H^- that are partly transmitted to I^+ and partly reflected to H^+ by the potential barrier.[7] For a graphical description see Fig. 3.11.

Moreover, we can also define another set of solutions which describes waves emanating from I^- and are partly transmitted to H^+ and partly reflected to I^+

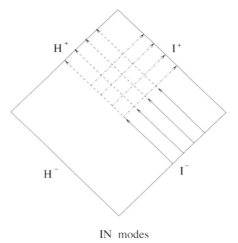

IN modes

Fig. 3.12 Graphical representation of the in modes. Starting from I^-, part of the wave is transmitted to H^+ and part is reflected to I^+.

$$f_l^{in}(r^*,w) \sim e^{-iwr^*} + R_l(w)e^{iwr^*} \quad r^* \to +\infty$$
$$f_l^{in}(r^*,w) \sim T_l(w)e^{-iwr^*} \quad r^* \to -\infty , \quad (3.172)$$

where $R_l(w)$ and $T_l(w)$ are the corresponding reflection and transmission coefficients (see Fig. 3.12). All these coefficients are not independent and

---
[7] For a rotating black hole these coefficients also depend on m.

in addition to the standard scattering relations

$$|t_l(w)|^2 + |r_l(w)|^2 = 1$$
$$|T_l(w)|^2 + |R_l(w)|^2 = 1 \,, \qquad (3.173)$$

they also obey

$$t_l(w) - T_l(w) = 0$$
$$R_l^*(w)t_l(w) + r_l(w)T_l^*(w) = 0 \,. \qquad (3.174)$$

These relations are obtained by evaluation of Wronskians at $r^* = \pm\infty$ (see, for instance, [Frolov and Novikov (1998)]).

3.4.2 Late-time basis to accommodate backscattering

We shall now explain the arguments leading to the correction (3.167) as a consequence of backscattering. The relevant mode basis for the problem in the collapsing spacetime are:

- "in" Fock space:

The "in" basis is naturally constructed, as usual, from the modes u_{wlm}^{in} defined by the condition at I^- [8]

$$u_{wlm}^{in}|_{I^-} \sim \frac{1}{\sqrt{4\pi w}} \frac{e^{-iwv}}{r} Y_{lm}(\theta,\varphi) \,. \qquad (3.175)$$

Therefore we can expand the field f as follows

$$f = \int_0^{+\infty} dw \sum_{lm} (a_{wlm}^{in} u_{wlm}^{in} + a_{wlm}^{in\dagger} u_{wlm}^{in*}) \,. \qquad (3.176)$$

- "out" Fock space:

The natural modes at I^+ are those that behave as

$$u_{wlm}^{out}|_{I^+} \sim \frac{1}{\sqrt{4\pi w}} \frac{e^{-iwu_{out}}}{r} Y_{lm}(\theta,\varphi) \,, \qquad (3.177)$$

---
[8] We maintain the description in terms of continuum-normalization modes for the "in" Fock space.

and vanish at H^+. Note that due to the backscattering of the potential barrier the modes u_{wlm}^{out} are nonvanishing at I^- when propagated backwards in time.

In order to describe particles going down to the black hole one needs to define modes at H^+. As we have already stressed there is no natural choice for them. We can, for instance, define the following modes

$$u_{wlm}^{down}|_{H^+} \sim \frac{1}{\sqrt{4\pi w}} \frac{e^{-iwv}}{r} Y_{lm}(\theta,\varphi) \, , \qquad (3.178)$$

and vanishing at I^+.

Since in the derivation of black hole radiation in the previous section a crucial point was the entire propagation of u_w^{out}, at late times, back to $v = v_H$, it will be convenient to replace u_{wlm}^{out} with the modes that, in a prolonged Schwarzschild eternal black hole, behave at H^- as the u^{out} modes defined in the previous section at $v = v_H$.[9] These modes verify

$$u_{wlm}^{up}|_{H^-} \sim \frac{1}{\sqrt{4\pi w}} \frac{e^{-iwu_{out}}}{r} Y_{lm}(\theta,\varphi) \, , \qquad (3.179)$$

and vanish at I^-. They correspond to the solutions f_l^{up} introduced in the previous subsection and can be written as

$$u_{wlm}^{up} = t_l(w) u_{wlm}^{out} + r_l(w) u_{wlm}^{down} \, . \qquad (3.180)$$

Note that when the potential is neglected $r_l(w) = 0$ "up" and "out" modes coincide.

Moreover, it is also convenient to replace the modes u_{wlm}^{down} with u_{wlm}^{int} considered in the previous section

$$u_{wlm}^{int} \equiv -\frac{1}{\sqrt{4\pi w}} \frac{e^{iw(v_H - 4M \ln \frac{v-v_H}{4M})}}{r} \theta(v-v_H) Y_{lm}(\theta,\varphi). \qquad (3.181)$$

It is important to stress that for our purposes the relevant modes are those close to v_H for which the effective frequency $w_{eff} = i\partial_v u^{int}/u^{int}$ is extremely large, so they are not sensitive to the potential barrier, and can propagate entirely from I^- to H^+, without being reflected to I^+.

---

[9] Due to the fact that the late time modes are highly blueshifted when traced back along the future horizon, the effect of the potential from v_H to H^- in the prolonged Schwarzschild spacetime can be neglected.

Given these modes one can construct the corresponding wave packets modes

$$u_{jnlm}^{up} = \frac{1}{\sqrt{\epsilon}} \int_{j\epsilon}^{(j+1)\epsilon} dw e^{2\pi i w n/\epsilon} u_{wlm}^{up}, \qquad (3.182)$$

and

$$u_{jnlm}^{int} = \frac{1}{\sqrt{\epsilon}} \int_{j\epsilon}^{(j+1)\epsilon} dw e^{2\pi i w n/\epsilon} u_{wlm}^{int}. \qquad (3.183)$$

An appropriate mode basis relevant for describing the late time radiation at I^+ (i.e., $n \to +\infty$) is made out essentially by the modes

$$u_{jnlm}^{up} \qquad u_{jnlm}^{int} \qquad n \gg 1. \qquad (3.184)$$

See Fig. 3.13.

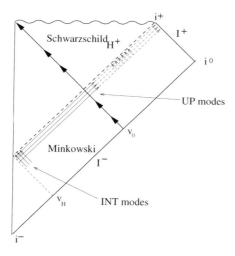

Fig. 3.13 The up modes propagate backward in time freely from v_0 to I^-, but suffer scattering in the propagation to I^+. The int modes, due to their large frequency, propagate freely from I^- to H^+.

The field f admits then the expansion ($n \gg 1$)

$$f = \sum_{jnlm} (a_{jnlm}^{up} u_{jnlm}^{up} + a_{jnlm}^{up\dagger} u_{jnlm}^{up*} + a_{jnlm}^{int} u_{jnlm}^{int} + a_{jnlm}^{int\dagger} u_{jnlm}^{int*} + ...). \qquad (3.185)$$

There are additional modes that enter in the expansion, but we have written explicitly only those relevant for the late time radiation. We have avoided those modes which are scattered by the potential and do not produce any mixing between positive and negative frequencies.

Proceeding now as usual we can define for very large n the corresponding Bogolubov coefficients (for simplicity we shall suppress the l, m indices in the modes, operators and states)

$$\beta_{jn,w'} = -(u_{jn}^{up}, u_{w'}^{in*}) \qquad \alpha_{jn,w'} = (u_{jn}^{up}, u_{w'}^{in}) \qquad (3.186)$$
$$\eta_{jn,w'} = -(u_{jn}^{int}, u_{w'}^{in*}) \qquad \gamma_{jn,w'} = (u_{jn}^{int}, u_{w'}^{in}) \ . \qquad (3.187)$$

We have maintained the same notation as in the case of no-backscattering. In fact, since the late time modes u_{jn}^{up} are a very good approximation to the old modes u_{jn}^{out}, in the absence of backscattering, at $v = v_H$, the evaluation of the above late time Bogolubov coefficients is *exactly the same* as in the computation of previous section. Therefore, the $|in\rangle$ state admits the late time expansion (very large n)

$$|in\rangle = \prod_{jn} \sqrt{1 - e^{-8\pi M w_j}} \sum_{N=0}^{+\infty} e^{-4\pi N M w_j} |N_{jn}^{up}\rangle \otimes |N_{jn}^{int}\rangle \ . \qquad (3.188)$$

The physical interpretation of this expression is as follows. It describes multiple pair creation with one quanta of the pair crossing the horizon, just after its formation, while the other quanta reaches I^+, with amplitude t_l, or is scattered back into the horizon, with amplitude r_l. In the case that the transmission amplitude is unity, u_{jn}^{up} becomes u_{jn}^{out} and $|N_{jn}^{up}\rangle$ turns out to be $|N_{jn}^{out}\rangle$. Then Eq. (3.188) reproduces the result of Eq. (3.159).

3.4.3 Thermal radiation and grey-body factors

To obtain the density matrix describing late time measurements at I^+ we have to trace over the quantum states at the horizon H^+. This can be done in two steps. First, we have trace over the $|N_{jn}^{int}\rangle$ states. This is easy, since the calculation is similar to that given in the previous section. So, to evaluate the expectation value of an operator O at I^+ we have

$$\langle in|O|in\rangle = \prod_{jn}(1 - e^{-8\pi M w_j}) \sum_{N=0}^{+\infty} e^{-8\pi N M w_j} \langle N_{jn}^{up}|O|N_{jn}^{up}\rangle \ . \qquad (3.189)$$

The state $|N_{jn}^{up}\rangle \equiv (N!)^{-1/2}(\hbar^{-1/2}a_{jn}^{up\dagger})^N(|out\rangle \otimes |down\rangle)$, where $|down\rangle$ is the vacuum state associated to the modes (3.178), is of the form

$$|N_{jn}^{up}\rangle = \frac{\hbar^{-N/2}}{\sqrt{N!}}\left(t_l(w_j)a_{jn}^{out\dagger} + r_l(w_j)a_{jn}^{down\dagger}\right)^N(|out\rangle \otimes |down\rangle)$$
$$= \frac{\hbar^{-N/2}}{\sqrt{N!}}\sum_{P=0}^{N}\binom{N}{P}t_l^P(w_j)r_l^{N-P}(w_j)(a_{jn}^{out\dagger})^P|out\rangle(a_{jn}^{down\dagger})^{N-P}|down\rangle \ . \tag{3.190}$$

Since the operator O is insensitive to the form of the states on the horizon we immediately get

$$\langle in|O|in\rangle = \prod_{jn}(1-e^{-8\pi Mw_j})\sum_{N=0}^{+\infty}e^{-8\pi NMw_j}\sum_{P=0}^{N}\frac{1}{N!}\binom{N}{P}^2 P!(N-P)!$$
$$\times |t_l(w_j)|^{2P}|r_l(w_j)|^{2(N-P)}\langle P_{jn}^{out}|O|P_{jn}^{out}\rangle \ . \tag{3.191}$$

Defining the so called *grey-body factor*

$$\Gamma_{wl} \equiv |t_l(w_j)|^2 \ , \tag{3.192}$$

and taking into account that $|r_l(w_j)|^2 = 1 - \Gamma_{wl}$, we can rewrite Eq. (3.191) in the form

$$\langle in|O|in\rangle = \prod_{jn}(1-e^{-8\pi Mw_j})\sum_{P=0}^{+\infty}\sum_{N=P}^{+\infty}\binom{N}{P}\Gamma_{wl}^P(1-\Gamma_{wl})^{(N-P)}$$
$$\times e^{-8\pi NMw_j}\langle P_{jn}^{out}|O|P_{jn}^{out}\rangle \ . \tag{3.193}$$

We can reorganize the sums and write

$$\sum_{P=0}^{+\infty}\sum_{N=P}^{+\infty}\binom{N}{P}\Gamma_{wl}^P(1-\Gamma_{wl})^{(N-P)}e^{-8\pi NMw_j} =$$
$$\sum_{P=0}^{+\infty}(\Gamma_{wl}e^{-8\pi Mw_j})^P\sum_{N=0}^{+\infty}\binom{N+P}{P}\left((1-\Gamma_{wl})e^{-8\pi Mw_j}\right)^N =$$
$$\sum_{P=0}^{+\infty}(\Gamma_{wl}e^{-8\pi Mw_j})^P\frac{1}{(1-(1-\Gamma_{wl})e^{-8\pi Mw_j})^{P+1}} \ . \tag{3.194}$$

Using this result we finally obtain

$$\langle in|O|in\rangle = \prod_{jn}(1-e^{-8\pi Mw_j})\sum_{P=0}^{+\infty}\frac{(\Gamma_{wl}e^{-8\pi Mw_j})^P}{(1-(1-\Gamma_{wl})e^{-8\pi Mw_j})^{P+1}}\langle P_{jn}^{out}|O|P_{jn}^{out}\rangle \ . \tag{3.195}$$

This expectation value can be rewritten in terms of a density matrix $\langle in|O|in\rangle = Tr\{\rho_{grey-body}O\}$, where

$$\rho_{grey-body} = \prod_{jn} \sum_{N=0}^{+\infty} P_{gb}(N_{jn}) |N_{jn}^{out}\rangle \langle N_{jn}^{out}| \, . \qquad (3.196)$$

The probability distribution $P_{gb}(N_{jn})$ is given by

$$P_{gb}(N_{jn}) = \frac{(1 - e^{-8\pi M w_j})}{1 - (1-\Gamma_{wl})e^{-8\pi M w_j}} \frac{(\Gamma_{wl} e^{-8\pi M w_j})^N}{(1 - (1-\Gamma_{wl})e^{-8\pi M w_j})^N} \, . \qquad (3.197)$$

This is just the thermal probability obtained in the previous section (with the assumption $\Gamma_{wl} = 1$) to radiate N particles in the given mode, modulated by the probability Γ_{wl} (i.e., the grey-body factor) that the emitted particle will reach infinity without being scattered back by the black hole potential barrier. The average number of particles can be derived from it

$$\langle in|N_{jn}|in\rangle = \sum_{N=0}^{+\infty} N P_{gb}(N_{jn}) = \frac{\Gamma_{wl}}{e^{8\pi M w_j} - 1} \, , \qquad (3.198)$$

recovering this way the announced correction (3.167) to the pure Planck spectrum.

3.4.4 Estimations for the luminosity

We can now return to our initial discussion and analyse the total luminosity L of a Schwarzschild black hole. At low frequencies $wM \ll 1$ the grey-body factor can be approximated as [Page (1976)]

$$\Gamma_{wl} \approx 16(wM)^{2l+2} \left(\frac{(l!)^3}{(2l)!(2l+1)!} \right)^2 \, . \qquad (3.199)$$

The main contribution is obtained for zero angular momenta $l = 0$

$$\Gamma_{w0} = 16 w^2 M^2 \, . \qquad (3.200)$$

Integration over the frequencies leads to the following approximate expression for the total luminosity[10]

$$L_{l=0} = \frac{1}{2\pi} \int_0^{+\infty} dw \frac{\hbar w 16 w^2 M^2}{e^{8\pi M w} - 1} = \frac{\hbar}{7680\pi M^2} \ . \quad (3.202)$$

The numerical calculation for the s-wave contribution to the luminosity gives [Balbinot et al. (2001)]

$$L_{l=0}^{numerical} \approx \frac{1.62\hbar}{7680\pi M^2} \ . \quad (3.203)$$

To get an estimation of the total luminosity due to all angular momenta we can consider the following approximation. At high frequencies $w \gg M$ the black hole behaves as a black sphere of effective radius $R_{eff} = 3\sqrt{3}M$ [Wald (1984)].[11] Taking into account that the allowed angular momenta for particles captured by the black hole verify $l \leq w R_{eff}$, we can approximate the grey-body factor as $\Gamma_{wl} = \theta(3\sqrt{3}wM - l)$. Using this approximation the sum in l in the expression (3.168) can be performed [DeWitt (1975)] and the result is

$$\sum_{l=0}^{+\infty}(2l+1)\Gamma_{wl} \approx 27\pi M^2 w^2 \ . \quad (3.204)$$

Integration over w gives

$$L \approx \frac{1.69\hbar}{7680\pi M^2} \ . \quad (3.205)$$

One can compare the above estimation with the numerical calculated value [Elster (1983)]

$$L^{numerical} \approx \frac{1.79\hbar}{7680\pi M^2} \ . \quad (3.206)$$

It is very interesting to note that the s-wave luminosity (3.203) is indeed the bulk of the Hawking radiation, as it represents the 90% of the total emitted radiation.

---

[10] One can appreciate the effects of the backscattering by noting that without it we would have obtained the result

$$L_{l=0} = \frac{1}{2\pi} \int_0^{+\infty} dw \frac{\hbar w}{e^{8\pi M w} - 1} = \frac{\hbar}{768\pi M^2} \quad (3.201)$$

which is ten times bigger.

[11] This is a consequence of the geometric optics approximation, which is valid for high frequencies. The factor $3\sqrt{3}$, instead of 2, is a consequence of the light bending effect. We shall explain in detail this approximation in Section 3.6.

3.4.4.1 Emission of massless and massive particles

As shown in Eq. (3.199) the grey-body factor decreases as the angular momenta increases. This also happens for the spin of the field considered. Although our discussion has been restricted to a massless scalar field, due to the universality of gravity, the Hawking derivation applies to any field. The only difference between the emission of different type of particles lies in the corresponding grey-body factors multiplying the universal Planckian thermal spectrum. A detailed investigation was performed in [Page (1976)]. It was found that, for a Schwarzschild black hole with $M >> 10^{17}g$ the luminosity is

$$L \approx (3.5 \times 10^{46}) \left(\frac{M}{1g}\right)^{-2} \frac{erg}{s} , \qquad (3.207)$$

of which 81.4% is in (massless) electron and muon neutrinos, 16.7% in photons and 1.9% in gravitons. Black holes of smaller masses, $5 \times 10^{14}g << M << 10^{17}$ can also emit thermal electron-positron pairs, in a proportion of 45%. In general massive particles of mass m are allowed if $k_B T_H > m$. More details can be found in [Page (1976)].

3.5 Importance of the Backreaction

It is worth to comment that a similar argument to the one leading to the justification of introducing backscattering, namely the necessity to have a finite luminosity when summing over l, can be used to justify the importance of *backreaction* effects of the radiated energy on the background geometry. The total energy radiated by the black hole is given by integrating the luminosity (energy emitted per unit time) over all the radiation time. If the backreaction is ignored, the black hole emits forever, and we get again a divergent result, but this time for the total energy. This is clearly in contradiction with the finite amount of energy contained inside the black hole. This means that one of the primary consequences of backreaction should be to provide a finite result for the total radiated energy, in parallel to the finite value of the luminosity provided by backscattering. This turns out to be a very difficult problem and one can only face it using suitable approximate schemes. This will be the main focus of Chapters 5–6.

3.6 Late-Time Independence on the Details of the Collapse

As we have already remarked the crucial physical feature that enforces the important result of the previous sections (i.e., the late-time thermal radiation produced by a black hole) is the very large redshift suffered by the wave that propagates from the proximity of the horizon to infinity. This is codified in the relation (3.74)

$$u_{out}(u_{in}) = -4M \ln \frac{v_H - u_{in}}{4Mc} , \qquad (3.208)$$

or, in terms of the coordinate v,

$$u_{out}(v) = -4M \ln \frac{v_H - v}{4Mc} , \qquad (3.209)$$

where c is a numerical constant $c = e^{-v_H/4M}$. This expression, as we will see in the following, turns out to be insensitive, up to the unimportant constant c, to the particular process of gravitational collapse producing the black hole. This is why the thermal radiation produced in forming a black hole is a *universal* property and does not depend on the details of the physical collapse.

3.6.1 The example of two shock waves

We shall illustrate this fact by considering a slight modification of the simple process considered in Section 3.3. Instead of having a single incoming shock wave we shall introduce two shock waves and replace (3.46) by

$$L(v) = M_1 \delta(v - v_i) + M_2 \delta(v - v_f) . \qquad (3.210)$$

Our spacetime splits then in three regions (see Fig. 3.14):
• A Minkowski vacuum region for $v < v_i$: "in" region

$$ds^2 = -dv du_{in} + r^2 d\Omega^2 . \qquad (3.211)$$

• An intermediate Schwarzschild region $v_i < v < v_f$ with mass M_1

$$ds^2 = -(1 - \frac{2M_1}{r})dv du + r^2 d\Omega^2 . \qquad (3.212)$$

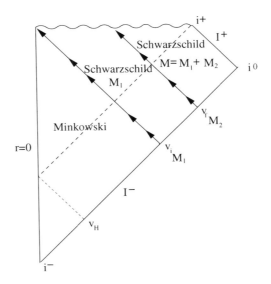

Fig. 3.14 Black hole of mass $M = M_1 + M_2$ formed by the collapse of two shock waves of masses M_1 and M_2.

• The final Schwarzschild black hole configuration $v > v_f$ with mass $M = M_1 + M_2$: "out" region

$$ds^2 = -(1 - \frac{2M}{r})dv du_{out} + r^2 d\Omega^2 \ . \tag{3.213}$$

We can determine the relation $u_{out} = u_{out}(u_{in})$ by matching the corresponding metrics. Continuity of the induced metrics along the lines $v = v_i$ and $v = v_f$ implies

$$r(v_i, u) = r(v_i, u_{in})$$
$$r(v_f, u_{out}) = r(v_f, u) \ . \tag{3.214}$$

The first equation allows to write

$$u = u_{in} - 4M_1 \ln \frac{|v_i - u_{in} - 4M_1|}{4M_1} \ , \tag{3.215}$$

and by differentiation we get

$$\frac{du}{du_{in}} = \frac{r(v_i, u)}{r(v_i, u) - 2M_1} \ , \tag{3.216}$$

and similarly, from the second of Eqs. (3.214) we obtain

$$\frac{du_{out}}{du} = \frac{r(v_f, u) - 2M_1}{r(v_f, u_{out}) - 2M} \; . \tag{3.217}$$

Combining the above two relations we arrive at

$$\frac{du_{out}}{du_{in}} = \frac{du_{out}}{du}\frac{du}{du_{in}} = \frac{r(v_f, u) - 2M_1}{r(v_f, u_{out}) - 2M}\frac{r(v_i, u)}{r(v_i, u) - 2M_1} \; . \tag{3.218}$$

For large u_{out} the value of the u coordinate in the intermediate region approaches the fixed value $u \approx u_H = v_f - 4M - 4M_1 \ln(\frac{M}{M_1} - 1)$. Therefore we can approximate

$$\frac{du_{out}}{du_{in}} \approx \frac{2M - 2M_1}{r(v_f, u_{out}) - 2M}\frac{r(v_i, u_H)}{r(v_i, u_H) - 2M_1} \; . \tag{3.219}$$

Since we also have

$$2M + 2M \ln(\frac{r(v_f, u_{out}) - 2M}{2M}) \approx \frac{v_f - u_{out}}{2} \; , \tag{3.220}$$

combining the above expressions we get

$$\frac{du_{out}}{du_{in}} \approx A e^{u_{out}/4M} \; , \tag{3.221}$$

where A is a numerical constant depending on M_1, M_2, v_i, v_f (i.e., on the details of the collapse)

$$A = \frac{M_2}{M}\frac{r(v_i, u_H)}{r(v_i, u_H) - 2M_1} e^{-(v_f - 4M)/4M} \; . \tag{3.222}$$

Integrating Eq. (3.221) we obtain

$$u_{out} = -4M \ln \frac{A(v_H - u_{in})}{4M} \; , \tag{3.223}$$

where v_H is now an integration constant.[12] We see immediately that this fits the expression (3.208) with $c = 1/A$.

Therefore the late-time results of the previous sections are still valid. The thermal radiation obtained depends only on the total mass of the final black hole, reproducing this way some sort of "no-hair" theorem. To finish the analysis of this example we shall remark the physical aspects underlying

---
[12]Note that for a single shock wave $v_H = v_f - 4M$.

this conclusion. As we have already stressed there is a divergent blueshift, of order

$$e^{(u_{out}-v_H)/4M} , \qquad (3.224)$$

in the wave propagation in the transition from I^+ to $r(v_f, u_{out})$. In addition there also exists a redshift in the propagation from $r(v_f, u_{out})$ to I^- (around $v \approx v_H$) passing through $r(v_i, u_H)$ and $r = 0$. This redshift is, in contrast, smaller than the blueshift. It is of order

$$A e^{v_H/4M} , \qquad (3.225)$$

and the net effect is still a divergent blueshift given by expression (3.221).[13] This is possible because we are in a dynamical situation. The travel of a wave from I^+ to I^- modifies the frequency only if the geometry is non-stationary. This is what we have in the formation of a black hole, and the divergent blueshift

$$\frac{du_{out}}{du_{in}} \sim e^{u_{out}/4M} , \qquad (3.226)$$

responsible for the thermal emission, is not modified by the details of the collapse. We have checked this fact explicitly, in this simple example, by matching the coordinates along the two shock wave lines. However, we can also perform the analysis based on the *geometric optics approximation*. The main advantage of this new point of view, leading to the same result, is that it has general validity and can be applied to a generic collapse.

3.6.2 *Geometric optics approximation*

We shall now reanalyse the example of the two shock waves by examining the law of wave propagation along the different spacetime regions. In the final Schwarzschild region, the wave propagation has already been studied in Section 3.4 and therefore we know that in the travel from I^+ to $(v = v_f, r \approx 2M)$ the wave suffers a very large blueshift. The main point is now to analyse the propagation of the wave from $v = v_f$, $r \approx 2M$ to I^-. The crucial modes u_{wlm}^{up}, when reaching $(v = v_f, r \approx 2M)$, are very well approximated by

$$u_{wlm}^{up} \approx \frac{1}{\sqrt{4\pi w}} \frac{e^{-iwu_{out}}}{r} Y_{lm}(\theta, \varphi) . \qquad (3.227)$$

---

[13] Note that with only one shock wave the finite redshift is trivial and we have only the infinite blueshift.

For an inertial free falling observer the measure of the effective frequency has to be performed with respect to its inertial null time. This null coordinate $\xi_{H^+}^-$ (see Section 2.4) is to a good approximation the Kruskal-type coordinate U defined as $U = -4Me^{-(u_{out}-v_H)/4M}$ (up to terms $O(U^2)$).[14] The presence of the constant v_H is irrelevant. So the approximate form of the wave is

$$u_{wlm}^{up} \approx \frac{1}{\sqrt{4\pi w}} \frac{e^{-iw(v_H - 4M \ln \frac{-U}{4M})}}{r} Y_{lm}(\theta, \varphi) \ . \tag{3.228}$$

The corresponding blueshift is then of order

$$\frac{du_{out}}{dU} = e^{(u_{out}-v_H)/4M} \ . \tag{3.229}$$

Due to this "ultra-high frequency" for $(u_{out} - v_H) \to +\infty$, the wave propagation throughout the remaining region to reach I^- can be performed by geometric optics. We remember that the geometric optics approximation is valid when the effective wavelength of the wave is very short compared to all the other length scales in the problem (see, for instance, [Misner et al. (1973)]), namely the typical radius of curvature of the spacetime and the typical length over which the amplitude of the wave varies. In this approximation, by writing $f = Re[ae^{iS}]$ the propagation of the waves is given, essentially, by the propagation of the phase of the waves. Light rays are defined to be the curves normal to surfaces of constant phase S. The wave vector k^μ is normal to these surfaces ($k_\mu = \partial_\mu S$). The wave equation ($\Box f = 0$) implies, in this approximation, that the wave vector is null

$$k^\mu k_\mu = 0 \ , \tag{3.230}$$

and also that it propagates along null geodesics

$$k^\nu \nabla_\nu k^\mu = 0 \ . \tag{3.231}$$

Moreover, the propagation equation for the wave amplitude a is

$$k^\mu \nabla_\mu a = -\frac{1}{2} (\nabla_\nu k^\nu) a \ . \tag{3.232}$$

The surfaces of constant phase are then governed by the law of propagation of the congruence of null geodesics which generates these surfaces. In our case we are interested in following the time evolution of the surfaces defined by $U = 0$ and $U = -\lambda \approx 0$ from $v = v_f$. The quantity λ is, very

---

[14]Only for the idealized situation $v_f \to -\infty$, which corresponds to H^-, the second order term disappears.

approximately, proportional to the affine distance between a point located at the future horizon, and a nearby point located in the null line γ, both lying in the radially ingoing null geodesic at $v = v_f$ (see Fig. 3.15).

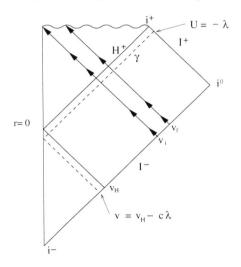

Fig. 3.15 The propagation law of waves in the geometric optics approximation for a Schwarzschild black hole formed by the collapse of two shock waves.

The detailed evolution, backward in time, of the radial geodesics that pass through these points depends on the specific matter that has produced the collapse (in our simple case we have two shock waves). However, the affine distance between both geodesics, which is given by the distance measured by locally inertial observers, remains bounded. This is equivalent to saying that the redshift suffered by the light rays in the propagation from v_f to I^- is always finite. Since at past null infinity the affine distance is given by the difference of the related advanced Eddington–Finkelstein coordinate, the distance between both geodesics there is $v_H - v$, so

$$\lambda = \frac{v_H - v}{c} . \qquad (3.233)$$

In other words, from the vicinity of the horizon at $v = v_f$ till I^- we have a finite redshift

$$\frac{dU}{dv} = \frac{1}{c} . \qquad (3.234)$$

Combining the above results we see how the geometric optics approximation agrees with the result of the previous subsection, obtained via the matching conditions. Then we have an expression for the phase of the wave at I^-

$$ae^{-iw(v_H - 4M \ln \frac{v_H - v}{4Mc})} Y_{lm}(\theta, \varphi) \,, \qquad (3.235)$$

where a is the amplitude for the radial part of the wave. We can also make use of the geometric optics approximation to evaluate the behavior of the amplitude a. It can be shown that from the basic equation (3.232) one can derive a conservation equation for the product $a^2 A$, where A is the cross-sectional area of the bundle of null rays surrounding a fiducial null ray. This can be seen ([Misner et al. (1973)] from the fact that $k^\mu \nabla_\mu A = (\nabla_\nu k^\nu) A$. Since all the geodesics are radial and the spacetime is spherically symmetric the cross-sectional area A is proportional to r^2. This means that the product ar is constant along the radial geodesics where geometric optics can be applied. The initial value of this product is

$$\frac{1}{\sqrt{4\pi w}}, \qquad (3.236)$$

so its final value at I^- should be the same. Combining the results for the phase and the amplitude we can find the explicit form of the wave at I^-

$$u^{up}_{wlm}|_{I^-} \approx \frac{1}{\sqrt{4\pi w}} \frac{e^{-iw(v_H - 4M \ln \frac{v_H - v}{4Mc})}}{r} Y_{lm}(\theta, \varphi) \,. \qquad (3.237)$$

We observe that the wave has not suffered any backscattering, in the way back from $v = v_f$, $r \approx 2M$ to I^-, as expected on the basis of geometric optics. All the backscattering is concentrated in the propagation from I^+ to the vicinity of the horizon in the static Schwarzschild region.

3.6.3 *General collapse*

We shall now discuss the situation of a general collapse. The considerations of the previous subsection make the analysis easy since the two basic ingredients in the propagation of a wave mode from I^+ to I^- can be equally applied in a generic spherically symmetric collapse. The wave propagation in the static Schwarzschild portion of the spacetime remains, obviously, unchanged. Moreover, the wave propagation through the center of the collapsing matter, the bouncing at $r = 0$ and the final return to past null infinity is governed by geometric optics. Therefore, the expressions for the wave propagation derived in the previous subsection are valid. All the dependence of the collapse is concentrated in the constant c which, in turn, produces an irrelevant phase factor in the Bogolubov coefficients. This implies that the basic results obtained for the late time radiation in the previous sections, using an extremely simple Vaidya spacetime, are still valid and represent the core of Hawking's remarkable result (see Fig. 3.16).

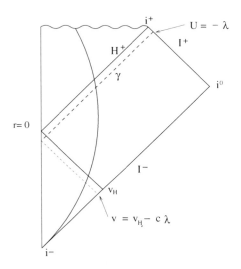

Fig. 3.16 The propagation law of waves in the geometric optics approximation for a Schwarzschild black hole formed by a general collapse.

We just want to note that the basic physical features used to derive the thermal radiation in the Vaidya spacetime play also a fundamental role to ensure the late-time independence of the result from the details of the collapse.

To complete the discussion on the general collapse it is interesting to briefly comment on the modifications to be added if the formed black hole is non-spherical and charged. Both cases are interesting also because we can see how the phenomena of spontaneous superradiance and Schwinger-type pair production are naturally recovered within Hawking's approach.

3.6.4 *Adding angular momentum and charge*

We shall now first describe briefly how the angular momenta of the black hole enters into the description of the Hawking effect. The first novelty is that the expansion of the scalar field in terms of spherical harmonics is no longer valid since we have only rotational symmetry with respect to one axis. Therefore the time and angular dependence of the modes of the wave equation in the Schwarzschild geometry $e^{-iwt}Y_{lm}(\theta,\phi)$ should be replaced by $e^{-iwt}e^{im\phi}$, where $\partial/\partial t$ and $\partial/\partial\phi$ are the asymptotic time translation and rotational Killing vectors. The dependence with respect to the radial and angular variables is governed by the so called Teukolsky equation (see, for instance [Frolov and Novikov (1998)]), which admits separation of variables. The angular equation leads to the spheroidal harmonics labeled by m and an additional index l, analogous to the total angular momenta of spherical harmonics. The main result concerns the behavior of the solutions for the radial equation. The standard relation satisfied by the transmission and reflection amplitudes in the Schwarzschild geometry $|t_l(w)|^2 = 1 - |r_l(w)|^2$, is now replaced by

$$(1 - \frac{m\Omega_H}{w})|t_{lm}(w)|^2 = 1 - |r_{lm}(w)|^2 . \qquad (3.238)$$

This relation shows clearly that the reflection coefficient can be, in modulo, greater than 1. An incoming wave can be amplified as it scatters off a rotating black hole. This effect is called *superradiance* and it happens for frequencies obeying the condition

$$w < m\Omega_H . \qquad (3.239)$$

In particular, the product $m\Omega_H$ should be positive to allow superradiant wave modes.

In addition to this, and since the event horizon is now a Killing horizon with respect to the Killing field $\partial/\partial t + \Omega_H \partial/\partial\phi$, one should replace w by $w - m\Omega_H$ in the Planck factor. This together with the substitution $|t_l(w)|^2$ by $(1 - m\Omega_H/w)|t_{lm}(w)|^2$, leads to the final result for the mean value of the particle number operator

$$\langle in|N_{jnlm}|in\rangle = (1 - \frac{m\Omega_H}{w_j})|t_{lm}(w_j)|^2 \frac{1}{e^{2\pi\kappa^{-1}(w_j - m\Omega_H)} - 1} , \qquad (3.240)$$

where κ is the surface gravity of the rotating black hole. Note that Hawking's temperature is still given by

$$T_H = \frac{\hbar\kappa}{2\pi k_B} , \qquad (3.241)$$

and that, as already stressed, the Planckian exponent is modified by the term $m\Omega_H$ which plays the role of a chemical potential. For very large black holes $M \to +\infty$ ($\kappa \to 0$) the pure thermal radiation is negligible and one finds that

$$\langle in|N_{jnlm}|in\rangle \to \frac{m\Omega_H - w_j}{w_j}|t_{lm}(w_j)|^2 > 0 \quad \text{for } w_j < m\Omega_H \quad (3.242)$$

$$\langle in|N_{jnlm}|in\rangle \to 0 \quad \text{for } w_j > m\Omega_H. \quad (3.243)$$

In this case the above result turns out to be the spontaneous quantum superradiant emission found by [Starobinsky (1973); Unruh (1974)]. It takes place even for eternal stationary configurations, without the requirement that the black hole has been formed by a gravitational collapse. The fundamental effect lies in the behavior of the wave amplitude and not in the frequency, as for the pure Hawking's effect. Note finally that particles are emitted preferably with angular momenta of the the same sign as the black hole itself, thus producing a net loss of the black hole angular momenta.

A similar analysis can be performed in the charged case. By considering the propagation of a charged field in a Reissner–Nordström black hole one finds

$$\langle in|N_{jnlm}|in\rangle = (1 - \frac{q\Phi_H}{w_j})|t_{lm}(w_j)|^2 \frac{1}{e^{2\pi\kappa^{-1}(w_j - q\Phi_H)} - 1}, \quad (3.244)$$

where q is the charge of the field and Φ_H is the electric potential at the horizon. The Hawking temperature is again given by the formula (3.241), where κ is now the surface gravity of the Reissner–Nordström black hole. The chemical potential-type term is represented by $q\Phi_H$ and implies that particles with the same sign of the charge as the black hole are preferentially emitted. For large black holes ($\kappa \to 0$) the above expression vanishes for $w > q\Phi_H$ but, for frequencies $w < q\Phi_H$ it reduces to a Schwinger-type pair production rate [Gibbons (1975)].

3.7 Black Hole Thermodynamics

One of the major arguments supporting, and giving confidence, to the surprising phenomena of thermal black hole radiance is that it allows to recover a complete agreement between the physical behavior of black holes and the laws of thermodynamics. As seen in Section 2.9, before Hawking's discovery of the thermal emission it was pointed out a close formal analogy between the classical laws of black holes mechanics and the laws of thermodynamics. The first signal of it was the "area law" theorem, followed by a series of theorems mimicking the zeroth and first laws of thermodynamics. This analogy was rather intriguing but the major obstacle to believe on it was that the thermodynamic temperature of a classical black hole is absolute zero. It absorbs particles but, according to the classical theory, it cannot emit. However, after Hawking's discovery that a black hole also emits particles, just as a black body at the temperature $T = \kappa/2\pi$, the formal analogy becomes physical. Quantum mechanics was a necessary and fundamental input to jump from the pure mathematical analogy to the physical interpretation. In fact, reestablishing all the physical constants, the relation between temperature and surface gravity requires Planck's constant

$$T_H = \frac{\hbar}{ck_B} \frac{\kappa}{2\pi} . \qquad (3.245)$$

According to our discussion of Section 2.9 this relation between temperature and surface gravity implies the following exact relation between entropy and area of a black hole

$$S_{BH} = \frac{k_B c^3}{G\hbar} \frac{A}{4} . \qquad (3.246)$$

Although Hawking's calculation on particle production of a black hole provides a clear physical understanding of the relation (3.245), the Eq. (3.246), called the Bekenstein–Hawking formula, is more elusive. A proper understanding of the entropy formula should require a direct derivation of it, based on first principles. The most direct way for this is by counting the (quantum) physical degrees of freedom associated to a black hole. In contrast to the explanation of Eq. (3.245), which only requires "quantum field theory on a curved background", the understanding of the Bekenstein–Hawking area law seems to need a complete quantum theory of gravity. We are still far from having such a theory, although considerable progress has been achieved in the last decades.

Despite this limitation one can provide strong physical reasons in favor of $A/4$ as the physical entropy of a black hole. The generalized second law proposed by Bekenstein is the key ingredient. As we have already stressed in Section 2.9, the ordinary second law of thermodynamics is violated if a black hole is present. Moreover, the area theorem goes into troubles if quantum effects are taken into account. The particle emission of a black hole must produce, by energy conservation, a decrease of the event horizon area. If a generalized entropy as

$$S' = S + \frac{k_B c^3}{G\hbar} \frac{A}{4} , \qquad (3.247)$$

where S represents the ordinary entropy of matter outside black holes, verifies the condition

$$\delta S' \geq 0 , \qquad (3.248)$$

as first suggested by Bekenstein, the identification of $A/4$ as the actual physical entropy of a black hole should be accepted. To investigate the validity of the above generalized second law a series of *gedanken experiments* have been considered in the literature. It seems that there is a general agreement on the validity of the generalized second law (see in this respect [Wald (1994)]). With this in mind we should then interpret the laws of black hole thermodynamics as just the laws of thermodynamics applied to ordinary matter in the presence of black holes. The Bekenstein–Hawking formula yields a huge amount of entropy. The quotient S_{BH}/k_B, which could represent the logarithm of the density of states, is proportional to the quotient of the area of the event horizon A and the Planck's area l_P^2

$$\frac{S_{BH}}{k_B} = \frac{A}{4l_P^2} . \qquad (3.249)$$

Since $l_P^2 \approx 10^{-66} cm^2$, the result for a black hole of a solar mass is enormous

$$\frac{S_{BH}}{k_B} \approx 10^{77} . \qquad (3.250)$$

It is much bigger than the ordinary entropy of the star that produces, by gravitational collapse, the black hole.[15] The fundamental problem is whether this entropy can be obtained via a statistical-mechanics derivation, as in ordinary physical systems. The black hole entropy would then correspond to the density of microstates consistent with the macrostate of the black holes, which, due to the no-hair theorem, are characterized by mass, charge, and angular momentum. Within superstring theory it has been possible to reproduce the Bekenstein–Hawking formula for extremal and near-extremal charged black holes [Strominger and Vafa (1996)].[16] Also in the Ashtekar's approach to quantum general relativity important results have been obtained [Rovelli (1996); Ashtekar *et al.* (1998)].[17] Another related and intriguing question is why the Bekenstein–Hawking formula is so

---

[15] For instance, the estimated entropy of the Sun is around $S_{Sun}/k_B \approx 10^{58}$.

[16] The string theory point of view also implies that the process of black hole formation and evaporation can be understood in terms of quantum decoherence [Myers (1997); Amati (2003)].

[17] Ambiguities related with the Immirzi parameter [Immirzi (1997)] have been discussed in [Garay and Mena-Marugán (2003)].

universal. Irrespective of the character of the black hole and of the field content of the theory, the area law formula applies. The microscopic explanation of it should be such that it applies equally to all possible types of black holes. This universality property, which is also shared by other physical features of black holes, like the universal thermal radiation, is rather puzzling.[18] This is why much of the recent research in this direction is based on fundamental symmetry principles which are expected to be shared by any theory of quantum gravity [Carlip (2000)].

We shall no longer insist on this issue. However it can serve to motivate the appearance of another puzzling consequence of black hole radiation, the so called *information loss paradox*, that we shall consider in the next section.

3.8 Physical Implications of Black Hole Radiance and the Information Loss Paradox

3.8.1 *Black hole evaporation*

The main feature (and also the main limitation) of the derivation of the Hawking effect is that it is based on the fixed background approximation. This is not in agreement with energy conservation, since the energy radiated by the black hole should be balanced by a corresponding decrease of its mass, just at the same rate at which energy is radiated out. This implies therefore a correction to the initial background geometry. However, for macroscopic black holes the temperature is very small $T_H \approx 10^{-7}(M_{sun}/M)^0 K$ and one can make the plausible assumption that the process of evaporation is quasistatic, being described very accurately by "thermal radiation" with temperature depending on the mass as $1/M$. Using the results described previously in Subsection 3.4.4 for the luminosity of a black hole one can estimate the lifetime of a (nonrotating) black hole by integrating the equation

$$\frac{dM}{dt} = -\beta \frac{m_P^3}{t_P} \frac{1}{M^2} , \qquad (3.251)$$

where the dimensionless constant β is of order 10^{-5}. This leads to the evolution law $M(t) = (M_0^3 - 3\beta \frac{m_P^3}{t_P} t)^{1/3}$, where M_0 is the initial mass. To

---
[18] Attempts to explain the origin of the Bekenstein–Hawking entropy have led to the idea of *holography* ['t Hooft (1993); Susskind (1995)].

get complete evaporation the process needs a large amount of time

$$\Delta t = \frac{t_P}{3\beta}\left(\frac{M_0}{m_P}\right)^3. \tag{3.252}$$

The above argument cannot be trusted when the black hole reaches the Planck mass $m_P \approx 2.2 \times 10^{-5} g$. However, since the rate of radiation grows as the mass of the black hole decreases, for most of the time Δt the black hole mass is much bigger than the Planck mass. Therefore, the above estimation is likely to provide the correct order of magnitude. For a black hole of solar mass or higher the lifetime exceeds, unfortunately, the age of the universe. The upper mass bound of a black hole that has undertaken a complete evaporation up to now is $M_0 \approx 5 \times 10^{14} g$. This (primordial) black hole cannot be formed by gravitational collapse, but rather it should be created in the early stages of the universe.

We should also note that the process of black hole evaporation, as described above, leads to a wild violation of the baryon number. Most of the energy radiated by a large black hole carries zero baryon number since $k_B T_H$ is much less than the mass of any baryon, until the end stages of the evaporation.

3.8.2 Breakdown of quantum predictability

There is an important, and perhaps dramatic, physical implication in the process of black hole formation and subsequent evaporation. This was first pointed out in [Hawking (1976)], and it is the so called *information loss paradox*. To explain it let us first simplify the scenario and ignore quantum effects. This means that we just consider the formation of a black hole from a gravitational collapse. The mere existence of an event horizon, and due to the no-hair theorem, prevents an external observer at infinity to know the full details of the star from which the black hole has been formed. The observer at future infinity cannot reconstruct all the data identifying the star. The horizon necessarily hides the data of matter that has crossed it and fell into the black hole. Part of the data has been radiated out before the black hole has settled down, but part remains inside the event horizon. An illustrative example is also in the Oppenheimer–Snyder model discussed in Subsection 2.1.1. For the exterior observer only one of the two parameters, R_0 and χ_0, characterizing the collapsed ball of dust is known, namely the mass $M = (R_0/2)\sin^3 \chi_0$.

However, and in general, the external observer can always argue that the remaining information or data is not lost, since it still lies inside the black hole. The only point is that it is not fully accessible in the external region. Although the exterior observer looses the possibility of reconstructing the past (breakdown of "postdictability"), there is no problem at all for predicting measurements at future infinity from the data at past infinity. In fact, the laws of classical physics allow to know, in principle, the emitted classical radiation at future infinity. Notice that a breakdown of classical predictability could happen if "naked singularities" were allowed. Then the classical evolution equations cannot determine what an observer can measure if a singularity is visible. The "cosmic censorship hypothesis" prevents, via the existence of the event horizon, this breakdown of classical predictability for observers outside the black holes.

Things get more complicated when quantum effects are considered. We know from the quantum uncertainty principle that we cannot know simultaneously the position and the momenta of a particle, in particular for those particles that are emitted by the collapsing body. As it is well known this implies a fundamental limitation for predicting physical measurements, but the laws of quantum mechanics tell us how to actually deal with physical systems. The evolution of a physical system, represented by a quantum state $|in\rangle$ in an initial Cauchy surface Σ_i in Minkowski space, is given by a unitary operator that maps it into a final state $|f\rangle$, at a final Cauchy surface Σ_f. Writing the initial state in terms of a fixed basis $|in\rangle = \sum_j c_j^{in} |\psi_j\rangle$ one can determine exactly the complex coefficients c_k^f of the final state $|f\rangle = \sum_k c_k^f |\psi_k\rangle$, out of the initial coefficients c_j^{in}. In this way we recover *predictability*, although in a different sense to that of classical theory.

However, things can get more complicated when the causal structure of spacetime is different to that of Minkowski space and a black hole is present. This is so because the Cauchy surface Σ_f can split into two surfaces $\Sigma_f = \Sigma_{int} \bigcup \Sigma_{ext}$, of which the first lies inside the black hole (we can take, as a limiting case, the event horizon as Σ_{int}) and the second one lies in the exterior region (we can also take future null infinity as Σ_{ext}). This is depicted in Fig. 3.17. We can repeat the above description, an initial state $|in\rangle$ evolves to a final state $|f\rangle$ in a unitary way. However, when physical measurements are restricted to the exterior of the black hole, and this is necessarily so for the outside observer, a new phenomena emerges.

We already know from the analysis of the previous sections that if $|in\rangle$ is the vacuum for an observer at past null infinity, this state evolves or equivalently, in the Heisenberg's picture, it can be regarded as a flux of

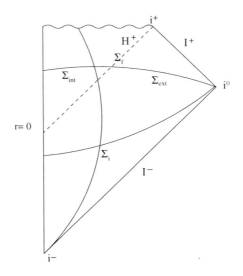

Fig. 3.17 Initial and final Cauchy surfaces in the Penrose diagram of a black hole.

pairs of entangled particles, one particle falling into the black hole and the other emitted to future null infinity. The outgoing flux forms, at late times, thermal radiation of uncorrelated particles. The final quantum state for the exterior observer cannot any longer be described as a (pure) state of the form $|f\rangle = \sum_k c_k^f |\psi_k\rangle$. We cannot know the value of the coefficients c_k^f. We can only know the probabilities for finding the states $|\psi_k\rangle$. In qualitative terms, we only know the probability distribution $P_{gb}(N_{jnlm})$ for detecting N particles in the state defined by the mode u_{jnlm}^{out}. We cannot define a pure final state for the outside region. The reason for this new limitation is entirely qualitative. It is the causal spacetime structure characteristic of a black hole the ultimate responsible of this effective *breakdown of predictability*.

To be clear, let us repeat the conventional argument in a generic way. If we expand a generic state $|f\rangle$ in the form

$$|f\rangle = \sum_{i,j} c_{i,j} |\psi_i\rangle_{int} \otimes |\psi_j\rangle_{ext} \qquad (3.253)$$

where $|\psi_j\rangle_{ext}$ and $|\psi_i\rangle_{int}$ are orthonormal basis in the external and internal region, a local measurement in the external region associated with an operator O will be of the form

$$\langle f|O|f\rangle = \sum_{i,j,i',j'} \langle\psi_{j'}|_{int}\langle\psi_{i'}|_{ext} c^*_{i',j'} O c_{i,j}|\psi_i\rangle_{ext}\psi_j\rangle_{int}$$

$$= \sum_{i,j,i',j'} c^*_{i',j'} c_{i,j} \langle\psi_{j'}|\psi_j\rangle_{int}\langle\psi_{i'}|O|\psi_i\rangle_{ext}$$

$$= \sum_{i,j,i'} c^*_{i',j} c_{i,j} \langle\psi_{i'}|O|\psi_i\rangle_{ext} = \sum_{i,j,i'} Tr\{c^*_{i',j} c_{i,j}|\psi_i\rangle\langle\psi_{i'}|_{ext} O\}$$

$$= Tr\{\rho O\} \qquad (3.254)$$

where we have introduced the density matrix ρ

$$\rho = \sum_{i,j,i'} c^*_{i',j} c_{i,j} |\psi_i\rangle\langle\psi_{i'}|_{ext} , \qquad (3.255)$$

obtained from $|f\rangle\langle f|$ by tracing over all the internal states. It is straightforward to see that the density matrix (3.255) is independent of the orthonormal basis $\{|\psi_j\rangle_{int}\}$ used for the internal region. Every unitary transformation relating two orthonormal basis for the internal region will leave unchanged ρ.

Let us now apply these considerations to our physical situation. The exterior observer, being confined to its own external region, will then describe his local measurements, as it has been shown in previous sections, in terms of the density matrix

$$\rho_{grey-body} = \prod_{jnlm} \sum_{N=0}^{+\infty} P_{gb}(N_{jnlm})|N^{out}_{jnlm}\rangle\langle N^{out}_{jnlm}| \qquad (3.256)$$

obtained by tracing over all the internal states. This reflects the conclusions of the above discussion in a very extreme way. The final state of the exterior observer is not only a mixed state, it is also made out of stochastic thermal radiation, which is the quickest way to loose quantum coherence.

In summary, as far as the external observer is concerned, the initial *pure vacuum state* $|in\rangle$ has evolved into a *mixed state* $\rho_{grey-body}$. For the external observer the correlations between the internal and external regions are lost due to the tracing operation, although the correlations do actually exist. From the exact knowledge of the density matrix (3.255) one cannot indeed reconstruct uniquely the initial state (adding internal states to $|f\rangle$ will produce the same density matrix) and we have lost postdiction, as already happens in the classical theory. Moreover, the initial state does not determine uniquely a final pure state for the external observer. Many possible final states are allowed, as it is expressed by the expansion of the density matrix. The quantum coherence in the expansion (3.253) is lost.

Despite the above considerations one can always claim that, although the information is effectively lost for the external observer, the quantum unitary evolution is preserved at a fundamental level. The exterior region is only part of the full system and correlations between the internal and external regions do exist. A "superobserver" covering the exterior region and the event horizon will describe indeed the final state as a pure state. Although the physical possibility of having such an observer is problematic by itself one can argue that the back hole will necessarily evaporate until, at least, reaching the Planck scale. When $M \approx m_P$ we do not know what happens but a natural possibility is that the black hole evaporates completely leaving nothing else but Hawking radiation (see Fig. 3.18).

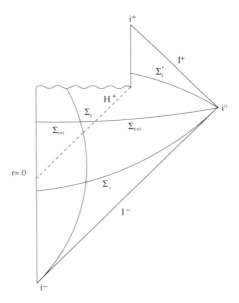

Fig. 3.18 Penrose diagram of a spacetime in which a black hole forms and completely evaporates.

This radiation is at first approximation thermal, although backreaction effects will make it deviate from thermality. Therefore the final state in the surface Σ'_f is necessarily a mixed state, even at a fundamental level. We cannot predict univocally the final quantum state of the system (i.e., that of the Hawking radiation). We can only assign probabilities $P(\chi_i)$ to possible final states $|\chi_i\rangle$ as given by a density matrix ρ, where $P(\chi_i)$ are the eigenvalues of ρ and $|\chi_i\rangle\langle\chi_i|$ the corresponding eigenstates [19]

$$\rho = \sum_i P(\chi_i)|\chi_i\rangle\langle\chi_i| \ . \tag{3.257}$$

This implies a fundamental *breakdown of quantum-mechanical predictability*. Another way of expressing this fact is saying that the transformation $|in\rangle \to \rho$ *is not unitary.* The correlations between the exterior and the interior region, that could have preserved purity (and they do in the fixed background approximation), are necessarily lost in the singularity when the black hole ceases to exist.

The above discussion is called the *information loss paradox* posed by macroscopic black holes to quantum mechanics. It is a paradox in the sense that on the basis of the principles of quantum mechanics and general relativity we are led to a conclusion that contradicts one of the fundamental rules of quantum mechanics itself: the unitary evolution of quantum states. This poses a major challenging problem to the interplay between quantum mechanics and general relativity, a very important aspect of fundamental physics. The resolution of this paradox is indeed so controversial that there is not a definitive answer. We shall now discuss the different possibilities that have emerged. The first one is the most radical viewpoint, advocated by Hawking in 1976, and it is just the acceptation of the breakdown of quantum predictability explained above. There are other alternatives aiming to avoid the apparent loss of quantum coherence.[20]

3.8.3 *Alternatives to restore quantum predictability*

a) The black hole disappears completely and radiation is sufficiently correlated to determine a pure state.

This possibility, strongly advocated in [Page (1980); 't Hooft (1990)] assumes that the standard semiclassical calculation of black hole radiance is not accurate enough to infer the evolution of an evaporating black hole even before reaching the Planck scale. During the time the black hole has radiated out most of its mass, the deviation from the uncorrelated thermal

---

[19]This decomposition is not unique in the case of degeneracy.
[20]For a review see [Page (1993); Giddings (1994); Thorlacius (1994); Strominger (1995)].

radiation obtained in the fixed background approximation could be such that it produces subtle correlations between quanta emitted at early and late times. Since the evaporation of macroscopic black holes is slow these correlations could be enough to ensure the purity of the final state. This is the most pragmatic possibility, since it is indeed how an ordinary hot body gets cooled down to an almost vanishing temperature. The emission of photons is initially thermal, and the purity of the state of the full system is ensured by correlations between photons and the atoms of the body. But these correlations are slowly transported to the photons emitted at late times. So, at the end we get strongly correlated radiation that can be described by a unique pure state.

Where is the trouble with this possibility? We can go back and reexamine the expansion (3.253). We can always choose the Cauchy surface $\Sigma_f = \Sigma_{int} \times \Sigma_{ext}$ to stay in regions of low curvature, to maintain confidence in the causal structure provided by the metric with backreaction corrections. In the proximities of curvature singularities, instead, one could expect that large quantum gravity corrections will appear. As we have already argued, ρ describes, in general, a non-pure state. The purity of the radiation is lost, unless the decomposition (3.253) factors out as

$$|f\rangle = |\chi\rangle_{int} \otimes |\chi\rangle_{ext} \qquad (3.258)$$

for some internal and external states $|\chi\rangle_{int}, |\chi\rangle_{ext}$. In this case ρ is just $|\chi\rangle\langle\chi|_{ext}$. Observe that the condition for purity $Tr\rho^2 = Tr\rho$ is automatically satisfied. If this is the case, the internal and external region are completely uncorrelated. If $O_1(x_{int})$ and $O_2(x_{ext})$ are a pair operators involving internal and external points, the correlation functions vanish immediately

$$\langle f|O_1(x_{int})O_2(x_{ext})|f\rangle - \langle f|O_1(x_{int})|f\rangle\langle f|O(x_{ext})|f\rangle = 0 \ . \qquad (3.259)$$

This possibility is however very exotic since the superposition principle for the possible initial states $|in\rangle$ is not consistent with the splitting of $|f\rangle$ as a tensor product of a pure state in the internal region and a pure state outside. If we consider a superposition of initial states $\sum_j \alpha_j |in_j\rangle$, they will evolve as

$$\sum_j \alpha_j |in_j\rangle \rightarrow \sum_j \alpha_j (|\chi_j\rangle_{int} \otimes |\chi_j\rangle_{ext}) \ , \qquad (3.260)$$

and will inevitably produce non-trivial correlations. The only way out is that all of the internal states $|\chi_j\rangle_{int}$ are actually the same state. So the internal region must be in a *unique* quantum state, unable to see all the information about the collapsing body, to permit the existence of a well defined pure state in the exterior region. It is very hard to admit the possibility of the above exotic situation. We know that, for a free falling observer, the event horizon is not a "singular" place that could cause such a "quantum bleaching" of the information about the initial state. Therefore we should conclude that the information falling into the black hole is lost forever.

There is also another strong reason leading to the same conclusion. The event horizon cannot be thought of as being a mere potential barrier preventing the transmission of energy (and information) from inside to outside. If this were the case, quantum tunneling could allow in principle the transfer of correlations. The event horizon, at least before reaching the Planck size, can maintain its classical meaning as a limiting surface to allow causal propagation. So, the possibility of radiating away all the correlations between the interior and exterior regions should cause a violation of macroscopic causality .

b) The correlations are maintained with a black hole "remnant" of Planck size, or radiated away at the final stages of the evaporation.

The difficulties of possibility a) are weaker when the black hole reaches the Planck mass. There the semiclassical approximation is expected to break down as well as the description in terms of the metric and the causal structure of spacetime, due to large quantum fluctuations. So when the black hole is of Planck size new possibilities can emerge, although they are necessarily more speculative since they require Planck-scale physics and quantum gravity. Nevertheless one can use "common sense" arguments to discuss these possibilities.

The easiest way to avoid information loss is to assume that Hawking radiation stops, leaving a stable remnant of Planck mass (see, for instance the reviews [Giddings (1994); Banks (1995)]). The remnant should store almost all the information of the initial state, since the information has not been radiated away previously, as stressed in the discussion of possibility a). Therefore the remnant should admit an infinite number of quantum states, all with mass of order m_P, to encode the information of the initial state (corresponding to the infinite possibilities to form the same black hole). But then we face a sharp disagreement with the statistical interpretation of the Bekenstein–Hawking entropy. If $e^{S(M)/k_B}$ measures the number of internal black hole states, a black hole of Planck size (i.e., the remnant) should have a few number of internal states. There is also a fundamental problem related to the existence of an infinite number of stable remnant species. Pair-production of these remnants can be generated in a non-stationary gravitational field, as the one given by the collapse of a large black hole. Although the probability of producing a given species of remnant is very small, due to the infinite degeneracy the total probability is then divergent.

A variant to the existence of stable remnants is that they cannot be eternal, but they will slowly evaporate. This opens the possibility of radiating the correlations in the final stages of the evaporation. However this alternative also encounters the difficulties of remnants. The black hole has a very large entropy to be radiated in the last stages of the evaporation, when it has only a small amount of energy. The radiation should be, at last, in the form of massless particles of long wavelength. This requires an extremely long period of evaporation, of order M^4 in Planck units. This is equivalent to have long-lived remnants, with the same problems as the stable remnants.

c) The correlations are maintained though a "baby universe".

A natural way to avoid that information is destroyed in the singularity is offered if quantum gravity effects prevent its existence. The singularity can be then replaced by a non-singular, closed "baby universe", causally disconnected to the external region of the black hole [Hawking (1988)]. The main idea is that the "baby universe" has enough room to contain the correlations with the Hawking radiation even after complete evaporation. Purity can be maintained for the whole system, made out of the external region and the baby universe, in parallel to the situation described in the fixed background approximation. The restriction of the initial pure state to either the baby universe or the external region produces a density matrix. For an observer in the exterior region the information is lost.

d) The information is encoded in "quantum hair".

At the classical level there are a few possibilities for hair [Bekenstein (1996)], and there are additional possibilities if the black hole is regarded as a quantum object. However, this does not seem to be an efficient way for storing all information (see [Wilczek (1998)] for a discussion).

e) None of the above possibilities. Either information loss occurs or more radical proposals are needed to avoid it.

In view of the difficulties with the above alternatives it may be quite possible that either the information is irremediably lost in the process of black hole formation and evaporation or more radical proposals are required. These opposites views can be exposed by the opinions of two leading authors:

- [Wald (1994)] "This evolution from a pure state to a mixed state is commonly regarded as corresponding to a serious breakdown of quantum theory. I do not share this view ... this phenomenon is entirely attributable to the failure of the "final" hypersurface, Σ'_f,[21] to be a Cauchy surface, so the "final state" does not provide a complete description of the "full state" of the field ... . It should be emphasized that there is not corresponding "breakdown" in any of the local laws governing the behavior of the quantum field."

Within this view the loss of quantum coherence is due to the disappearance of the infalling quanta into the black hole singularity, and not to the breakdown of ordinary evolution laws of physics. Nevertheless it suggests that the loss of quantum coherence of black holes could be spread out to other parts of physics, via "virtual" black holes contributions. In fact, Hawking proposed [Hawking (1982)] that the laws of physics reflect such a breakdown of quantum predictability. He suggested to replace the unitary evolution operator $U = e^{-iHt/\hbar}$ by an evolution operator \$ acting linearly on density matrices

$$\rho \to \$\rho \ . \tag{3.261}$$

The usual evolution equation for ρ is then modified by new non-unitary terms $\dot\rho = -i[H, \rho] + \dots$. However, the conflict with ordinary physics could be then even greater since a nonunitary evolution implies violations of local energy conservation [Banks et al. (1984)].

- [Preskill (1992)] " ... As I have pondered this puzzle, it has come to seem less and less likely to me that the accepted principles of quantum mechanics and relativity can be reconciled with the phenomena of black hole evaporation. In other words, I have come to believe more and more (only 15 years after Hawking) that the

---
[21]See Fig. 3.18.

accepted principles lead to a truly paradoxical conclusion ... Conceivably, the puzzle of black hole evaporation portends a scientific revolution as sweeping as that that led to the formulation of quantum theory in the early 20th century."

Along with the philosophy expressed by the above opinions a somewhat radical proposal has been put forward to bypass the strong arguments against information retrieval. It was pointed out by ['t Hooft (1985); Jacobson (1991)] that due to the infinite redshift at the horizon, the derivation of the Hawking radiation involves modes which have arbitrarily large *transplanckian* frequency there with respect to a free falling observer. Therefore one could expect a breakdown of the fixed background approximation due to large gravitational interactions. This fact could lead potentially to large deviations from the usual semiclassical intuition.[22] This partially motivated the introduction of the so called *black hole complementarity principle* ['t Hooft (1990); Susskind et al. (1993); Stephens et al. (1994); Susskind and Uglum (1995)], which we are going to explain briefly.

The logic of the argument to rule out that information comes out with the Hawking radiation lies in the partition of the late time Cauchy surface Σ_f into two portions, Σ_{int} and Σ_{ext}. The first one lies in the interior region of the black hole and the second one outside it. Both surfaces can be chosen to be in regions of weak curvature. Since the points of Σ_f are spacelike separated, and this is true in particular for points $x_{int} \in \Sigma_{int}$ and $x_{ext} \in \Sigma_{ext}$, the Fock space of states on the surface Σ_f factorizes into a tensor product [23]

$$F(\Sigma_f) = F(\Sigma_{int}) \otimes F(\Sigma_{ext}) . \qquad (3.262)$$

The impossibility of replicating quantum information in two separate sets of commuting degrees of freedom implies the loss of quantum coherence. If the information of the infalling quanta is coded in the Fock space $F(\Sigma_{int})$ it cannot be, simultaneously, recorded in the Fock space $F(\Sigma_{ext})$, and the other way around, otherwise a violation of the superposition principle would take place, as we have already explained earlier.

The black hole complementarity principle states that simultaneous physical measurements, inside and outside the black hole, cannot be performed,

---

[22]It must be mentioned, however, that as we will see in Chapter 5 the stress tensor containing the Hawking radiation at infinity is finite on the horizon despite the presence of these large frequencies.
[23]The one-particle Hilbert spaces factorize as $H(\Sigma_f) = H(\Sigma_{int}) \oplus H(\Sigma_{ext})$.

in analogy with the uncertainty principle for the position and momenta of a particle in ordinary quantum mechanics. This means in practice that the statement (3.262) no longer applies and invalidates the above argument to rule out the possibility of radiating away all the information of the initial quantum state. This situation has a resemblance with the "shocking" historical situation involving the description of gravitational collapse. From the point of view of a free falling observer the collapsing body crosses the Schwarzschild radius in a finite proper time, but from the point of view of an external asymptotic observer the collapse takes place in an infinite amount of time. These two descriptions are apparently in contradiction, but when we go further into the problem we realize that they are indeed complementary. Although there are some indications [Schoutens et al. (1993); Kiem et al. (1995)] that this could be also the case for the descriptions of the evaporation process itself, further investigations are required.

A new approach to the problem is offered by Maldacena's AdS/CFT correspondence, which conjectures a holographic equivalence between a theory containing gravity in Anti-de Sitter space and a conformal field theory living in its boundary at infinity [Aharony et al. (2000)]. One can either consider a black hole in AdS space and study its evolution in terms of the states of the boundary unitary CFT [Lowe and Thorlacius (1999)] or, in the other way around, by considering the black hole in the boundary and describing the effects of the evaporation in terms of classical physics in AdS [Tanaka (2003); Emparan et al. (2002)].

Recently, a new, but still preliminary, proposal has been advocated by [Hawking (2004)]. It involves quantum gravity arguments in the so called Euclidean approach, which means working with positive definite metrics and go back to the Lorentzian signature, by Wick rotation, at the end. Non-trivial topologies in the Euclidean regime correspond to Lorentzian configurations with horizons. It is claimed that, at least for an asymptotically Anti-de Sitter space, only topologically trivial metrics matter in evaluating the amplitude connecting initial and final states, implying that the evolution is unitary. The role of the classical horizon is diminished, as in the complementarity principle, due to the quantum uncertainty. Only the values of the fields at the asymptotic regions are actual observables.

These far reaching proposals are expected to contribute significantly towards the resolution of the information loss paradox. Nevertheless, it will also be crucially important to understand the relation of these approaches with the semiclassical approximation,[24] in order to have a full and intuitive picture of the black hole evaporation process.

---

[24] More concretely, in limit where one has a huge number N of matter fields so that one can "safely" neglect the gravitational fluctuations. This is the approach we shall take in Chapter 6.

Chapter 4

Near-Horizon Approximation and Conformal Symmetry

In this chapter we shall clarify many aspects of the Hawking radiation effect by considering mainly flat space physics. This way of looking at the problem lies in the heart of the equivalence principle and proves to be very useful in the understanding of black hole radiation. A uniformly accelerated observer in Minkowski space cannot distinguish the force producing his non-inertial trajectory from a fictitious gravitational field. Moreover, this observer experiences the presence of a horizon in the same way as an exterior observer in the Schwarzschild geometry. Therefore a natural question arises: does the fictitious gravitational field, and the associated (observer-dependent) horizon, produce particles in analogy with the Hawking effect? The answer is positive and the associated phenomena [Fulling (1973); Davies (1975); Unruh (1976)] is usually called the *Unruh effect*. We shall explain the above analogy in the context of the so called *near-horizon approximation*. This is a way to understand better, and from new viewpoints, the Hawking effect.

We shall first describe the more standard (static) picture of Rindler space arising from the near-horizon approximation of the Schwarzschild metric and the emergence of the *infinite-dimensional conformal symmetry* of matter fields in the vicinity of the horizon. Then we shall analyse the physical consequences in dealing with a dynamical situation mimicking the gravitational collapse. We shall show in detail how the Hawking radiation and its properties can be understood in terms of the transformation law properties of the correlation functions of matter under spacetime conformal transformations. The Unruh effect will be explained in this context. We shall also consider extremal and near-extremal black holes in the vicinity of the horizon. This way the Anti-de Sitter space will emerge naturally in our considerations.

Finally, we shall illustrate all these features using a simple toy model which reproduces the properties of black hole radiance in flat spacetime. This is the so called *moving-mirror* analogy [Fulling and Davies (1976); Davies and Fulling (1977)].

4.1 Rindler Space

The introduction of the Rindler geometry can be motivated, from the physical point of view, from the important observation that the $(t-r)$-part of the Schwarzschild metric[1]

$$ds^2_{(4)} = -(1 - \frac{2M}{r})dt^2 + \frac{dr^2}{(1 - \frac{2M}{r})} + r^2 d\Omega^2 \qquad (4.1)$$

is well approximated, around the horizon, by flat space. Introducing the coordinate $0 < x << 2M$ defined by

$$r = 2M + \frac{x^2}{8M}, \qquad (4.2)$$

and expanding the Schwarzschild metric around $2M$

$$1 - \frac{2M}{r} = \frac{x^2}{16M^2} + O((\frac{x}{M})^4) \qquad (4.3)$$

one gets

$$ds^2_{(4)} \sim -(\kappa x)^2 dt^2 + dx^2 + (2M)^2 d\Omega^2, \qquad (4.4)$$

where the constant κ coincides with the surface gravity of the Schwarzschild metric ($\kappa = 1/4M$, see later for a justification). The first two terms of the above metric define a two-dimensional flat spacetime known as *Rindler space* [Rindler (1966); Rindler (2001)]. The last term describes a two-sphere with radius $2M$. It is easy to see that the apparent singularity at $x = 0$ can be eliminated introducing the null Minkowskian coordinates (U, V)

$$U = -xe^{-\kappa t}$$
$$V = +xe^{\kappa t}, \qquad (4.5)$$

with respect to which the (two-dimensional) Rindler metric is[2]

$$ds^2 = -dU dV. \qquad (4.6)$$

---
[1] We introduce the subscrip (4) to refer to a four-dimensional metric.
[2] One can introduce two extra flat coordinates and construct the four-dimensional Rindler space. These two additional coordinates are not relevant for our considerations.

Therefore, Rindler space is just the right wedge $x > 0$ ($V > 0, U < 0$) of Minkowski space. The left wedge ($V < 0, U > 0$) of Minkowski space is also isomorphic to it. This is represented in Fig. 4.1.

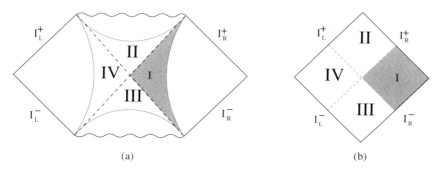

Fig. 4.1 (a) Near-horizon region of the Schwarzschild black hole. The shaded region I is mapped to the corresponding Rindler wedge I of the full Minkowski space in (b).

It is convenient to introduce now a new coordinate ξ defined by the relation $x = \kappa^{-1} e^{\kappa \xi}$. The metric in the (t, ξ) coordinates reads

$$ds^2 = e^{2\kappa\xi}(-dt^2 + d\xi^2) \ . \tag{4.7}$$

The associated null coordinates $v \equiv t + \xi$, $u \equiv t - \xi$ are related to (U, V) by

$$U = -\kappa^{-1} e^{-\kappa u}$$
$$V = +\kappa^{-1} e^{\kappa v} \ , \tag{4.8}$$

for the right wedge, and there

$$ds^2 = -e^{\kappa(v-u)} du\, dv \ . \tag{4.9}$$

For the left wedge the relation is $U = +\kappa^{-1} e^{-\kappa u}$, $V = -\kappa^{-1} e^{\kappa v}$ leading again to Eq. (4.9).

It is easy to see that the metric (4.9) corresponds to the radial part of (4.1) expressed in null Eddington–Finkelstein coordinates $t \pm r^*$ in the limit $r \to 2M$. The coordinate ξ plays the role of the tortoise coordinate r^* in the near-horizon limit

$$\xi = 2M \ln \frac{x^2}{16M^2} = 2M \ln \frac{r - 2M}{2M} \sim r^* \ . \tag{4.10}$$

Moreover, the coordinates (U, V) are essentially the near-horizon limit of the null Kruskal coordinates, regular at the horizon, of the Schwarzschild geometry (this is why we use the same notation):

$$-UV = 8M(r - 2M)e^{r/2M} \sim 8Me(r - 2M) = ex^2,$$
$$U/V = -e^{-t/2M} . \qquad (4.11)$$

The latter formulae should not be a surprise since, as we have shown in Chapter 2, the Kruskal coordinates of the Schwarzschild geometry are naturally associated with the locally inertial coordinates at the horizons. In the near-horizon approximation they turn out to be the global inertial coordinates of the resulting flat geometry.

The physical meaning of the Rindler coordinates (t, x) can be deduced from the relations

$$T \equiv \frac{V + U}{2} = x \sinh \kappa t ,$$
$$X \equiv \frac{V - U}{2} = x \cosh \kappa t , \qquad (4.12)$$

where T and X are the two-dimensional Minkowski coordinates. For fixed x the above hyperbolic curve $(-T^2 + X^2 = x^2)$ defines the trajectory of a uniformly accelerated observer, with proper acceleration $a \equiv (a^\mu a_\mu)^{1/2} = 1/x$, where $a^\mu = \frac{d}{d\tau} u^\mu$. This is depicted in Fig. 4.2.

The acceleration of the observer at fixed x (or ξ) is what is required to counterbalance the gravitational field and keep the observer at constant r (or r^*). This is the manifestation of the equivalence principle. Clearly this is possible only for $x > 0$, because when $x = 0$, i.e., on the horizon, the proper acceleration becomes divergent. Nevertheless, the acceleration, as measured by an observer whose proper time is "t" (this corresponds to the asymptotic observer at infinity in the Schwarzschild geometry), $\bar{a}^\mu = \frac{d}{dt} u^\mu$, is not divergent and the quantity $\bar{a} \equiv (\bar{a}^\mu \bar{a}_\mu)^{1/2}$ is constant and coincides with the surface gravity of the Schwarzschild black hole: $\bar{a} = \kappa$.

4.2 Conformal Symmetry, Stress Tensor and Particle Number

We can also analyze the near-horizon behavior of the four-dimensional matter scalar field $f(x^\mu)$ considered in the previous chapter for the Hawking effect. As already stressed, since the Schwarzschild background is spheri-

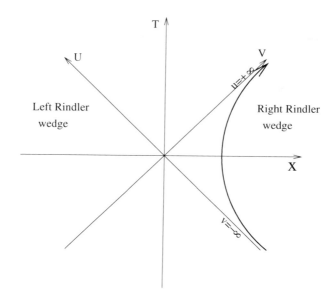

Fig. 4.2 Hyperbolic trajectory of a uniformly accelerated observer.

cally symmetric we can expand the field in spherical harmonics

$$f(x^\mu) = \sum_{l,m} \frac{f_l(t,r)}{r} Y_{lm}(\theta, \varphi) \ . \tag{4.13}$$

The four-dimensional Klein–Gordon equation for $f(x^\mu)$ is then converted into a two-dimensional wave equation, for each angular momentum component $f_l(t,r)$,

$$\left(-\frac{\partial^2}{\partial t^2} + \frac{\partial^2}{\partial r^{*2}} - V_l(r)\right) f_l(t,r) = 0 \ , \tag{4.14}$$

with the potential $V_l(r) = \left(1 - \frac{2M}{r}\right)\left[l(l+1)/r^2 + 2M/r^3\right]$. In the near-horizon limit $r \to 2M$ ($r^* \to -\infty$) the potential vanishes and Eq. (4.14) becomes the two-dimensional free wave equation

$$\left(-\frac{\partial^2}{\partial t^2} + \frac{\partial^2}{\partial \xi^2}\right) f_l(t,\xi) = 0 \ , \tag{4.15}$$

which in null Rindler coordinates (u, v) reads:[3]

$$\partial_u \partial_v f_l = 0 . \qquad (4.16)$$

From now on we shall omit the index l referring to the angular momenta. Our considerations will be valid for each l wave component.[4]

Equation (4.16) is very important since it exhibits the emergence of the two-dimensional conformal invariance[5] of the matter field in the vicinity of the horizon. Under arbitrary spacetime transformations maintaining the conformal form of the metric $ds^2 = -e^{2\rho} dx^+ dx^-$

$$x^\pm \to y^\pm = y^\pm(x^\pm) \qquad (4.17)$$

we have

$$e^{2\rho} dx^+ dx^- \to e^{2\rho} \frac{dx^+}{dy^+} \frac{dx^-}{dy^-} dy^+ dy^- \qquad (4.18)$$

and the wave equation

$$\partial_+ \partial_- f = 0 \qquad (4.19)$$

is kept invariant.[6] At the quantum level the implications of this symmetry have far reaching consequences. We shall now analyze this point in detail. Canonical quantization can be performed immediately expanding the field into positive and negative frequency orthonormal modes

$$f = \sum_i \left(\vec{a}_i u_i(x^-) + \vec{a}^\dagger_i u_i^*(x^-) + \overleftarrow{a}_i v_i(x^+) + \overleftarrow{a}^\dagger_i v_i^*(x^+) \right) . \qquad (4.20)$$

The scalar product for the left-moving sector is

$$(f_1(x^+), f_2(x^+)) = -i \int dx^+ (f_1 \partial_+ f_2^* - f_2^* \partial_+ f_1) , \qquad (4.21)$$

---

[3]In terms of the Rindler coordinates the potential behaves as $V(r) \approx Ae^{2\kappa\xi} \equiv Ae^{\kappa(v-u)}$ and only in the limit $v - u \to -\infty$ one recovers the above free wave equation.

[4]However, as we already noted in the previous chapter, the backscattering effects in the propagation from $r = 2M$ to $r = +\infty$ select the s wave component as the responsible for the main contribution to the Hawking effect.

[5]For a detailed account of this symmetry see, for instance, [Alvarez-Gaumé et al. (1989); Ginsparg (1990); Di Francesco et al. (1997)].

[6]The corresponding action functional leading to the wave equation is also kept invariant and therefore the conformal transformations are indeed symmetries in the strict sense.

and a similar expression for the right-moving sector. The modes $u_i(x^-)$ and $v_i(x^+)$ have positive frequency with respect to the timelike coordinate $(x^+ + x^-)/2$ and represent outgoing and ingoing waves, respectively. They are wave packet modes, constructed from appropriate superposition of plane wave modes $(4\pi w)^{-1/2} e^{-iwx^-}$, $(4\pi w)^{-1/2} e^{-iwx^+}$ (see Subsection 3.3.2). One can construct the Fock space from the non-vanishing commutation relations

$$[\vec{a}_i, \vec{a}_j^\dagger] = \hbar \delta_{ij} ,\qquad (4.22)$$

$$[\overleftarrow{a}_i, \overleftarrow{a}_j^\dagger] = \hbar \delta_{ij} .\qquad (4.23)$$

The vacuum state $|0_x\rangle$ is defined by

$$\vec{a}_i |0_x\rangle = 0, \quad \overleftarrow{a}_i |0_x\rangle = 0 ,\qquad (4.24)$$

and the excited states can be obtained by the application of creation operators $\vec{a}_i^\dagger, \overleftarrow{a}_i^\dagger$ out of the vacuum.

As usual in the quantum theory, the generators of spacetime symmetries are constructed in terms of the stress tensor operator. In null coordinates the only nonvanishing components of the classical stress tensor of the planar theory emerging in the near-horizon approximation are $T_{\pm\pm}$. The component T_{+-} vanishes identically. Our aim is to analyse the quantum counterparts of these two null components of the stress tensor. To this end we shall first analyse the basic correlation functions of the planar scalar field f. It is also convenient, for this purpose, to expand the field in terms of pure plane wave basis

$$f(t,r) = \int_0^\infty \frac{dw}{\sqrt{4\pi w}} (\vec{a}_w e^{-iwx^-} + \overleftarrow{a}_w e^{-iwx^+} + \vec{a}_w^\dagger e^{iwx^-} + \overleftarrow{a}_w^\dagger e^{iwx^+}) .$$
$$(4.25)$$

Using the commutation relations $[\vec{a}_w, \vec{a}_{w'}^\dagger] = \hbar \delta(w-w')$, $[\overleftarrow{a}_w, \overleftarrow{a}_{w'}^\dagger] = \hbar \delta(w-w')$ we easily get

$$\langle 0_x | f(x) f(x') | 0_x \rangle = \hbar \int_0^\infty \frac{dw}{4\pi w} \left(e^{-iw(x^- - x'^-)} + e^{-iw(x^+ - x'^+)} \right) . \qquad (4.26)$$

This two-point function is infrared divergent. This can be cured by introducing a small cut-off λ for the frequency and replace Eq. (4.26) by

$$\langle 0_x | f(x) f(x') | 0_x \rangle = \hbar \int_\lambda^\infty \frac{dw}{4\pi w} \left(e^{-iw(x^- - x'^-)} + e^{-iw(x^+ - x'^+)} \right) . \qquad (4.27)$$

We get

$$\langle 0_x|f(x)f(x')|0_x\rangle = -\frac{\hbar}{4\pi}\left(2\gamma + \ln \lambda^2|(x^+ - x'^+)(x^- - x'^-)|\right), \quad (4.28)$$

where γ is the Euler constant. The ambiguity inherent to the cut-off disappears when one considers, instead of the two-point function for the field f, the correlations for the fields $\partial_\pm f \equiv \partial_{x^\pm} f(x^\pm)$. We have then

$$\langle 0_x|\partial_\pm f(x^\pm)\partial_\pm f(x'^\pm)|0_x\rangle = -\frac{\hbar}{4\pi}\frac{1}{(x^\pm - x'^\pm)^2}, \quad (4.29)$$

and

$$\langle 0_x|\partial_+ f(x^+)\partial_- f(x'^-)|0_x\rangle = 0. \quad (4.30)$$

We open a brief parenthesis to note that this result serves to illustrate the concept of the "primary field" in Conformal Field Theory (see, for instance [Di Francesco et al. (1997)]). It is a field Φ that, under conformal transformations $x^\pm \to y^\pm = y^\pm(x^\pm)$, transforms according to

$$\Phi(x^+,x^-) \to \left(\frac{dy^+}{dx^+}\right)^{h^+}\left(\frac{dy^-}{dx^-}\right)^{h^-}\Phi(y^+(x^+),y^-(x^-)), \quad (4.31)$$

where h^+, h^- are real numbers called conformal weights. Moreover, the correlation functions between primary fields $\Phi_j, j = 1,...,n$, should satisfy the following rule:

$$\langle 0_x|\Phi_1(x_1)\cdots\Phi_n(x_n)|0_x\rangle = \left|\frac{dy^+}{dx^+}\right|^{h_1^+}_{x^+=x_1^+}\left|\frac{dy^-}{dx^-}\right|^{h_1^-}_{x^-=x_1^-}\cdots$$
$$\left|\frac{dy^+}{dx^+}\right|^{h_n^+}_{x^+=x_n^+}\left|\frac{dy^-}{dx^-}\right|^{h_n^-}_{x^-=x_n^-}\langle 0_x|\Phi_1(y_1)\cdots\Phi_n(y_n)|0_x\rangle, \quad (4.32)$$

where $h_i^\pm$ are the conformal weights of the field Φ_i. In our case the fields $\partial_+ f (\partial_- f)$ are primary with conformal weights $h^+ = 1$, $h^- = 0$ ($h^+ = 0$, $h^- = 1$). We have then, under conformal transformations,

$$\langle 0_x|\partial_\pm f(y^\pm)\partial_\pm f(y'^\pm)|0_x\rangle = -\frac{\hbar}{4\pi}\frac{\frac{dx^\pm}{dy^\pm}(y^\pm)\frac{dx^\pm}{dy^\pm}(y'^\pm)}{(x^\pm(y^\pm) - x'^\pm(y^\pm))^2}. \quad (4.33)$$

We should note here that the field f is not primary, indeed the corresponding correlation functions do not fit the above general expression. In contrast, the fields $\partial_\pm f$ have good transformation properties under conformal change of coordinates.

4.2.1 The normal ordered stress "tensor" operator

A quantum operator for the stress tensor can be constructed by taking normal ordering of the two-point function (4.29)

$$: \partial_\pm f(x^\pm) \partial_\pm f(x'^\pm) := \partial_\pm f(x^\pm) \partial_\pm f(x'^\pm) - \langle 0_x | \partial_\pm f(x^\pm) \partial_\pm f(x'^\pm) | 0_x \rangle$$
$$= \partial_\pm f(x^\pm) \partial_\pm f(x'^\pm) + \frac{\hbar}{4\pi} \frac{1}{(x^\pm - x'^\pm)^2}, \qquad (4.34)$$

and then the coincidence-point limit

$$: T_{\pm\pm}(x^\pm) := \lim_{x'^\pm \to x^\pm} \partial_\pm f(x^\pm) \partial_\pm f(x'^\pm) + \frac{\hbar}{4\pi} \frac{1}{(x^\pm - x'^\pm)^2}. \qquad (4.35)$$

The crucial point arises when we analyse what happens when one performs an arbitrary conformal transformation

$$x^\pm \to y^\pm = y^\pm(x^\pm). \qquad (4.36)$$

In terms of the new coordinates $y^\pm$ one can consider the alternative mode expansion

$$f = \sum_j \left(\vec{b}_j \tilde{u}_j(y^-) + \vec{b}_j^\dagger \tilde{u}_j^*(y^-) + \overleftarrow{b}_j \tilde{v}_j(y^+) + \overleftarrow{b}_j^\dagger \tilde{v}_j^*(y^+) \right), \qquad (4.37)$$

where the modes $\tilde{u}_j(y^-)$ and $\tilde{v}_j(y^+)$ are wave packets constructed from superposition of plane wave modes $(4\pi w)^{-1/2} e^{-iwy^-}$, $(4\pi w)^{-1/2} e^{-iwy^+}$. The corresponding quantization, with the time choice $(y^+ + y^-)/2$, leads to the new vacuum state $|0_y\rangle$ and to an alternative definition of normal ordering. The quantum normal ordered operator is then defined as

$$: T_{\pm\pm}(y^\pm) := \lim_{y'^\pm \to y^\pm} \partial_\pm f(y^\pm) \partial_\pm f(y'^\pm) - \langle 0_y | \partial_\pm f(y^\pm) \partial_\pm f(y'^\pm) | 0_y \rangle$$
$$= \lim_{y'^\pm \to y^\pm} \partial_\pm f(y^\pm) \partial_\pm f(y'^\pm) + \frac{\hbar}{4\pi} \frac{1}{(y^\pm - y'^\pm)^2}. \qquad (4.38)$$

We shall now see how to relate $: T_{\pm\pm}(y^\pm) :$ with $: T_{\pm\pm}(x^\pm) :$. Taking into account that

$$: \partial_\pm f(y^\pm) \partial_\pm f(y'^\pm) := \frac{dx^\pm}{dy^\pm}(y^\pm) \frac{dx^\pm}{dy^\pm}(y'^\pm) \partial_\pm f(x^\pm) \partial_\pm f(x'^\pm)$$
$$+ \frac{\hbar}{4\pi} \frac{1}{(y^\pm - y'^\pm)^2}, \qquad (4.39)$$

we easily get

$$: T_{\pm\pm}(y^\pm) := \left(\frac{dx^\pm}{dy^\pm}(y^\pm)\right)^2 : T_{\pm\pm}(x^\pm) :$$
$$- \frac{\hbar}{4\pi}\left(\lim_{y'^\pm \to y^\pm} \frac{\frac{dx^\pm}{dy^\pm}(y^\pm)\frac{dx^\pm}{dy^\pm}(y'^\pm)}{(x^\pm(y^\pm) - x^\pm(y'^\pm))^2} - \frac{1}{(y^\pm - y'^\pm)^2}\right). \quad (4.40)$$

The limit inside the bracket can be evaluated by performing a Taylor expansion of the conformal transformation $x^\pm(y^\pm)$ (i.e., the inverse of $y^\pm(x^\pm)$)

$$x^\pm(y'^\pm) = x^\pm(y^\pm) + \frac{dx^\pm}{dy^\pm}(y^\pm)(y'^\pm - y^\pm) + \frac{1}{2!}\frac{d^2x^\pm}{dy^{\pm 2}}(y^\pm)(y'^\pm - y^\pm)^2$$
$$+ \frac{1}{3!}\frac{d^3x^\pm}{dy^{\pm 3}}(y^\pm)(y'^\pm - y^\pm)^3 + \ldots . \quad (4.41)$$

The limit is finite and, after some algebra, we obtain

$$: T_{\pm\pm}(y^\pm) := \left(\frac{dx^\pm}{dy^\pm}\right)^2 : T_{\pm\pm}(x^\pm) : - \frac{\hbar}{24\pi}\{x^\pm, y^\pm\}, \quad (4.42)$$

where

$$\{x^\pm, y^\pm\} = \frac{d^3x^\pm}{dy^{\pm 3}}\bigg/\frac{dx^\pm}{dy^\pm} - \frac{3}{2}\left(\frac{d^2x^\pm}{dy^{\pm 2}}\bigg/\frac{dx^\pm}{dy^\pm}\right)^2 \quad (4.43)$$

is the Schwarzian derivative.

We notice that normal ordering breaks the classical covariant transformation law under conformal transformations

$$T_{\pm\pm}(y^\pm) = \left(\frac{dx^\pm}{dy^\pm}\right)^2 T_{\pm\pm}(x^\pm), \quad (4.44)$$

and modifies this expression by the anomalous transformation law (4.42), usually referred to as the *Virasoro anomaly*. The reason is that normal ordering requires a selection of modes, and therefore of coordinates and, in particular, of time. For instance, $: T_{\pm\pm}(x^\pm) :$ can be defined from the plane wave modes $(4\pi w)^{-1/2}e^{-iwx^\pm}$ and $: T_{\pm\pm}(y^\pm) :$, instead, from the modes $(4\pi w)^{-1/2}e^{-iwy^\pm}$.

4.2.2 The SO(d,2) conformal group and Möbius transformations

To grasp the physical meaning of the above results it is useful to consider their infinitesimal version. Under infinitesimal conformal transformations

$x^\pm \to y^\pm = x^\pm + \epsilon^\pm(x^\pm)$ Eq. (4.42) can be reexpressed as

$$\delta_{\epsilon^\pm} : T_{\pm\pm} := \epsilon^\pm \partial_\pm : T_{\pm\pm} : +2\partial_\pm \epsilon^\pm : T_{\pm\pm} : -\frac{\hbar}{24\pi}\partial_\pm^3 \epsilon^\pm . \qquad (4.45)$$

It is easy to see that the infinitesimal transformations with no contribution to the Virasoro anomaly ($\partial_\pm^3 \epsilon^\pm = 0$), i.e.,

$$\epsilon^\pm(x^\pm) = \epsilon_{-1}^\pm + \epsilon_0^\pm x^\pm + \epsilon_{+1}^\pm (x^\pm)^2 , \qquad (4.46)$$

where $\epsilon_{-1}^\pm, \epsilon_0^\pm, \epsilon_{+1}^\pm$ are arbitrary constants, generate the Lie algebra of the $SO(2,2)$ group. It is nothing else but the restriction to $d = 2$ of the higher dimensional conformal group $SO(d, 2)$ defined by those spacetime transformations leaving the Minkowskian metric invariant up to a global scaling (Weyl transformation) [Di Francesco et al. (1997)].[7]

To be clear we shall devote a few lines to explain the conformal group in an arbitrary Minkowskian spacetime R^d of dimension d. The spacetime transformations $x \to x'$ that change the Minkowskian metric by a Weyl transformation

$$g'_{\mu\nu}(x') = \Omega^2(x)g_{\mu\nu}(x) \qquad (4.47)$$

can be classified and turn out to be:

- Translations:
$$x'^\mu = x^\mu + a^\mu ; \qquad (4.48)$$

- Lorentz transformations:
$$x'^\mu = \Lambda^\mu{}_\nu x^\nu ; \qquad (4.49)$$

- Dilatations:
$$x'^\mu = \lambda x^\nu, \quad \lambda \in R^+; \qquad (4.50)$$

- Special conformal transformations:
$$x'^\mu = \frac{x^\mu + x^2 b^\mu}{1 + 2bx + b^2 x^2} \quad b^\mu \in R^d . \qquad (4.51)$$

These transformations generate the group $SO(d, 2)$.

---
[7]See [de Azcárraga and Izquierdo (1995)] for a more general viewpoint.

4.2.2.1 Infinitesimal transformations

It is not difficult to determine the corresponding infinitesimal transformations and, for $d = 2$ and in null coordinates, they are of the form (ϵ is an infinitesimal parameter):

- Translations:
$$\delta x^+ = \epsilon \qquad \delta x^- = 0 , \qquad (4.52)$$

or

$$\delta x^- = \epsilon \qquad \delta x^+ = 0 ; \qquad (4.53)$$

- Lorentz transformations:
$$\delta x^\pm = \pm \epsilon x^\pm ; \qquad (4.54)$$

- Dilatations:
$$\delta x^\pm = \epsilon x^\pm ; \qquad (4.55)$$

- Special conformal transformations:
$$\delta x^+ = \epsilon (x^+)^2 \qquad \delta x^- = 0 , \qquad (4.56)$$

or

$$\delta x^- = \epsilon (x^-)^2 \qquad \delta x^+ = 0 . \qquad (4.57)$$

It is now clear that, in Eq. (4.46), $\epsilon^\pm_{-1}$ generate translations of $x^\pm$; $(\epsilon^+_0, \epsilon^-_0)$ with $\epsilon^+_0 = -\epsilon^-_0$ and $\epsilon^+_0 = \epsilon^-_0$ generate Lorentz transformations and dilatations, respectively, and $\epsilon^\pm_{+1}$ are the generators of the so called special conformal transformations. The finite form of these transformations are the *Möbius transformations*:

$$x^\pm \to x'^\pm = \frac{a^\pm x^\pm + b^\pm}{c^\pm x^\pm + d^\pm} \qquad (4.58)$$

where $a^\pm d^\pm - b^\pm c^\pm = 1$. This is the group $(SL(2, R) \otimes SL(2, R))/Z_2 \approx SO(2, 2)$. The quotient by Z_2 is due to the fact that Eq. (4.58) is not modified by reversing the sign of the constants $a^\pm, b^\pm, c^\pm, d^\pm$. It is easy to check that the Schwarzian derivative of the Möbius transformations (4.58) vanishes. They are the only transformations with this property.

4.2.3 The particle number operator

We shall see now how the two-point correlation function $\partial_{\pm}f(y^{\pm})\partial_{\pm}f(y'^{\pm})$ also serves to construct the particle number operator [Fabbri et al. (2004)]. To achieve this we shall consider the normal ordered operator $:\partial_{\pm}f(y^{\pm})\partial_{\pm}f(y'^{\pm}):$, in particular, its explicit form in terms of creation and annihilation operators. For simplicity we restrict our considerations only to the outgoing sector (similar results can be obtained for the ingoing sector)

$$\langle 0_x | : \partial_{y^-}f(y^-)\partial_{y^-}f(y'^-) : |0_x\rangle =$$
$$\sum_{ji}\left\{\langle 0_x|\vec{b}_j^\dagger \vec{b}_i|0_x\rangle\left(\partial_{y^-}\tilde{u}_i(y^-)\partial_{y^-}\tilde{u}_j^*(y'^-) + \partial_{y^-}\tilde{u}_j^*(y^-)\partial_{y^-}\tilde{u}_i(y'^-)\right)\right.$$
$$\left. + \left(\langle 0_x|\vec{b}_j\vec{b}_i|0_x\rangle \partial_{y^-}\tilde{u}_j(y^-)\partial_{y^-}\tilde{u}_i(y'^-) + c.c.\right)\right\} . \quad (4.59)$$

Now, instead of taking the limit $y^- \to y'^-$, as in the construction of the stress tensor, we shall perform the following transform

$$\int_{-\infty}^{+\infty} dy^- dy'^- \tilde{u}_k(y^-)\tilde{u}_{k'}^*(y'^-)\langle 0_x | : \partial_{y^-}f(y^-)\partial_{y^-}f(y'^-) : |0_x\rangle . \quad (4.60)$$

We can evaluate this expression in terms of the particle number operator. To this end we shall use Eq. (4.59) together with the relations

$$\begin{aligned}(\tilde{u}_i, \tilde{u}_j) &= -2i\int_{-\infty}^\infty dy^- \tilde{u}_i \partial_{y^-}\tilde{u}_j^* = \delta_{ij}\\ (\tilde{u}_i^*, \tilde{u}_j^*) &= -2i\int_{-\infty}^\infty dy^- \tilde{u}_i^* \partial_{y^-}\tilde{u}_j = -\delta_{ij}\\ (\tilde{u}_i, \tilde{u}_j^*) &= -2i\int_{-\infty}^\infty dy^- \tilde{u}_i \partial_{y^-}\tilde{u}_j = 0 .\end{aligned} \quad (4.61)$$

The result is as follows

$$\int_{-\infty}^{+\infty} dy^- dy'^- \tilde{u}_k(y^-)\tilde{u}_{k'}^*(y'^-)\ \langle 0_x | : \partial_{y^-}f(y^-)\partial_{y^-}f(y'^-) : |0_x\rangle =$$
$$\frac{1}{4}\langle 0_x|\vec{b}_k^\dagger \vec{b}_{k'}|0_x\rangle . \quad (4.62)$$

We then immediately get an expression for the expectation value of the particle number operator $\vec{N}_k = \hbar^{-1}\vec{b}_k^\dagger \vec{b}_k$ associated to the right-moving mode k :

$$\langle 0_x|\vec{N}_k|0_x\rangle = \frac{4}{\hbar}\int_{-\infty}^{+\infty} dy^- dy'^- \tilde{u}_k(y^-)\tilde{u}_k^*(y'^-) \times$$
$$\langle 0_x | : \partial_{y^-}f(y^-)\partial_{y^-}f(y'^-) : |0_x\rangle. \quad (4.63)$$

Taking into account Eq. (4.39) we obtain

$$\langle 0_x|\vec{N}_k|0_x\rangle = -\frac{1}{\pi}\int_{-\infty}^{+\infty} dy^- dy'^- \tilde{u}_k(y^-)\tilde{u}_k^*(y'^-) \times \qquad (4.64)$$

$$\left[\frac{dx^-(y^-)}{dy^-}\frac{dx^-(y'^-)}{dy^-}\frac{1}{(x^-(y^-)-x^-(y'^-))^2} - \frac{1}{(y^- - y'^-)^2}\right].$$

This expression has a nice physical interpretation. The production of quanta, as measured by an observer with coordinates $y^\pm$, is clearly associated to the deviation of the correlations $\langle 0_x|\partial_{y^-}f(y^-)\partial_{y^-}f(y'^-)|0_x\rangle$ from their corresponding values in the vacuum $|0_y\rangle$. Moreover, the correlations contributing to the production of quanta in the mode k are those supported at the set of points y^- and y'^- where the mode is located. This is more clear when one introduces finite-normalization wave packet modes, instead of the usual plane wave modes. For wave packets peaked about $y^- = 2\pi n/\epsilon$ with width $2\pi/\epsilon$ the main contribution to $\langle 0_x|\vec{N}_{jn}|0_x\rangle$ comes from correlations of range similar to the support of the wave packet and around the referred point.

We notice that the difference of two-point functions in Eq. (4.64) at $y^- = y'^-$ is not singular. In fact, for $y^- = y'^- + \epsilon$ and $|\epsilon| \ll 1$, it is proportional to

$$-4\pi\langle 0_x|:T_{--}(y^-):|0_x\rangle - 2\pi\frac{d}{dy^-}\langle 0_x|:T_{--}(y^-):|0_x\rangle\epsilon + O(\epsilon^2), \qquad (4.65)$$

which clearly shows a smooth behavior at the coincident-point limit. Moreover, using plane wave modes the expression (4.64) turns out to be

$$\langle 0_x|\vec{N}_w|0_x\rangle = -\frac{1}{4\pi^2 w}\int_{-\infty}^{+\infty} dy^- dy'^- e^{-iw(y^- - y'^-)} \times$$

$$\left(\frac{dx^-}{dy^-}(y^-)\frac{dx^-}{dy^-}(y'^-)\frac{1}{(x^- - x'^-)^2} - \frac{1}{(y^- - y'^-)^2}\right). \qquad (4.66)$$

It is now interesting to evaluate the following quantity

$$\int_0^\infty dw \langle 0_x|\vec{N}_w|0_x\rangle \hbar w = -\frac{\hbar}{4\pi^2}\int_0^\infty dw \int_{-\infty}^{+\infty} dy^- dy'^- e^{-iw(y^- - y'^-)}$$

$$\left(\frac{dx^-}{dy^-}(y^-)\frac{dx^-}{dy^-}(y'^-)\frac{1}{(x^- - x'^-)^2} - \frac{1}{(y^- - y'^-)^2}\right). \qquad (4.67)$$

The integral over w can be easily evaluated and we obtain

$$-\frac{\hbar}{4\pi}\int_{-\infty}^{+\infty}dy^-dy'^-\delta(y^- - y'^-)\left(\frac{\frac{dx^-}{dy^-}(y^-)\frac{dx^-}{dy^-}(y'^-)}{(x^- - x'^-)^2} - \frac{1}{(y^- - y'^-)^2}\right) . \quad (4.68)$$

The delta function now involves to evaluate the coincidence limit of the integrand giving (see (4.65))

$$\int_0^\infty dw\langle 0_x|\vec{N}_w|0_x\rangle\hbar w = \int_{-\infty}^{+\infty}dy^-\langle 0_x|:T_{--}(y^-):|0_x\rangle . \quad (4.69)$$

This quantity is nothing else but the total energy, as one expects on physical grounds.

4.3 Radiation in Rindler Space: Hawking and Unruh Effects

In this section we shall consider the process involving the formation of a Schwarzschild black hole in a dynamical Vaidya spacetime, widely studied in detail in Chapter 3, in the near-horizon approximation. In this limit, the process is viewed as the dynamical formation of Rindler space out of Minkowski. Moreover, we will consider the simplified situation where the formation of Rindler space takes place along an ingoing null surface located at some $v = v_0$. The full solution to this problem is given by the metric

$$ds_{(4)}^2 = -(1 - \frac{2M(v)}{r})dv^2 + 2drdv + r^2d\Omega^2 , \quad (4.70)$$

where

$$M(v) = M\Theta(v - v_0) . \quad (4.71)$$

This corresponds to a classical four-dimensional stress tensor $T_{vv}^{(4)} = \frac{M}{4\pi r^2}\delta(v - v_0)$. In the region $v < v_0$ the relevant part of the metric can be cast in the double null form

$$ds^2 = -du_{in}dv \quad (4.72)$$

via the introduction of the outgoing null coordinate $u_{in} = v - 2r$. In the black hole region $v > v_0$, expanding r around the Schwarzschild radius

$r = 2M + \frac{x^2}{8M}$ we obtain, in the leading order approximation,

$$ds^2 \sim -\frac{x^2}{16M^2}dv^2 + \frac{x}{2M}dvdx . \tag{4.73}$$

Introduction of the null coordinate

$$u = v - 8M \ln \frac{x}{4M} \tag{4.74}$$

leads to the metric

$$ds^2 = -e^{\kappa(v-u)}dvdu , \tag{4.75}$$

in agreement with Eq. (4.9). The near-horizon description of the process is given in Fig. 4.3.

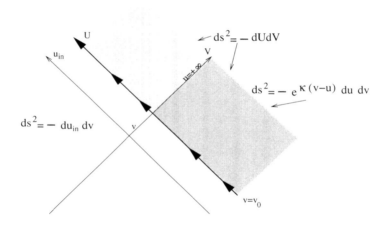

Fig. 4.3 Near-horizon geometry modeling the collapse of a Schwarzschild black hole by a single shock wave.

It is crucial to relate the coordinates u_{in} and u along the shock wave (the v coordinate remains unchanged). We then have, imposing the continuity of the radial coordinate r, $r(u_{in}, v_0) = r(u, v_0)$

$$u_{in} = v_0 - 2r = v_0 - 4M - \frac{x(u,v_0)^2}{4M} \tag{4.76}$$

leading to

$$u_{in} = v_0 - 4M - \kappa^{-1}e^{\kappa(v_0-u)} . \tag{4.77}$$

4.3.1 The Hawking effect

Now let us consider a matter field initially prepared in the Minkowski ground state $|in\rangle$ in the Minkowskian "in" region $v < v_0$. The region $v > v_0$ has, as natural vacuum state, the Rindler state $|0_R\rangle$, associated to the positive frequency modes

$$\vec{X}_w^R \equiv \frac{1}{\sqrt{4\pi w}} e^{-iwu} ,$$

$$\overleftarrow{X}_w^R \equiv \frac{1}{\sqrt{4\pi w}} e^{-iwv} . \qquad (4.78)$$

How does a Rindler observer, with coordinates (u, v), perceive the initial Minkowski state $|in\rangle$? To answer this question we shall employ the relation (4.77). This allows to compute $\langle in| : T_{uu} : |in\rangle$ using the anomalous transformation law for the normal ordered stress tensor and taking into account that

$$\langle in| : T_{u_{in}u_{in}} : |in\rangle = 0 . \qquad (4.79)$$

Therefore we have

$$\langle in| : T_{uu}(u) : |in\rangle = -\frac{\hbar}{24\pi} \left(\frac{d^3 u_{in}}{du^3} / \frac{du_{in}}{du} - \frac{3}{2} \left(\frac{d^2 u_{in}}{du^2} / \frac{du_{in}}{du} \right)^2 \right) . \qquad (4.80)$$

The result is a constant, namely

$$\langle in| : T_{uu} : |in\rangle = \frac{\hbar \kappa^2}{48\pi} . \qquad (4.81)$$

For the ingoing sector, due to the fact that v is always the same before and after the shock wave the result is trivial

$$\langle in| : T_{vv} : |in\rangle = 0 . \qquad (4.82)$$

The non-vanishing flux $\langle in| : T_{uu}(u) : |in\rangle$ reflects the phenomena of particle production. This result has a clear physical interpretation in the near-horizon approximation we are using: this is just the realization of the Hawking effect for the Schwarzschild black hole, where $\kappa = 1/4M$ giving $\langle in| : T_{uu} : |in\rangle = \hbar/768\pi M^2$. The quantity $\langle in| : T_{uu}(u) : |in\rangle$ represents indeed the energy flux radiated by the black hole at late times, i.e., that propagates along outgoing null lines to I^+ in the limit when these null lines approach the future event horizon. In the propagation from $r \approx 2M$ to $r = +\infty$, this flux will decrease due to backscattering. Since the flux has been calculated in the coordinates (u, v), which are asymptotically inertial at I^+, the quantity $\langle in| : T_{uu}(u) : |in\rangle$ is nothing else but the Hawking flux of radiation when backscattering is neglected.

A more detailed analysis of the properties of the radiation emitted involves the calculation of the expectation value of the outgoing particle number operator $\langle in|\vec{N}_k|in\rangle$ [Fabbri et al. (2004)]. In our case, Eq. (4.64), with the relation (4.77), turns out to be

$$\langle in|\vec{N}_k|in\rangle = -\frac{1}{\pi}\int_{-\infty}^{+\infty} dudu' \tilde{u}_k(u)\tilde{u}_k^*(u')\left[\frac{\kappa^2 e^{-\kappa(u-u')}}{(e^{-\kappa(u-u')}-1)^2} - \frac{1}{(u-u')^2}\right]. \tag{4.83}$$

This integral can be rewritten as

$$\langle in|\vec{N}_k|in\rangle = \frac{1}{\pi}\int_{-\infty}^{+\infty} dudu' [\partial_u \tilde{u}_k(u)\tilde{u}_k^*(u')\frac{\kappa}{e^{-\kappa(u-u')}-1} + \tilde{u}_k(u)\tilde{u}_k^*(u')\frac{1}{(u-u')^2}]. \tag{4.84}$$

Taking into account the form of the wave packet mode $\tilde{u}_k(u) \equiv u_{jn}^{out}(u)$

$$u_{jn}^{out}(u) = \frac{1}{\sqrt{\epsilon}}\int_{j\epsilon}^{(j+1)\epsilon} dw e^{2\pi iwn/\epsilon}\frac{e^{-iwu}}{\sqrt{4\pi w}}, \tag{4.85}$$

we obtain $(z = u - u')$

$$\langle in|\vec{N}_{jn}|in\rangle = \frac{1}{2\pi}\int_{-\infty}^{+\infty} dz e^{-izw_j}\left[\frac{-i\kappa}{e^{-\kappa z}-1} + \frac{1}{w_j z^2}\right], \tag{4.86}$$

where we have assumed that the frequencies of the wave packet vary in a small interval around $w_j \approx (j+1/2)\epsilon$. Evaluation of the integral using the identity

$$\sum_{n=1}^{\infty}\frac{1}{n^2+w_j^2/\kappa^2} = -\frac{\kappa^2}{2w_j^2} + \frac{\pi\kappa}{2w_j}\coth(\pi w_j/\kappa) \tag{4.87}$$

leads to

$$\langle in|\vec{N}_{jn}|in\rangle = \frac{1}{e^{2\pi w_j \kappa^{-1}}-1}, \tag{4.88}$$

which coincides with the Planckian spectrum of radiation at the temperature $k_B T = \frac{\kappa\hbar}{2\pi}$.

4.3.1.1 Correlation functions and thermal radiation

To properly show that we have indeed thermal radiation one should check that every possible expectation value corresponds to that predicted by a thermal quantum state. We shall now see that the constant flux (4.81) is associated to a thermal density matrix of outgoing flux:

$$\vec{\rho} = \prod_w \left(1 - e^{-2\pi w\kappa^{-1}}\right) \sum_{N=0}^{+\infty} e^{-2\pi N w\kappa^{-1}} |\vec{N}_w\rangle\langle\vec{N}_w| , \qquad (4.89)$$

where $|\vec{N}_w\rangle$ is the state in the Fock space with N outgoing particles of frequency w. Therefore we shall compute the expectation value $Tr[: T_{uu} : \vec{\rho}]$

$$Tr[: T_{uu} : \vec{\rho}] = \int_0^\infty dw \sum_{N=1}^{+\infty} \langle \vec{N}_w | : T_{uu} : \vec{\rho} |\vec{N}_w\rangle . \qquad (4.90)$$

To evaluate this quantity we insert a complete set of states and taking into account that $\vec{\rho}$ is diagonal

$$\langle \vec{N}_w | \vec{\rho} | \vec{N'}_{w'} \rangle = (1 - e^{-2\pi w\kappa^{-1}}) e^{-2N\pi w\kappa^{-1}} \delta_{NN'} \delta(w - w') , \qquad (4.91)$$

we find that

$$Tr[: T_{uu} : \vec{\rho}] = \int_0^\infty dw \sum_{N=0}^{+\infty} \langle \vec{N}_w | : T_{uu} : |\vec{N}_w\rangle (1 - e^{-2\pi w\kappa^{-1}}) e^{-2N\pi w\kappa^{-1}} . \qquad (4.92)$$

Now using that

$$\langle \vec{N}_w | : T_{uu} : |\vec{N}_w\rangle = \frac{\hbar N w}{2\pi} , \qquad (4.93)$$

and

$$\sum_{N=0}^\infty N e^{-2N\pi w\kappa^{-1}} = \frac{e^{-2\pi w\kappa^{-1}}}{(1 - e^{-2\pi w\kappa^{-1}})^2} , \qquad (4.94)$$

the result is

$$Tr[: T_{uu} : \vec{\rho}] = \frac{\hbar}{2\pi} \int_0^\infty dw \frac{w e^{-2\pi w\kappa^{-1}}}{(1 - e^{-2\pi w\kappa^{-1}})} = \frac{\hbar \kappa^2}{48\pi} \qquad (4.95)$$

in agreement with (4.81).

We have shown that the mean value of the particle number operator, as well as of the normal ordered stress tensor, agree with the thermal results.

However, as we already remarked in Chapter 3, this is not enough to ensure the exact thermal nature of the radiation. In the analysis of the Hawking effect of Chapter 3 we have shown that the expectation values of different products of *particle number operators* are consistent with thermal radiation. Here we shall follow a different route to prove this [Olmo (2003)]. We shall check that the *two-point function* $\langle in|\partial_u f(u)\partial_u f(u')|in\rangle$ corresponds to thermal radiation at the temperature $k_B T = \frac{\hbar\kappa}{2\pi}$. Taking into account that the two-point correlation function of the field ∂f transforms as

$$\langle in|\partial_u f(u)\partial_u f(u')|in\rangle = \frac{du_{in}}{du}(u)\frac{du_{in}}{du}(u')\langle in|\partial_{u_{in}} f(u_{in})\partial_{u_{in}} f(u'_{in})|in\rangle \tag{4.96}$$

we find

$$\langle in|\partial_u f(u)\partial_u f(u')|in\rangle = \frac{-\hbar\kappa^2}{16\pi}\frac{1}{\sinh^2\kappa(u-u')/2} . \tag{4.97}$$

We can now easily verify that this two-point function can be reproduced as an expectation value with respect to $\vec{\rho}$

$$\langle in|\partial_u f(u)\partial_u f(u')|in\rangle = Tr[\partial_u f(u)\partial_u f(u')\vec{\rho}] . \tag{4.98}$$

In analogy with the previous calculation for the normal ordered stress tensor we have to evaluate

$$Tr[\partial_u f(u)\partial_u f(u')\vec{\rho}] = \int_0^\infty dw \sum_{N=1}^\infty \langle \vec{N}_w|\partial_u f(u)\partial_u f(u')\vec{\rho}|\vec{N}_w\rangle$$
$$+ \langle 0_R|\partial_u f(u)\partial_u f(u')|0_R\rangle . \tag{4.99}$$

Taking into account that $\vec{\rho}$ is diagonal and that

$$\langle \vec{N}_w|\partial_u f(u)\partial_u f(u')|\vec{N}_w\rangle = \frac{\hbar N w}{4\pi}[e^{iw(u-u')} + e^{-iw(u-u')}] , \tag{4.100}$$

we get

$$Tr[\partial_u f(u)\partial_u f(u')\vec{\rho}] = -\frac{\hbar}{4\pi(u-u')^2}$$
$$+ \frac{\hbar}{4\pi}\int_0^\infty dw w[e^{iw(u-u')} + e^{-iw(u-u')}]\frac{e^{-2\pi w\kappa^{-1}}}{1-e^{-2\pi w\kappa^{-1}}} . \tag{4.101}$$

The second term can be rewritten as

$$\frac{\hbar}{4\pi}\sum_{N=0}^\infty \int_0^\infty dww[e^{iw(u-u')-2\pi w\kappa^{-1}(N+1)} + c.c.] , \tag{4.102}$$

which can be then transformed into

$$\frac{\hbar\kappa^2}{4\pi}\sum_{N=1}^{\infty}\frac{1}{(2\pi N-i\kappa(u-u'))^2}+c.c.\,. \qquad (4.103)$$

Summing up the vacuum contribution in Eq. (4.101), and using the identity

$$\sum_{N=-\infty}^{+\infty}\frac{1}{(N-ix)^2}=-\frac{\pi^2}{\sinh^2\pi x} \qquad (4.104)$$

we finally get

$$Tr[\partial_u f(u)\partial_u f(u')\vec{\rho}]=\frac{-\hbar\kappa^2}{16\pi}\frac{1}{\sinh^2\kappa(u-u')/2}\,, \qquad (4.105)$$

in agreement with Eq. (4.97).

Since in our near-horizon limit we are dealing with a free field theory the two-point function determines all the higher-order point functions and so this completes the proof that we have indeed an outgoing thermal state described by $\vec{\rho}$.

4.3.2 Radiation through the horizon

Until now we have introduced two different vacuum states: before v_0 the Minkowski ground state $|in\rangle$ is the natural one. We have seen that this vacuum state is perceived, in the region $v>v_0$ and for the Rindler observer, as a no-particle state in the incoming sector, but a thermal one in the outgoing sector. The natural state for this region is the Rindler vacuum $|0_R\rangle$. However, since this region is also flat we can also define a new state which is the Minkowski state $|0_M\rangle$ for the Minkowskian coordinates

$$\begin{aligned}U&=-\kappa^{-1}e^{-\kappa u},\\V&=\kappa^{-1}e^{\kappa v}\,.\end{aligned} \qquad (4.106)$$

It is interesting to investigate how the initial Minkowski ground state $|in\rangle$ is seen by a Minkowskian observer using coordinates U, V (and normal ordering with respect to the vacuum state $|0_M\rangle$). These coordinates are the analog, in the near-horizon approximation, of the locally inertial coordinates which characterize a free falling observer at the horizon. The relation between these coordinates and (u_{in}, v) is

$$u_{in} = v_0 - 4M + U ,$$
$$v = \kappa^{-1} \ln \kappa V . \qquad (4.107)$$

The trivial relation between the outgoing null coordinates u_{in} and U immediately implies, unlike what happens for the Rindler observer, that

$$\langle in| : T_{UU} : |in\rangle = 0 . \qquad (4.108)$$

It is the ingoing sector which now contains interesting physical information. Indeed, evaluation of the Schwarzian derivative

$$\langle in| : T_{VV} : |in\rangle = -\frac{\hbar}{24\pi} \left(\frac{d^3v}{dV^3} / \frac{dv}{dV} - \frac{3}{2} \left(\frac{d^2v}{dV^2} / \frac{dv}{dV} \right)^2 \right) \qquad (4.109)$$

gives the result

$$\langle in| : T_{VV} : |in\rangle = -\frac{\hbar}{48\pi} \frac{1}{V^2} . \qquad (4.110)$$

The emergence of a *negative influx of radiation*[8] is, in the black hole near-horizon picture underlying our construction, the counterpart of the Hawking effect, which causes the black hole to shrink due to the evaporation.

4.3.3 The Unruh effect

The phenomena of particle production in Rindler space has been explained so far only as a way to provide a simplified derivation of the Hawking effect. The near-horizon limit has allowed the required simplification to model the real curved spacetime geometry in terms of a flat geometry. Despite this simplification, the scenario used maintains all the crucial physical ingredients. The formation of the Rindler wedge from an initial Minkowski space, with its natural vacuum state $|in\rangle$, provides the necessary dynamics. The Hawking effect answers the question of how the state $|in\rangle$ is perceived by the Rindler observer. However, as already stressed in the previous subsection, since the "out" region is flat one can naturally consider the Minkowskian coordinates (U, V) and the corresponding vacuum state $|0_M\rangle$. A natural question, not necessarily tied to the original Schwarzschild geometry, emerges. How is the Minkowski vacuum $|0_M\rangle$, not $|in\rangle$, perceived by a uniformly accelerated Rindler observer?

---

[8]We should remark at this point that the emergence of negative energies can arise in quantum theory due to quantum coherence. See, for instance, [Ford (1997)].

A simple calculation gives

$$\langle 0_M | : T_{uu} : | 0_M \rangle = \frac{\hbar \kappa^2}{48\pi},$$
$$\langle 0_M | : T_{vv} : | 0_M \rangle = \frac{\hbar \kappa^2}{48\pi}. \qquad (4.111)$$

It can be shown, using the same arguments as for the Hawking effect, that with respect to a Rindler observer the state $|0_M\rangle$ corresponds to a *thermal bath of radiation* (both in the left and right moving sectors) at the temperature $k_B T = \hbar \kappa / 2\pi$. It is then described by the following thermal density matrix ρ

$$\rho = \overleftarrow{\rho} \otimes \overrightarrow{\rho}, \qquad (4.112)$$

where $\overrightarrow{\rho}$ is given in (4.89) and $\overleftarrow{\rho}$ is the analogous thermal density matrix for the left-moving sector. This is the *Fulling–Davies–Unruh effect*, which is usually called the *Unruh effect*. It is a flat spacetime effect and does not require any dynamical background. It expresses the relation between two different ways of quantizing in Minkowski space, as first stressed in [Fulling (1973)]: the natural quantization, associated to the Minkowskian coordinates, and the quantization associated to a uniformly accelerating observer. That the relation between these two different quantizations is characterized by a temperature $k_B T = \hbar \kappa / 2\pi$ was first pointed out in [Davies (1975)]. Later, [Unruh (1976)] clarified operationally the physical meaning by introducing "particle detectors": any uniformly accelerated particle detector measuring the state of the field in terms of Rindler particles will describe the Minkowski vacuum as a thermal state.[9] This effect, as the Hawking effect, is very tiny. The Unruh temperature, in terms of the acceleration $\bar{a} \equiv \kappa$ (see Section 4.1) is given by

$$T/1°K \approx \bar{a}/10^{21} \ cm \ sec^{-2}. \qquad (4.113)$$

Let us analyse in more detail the reason why the *pure* Minkowski vacuum state is described as a *mixed* state, i.e., using the thermal density matrix ρ, by a uniformly accelerating observer. The crucial point is that the notion of a quantum state has a global character. It is defined in a region of the spacetime where the selected modes for the quantization are well defined. In Minkowski space the natural modes, defined everywhere, are defined

---
[9] More details and subtle aspects can be found in [Unruh and Wald (1984)] and [Sciama et al. (1981)].

using Minkowskian coordinates (U, V)

$$\vec{X}_w^M \equiv \frac{1}{\sqrt{4\pi w}} e^{-iwU} ,$$

$$\overleftarrow{X}_w^M \equiv \frac{1}{\sqrt{4\pi w}} e^{-iwV} . \qquad (4.114)$$

They have associated the standard Minkowski vacuum state $|0_M\rangle$. However, when the quantum state is *restricted* to a region of the spacetime it becomes, in general, a *mixed state* (see, for instance, [Wald (1994)]). There is a simple and intuitive reason for this. When the state is restricted to a region the existing correlations with the exterior are lost. This fact is reflected in the mixed nature of the state obtained in this way. For instance, if we restrict the vacuum $|0_M\rangle$ to the region $U < 0$ we loose the correlations

$$\langle 0_M | \partial_U f(U < 0) \partial_U f(U' > 0) | 0_M \rangle = -\frac{\hbar}{4\pi} \frac{1}{(U - U')^2} . \qquad (4.115)$$

Moreover, since the coordinate U has been restricted and it has no more the standard range $-\infty < U < +\infty$, the modes $(4\pi w)^{-1/2} e^{-iwU}$ are no longer the natural ones for quantizing the restricted region. We need to introduce a new coordinate u with the standard range $-\infty < u < +\infty$. A simple choice for this is the Eddington–Finkelstein type coordinate u:

$$U = -\kappa^{-1} e^{-\kappa u} . \qquad (4.116)$$

Although this is not the only possibility, it is the simplest one. The vacuum-type correlation function in the restricted region $U < 0$

$$\langle 0_M | \partial_U f(U < 0) \partial_U f(U' < 0) | 0_M \rangle = -\frac{\hbar}{4\pi} \frac{1}{(U - U')^2} , \qquad (4.117)$$

becomes, when expressed in terms of the unrestricted coordinate u,

$$\langle 0_M | \partial_u f(u) \partial_u f(u') | 0_M \rangle = \frac{-\hbar \kappa^2}{16\pi} \frac{1}{\sinh^2 \kappa(u - u')/2} . \qquad (4.118)$$

In the quantization defined by the modes $(4\pi w)^{-1/2} e^{-iwu}$ the above correlation function can only be reproduced if the restricted vacuum state is indeed a mixed (thermal) state. It becomes mixed because of the restriction and thermal since we have introduced the Eddington–Finkelstein type coordinate to account for the restriction.[10]

---

[10]Other type of coordinate adapted to the restriction would have produced a different type of mixed state.

If we also restrict the coordinate V to $V > 0$, we are forced to introduce a new coordinate adapted to the restriction. The simplest choice is again the Eddington–Finkelstein type coordinate v:

$$V = \kappa^{-1} e^{\kappa v} . \tag{4.119}$$

The vacuum-type correlation function in the restricted region $V > 0$

$$\langle 0_M | \partial_U f(V>0) \partial_U f(V'>0) | 0_M \rangle = -\frac{\hbar}{4\pi} \frac{1}{(V-V')^2} \tag{4.120}$$

is then transformed into

$$\langle 0_M | \partial_v f(v) \partial_v f(v') | 0_M \rangle = \frac{-\hbar \kappa^2}{16\pi} \frac{1}{\sinh^2 \kappa (v-v')/2} . \tag{4.121}$$

Therefore, in the quantization of the region $(U<0, V>0)$, with the modes $(4\pi w)^{-1/2} e^{-iwu}$ and $(4\pi w)^{-1/2} e^{-iwv}$, the above correlation functions identify the restricted state as

$$|0_M\rangle|_{(U<0,V>0)} \Leftrightarrow \rho . \tag{4.122}$$

The use of the modes $(4\pi w)^{-1/2} e^{-iwu}$ and $(4\pi w)^{-1/2} e^{-iwv}$ in the restricted region allows to physically interpret the result since these are the natural modes of a uniformly accelerating observer in the (right) Rindler wedge.

Similar arguments can be equally applied if one restricts the Minkowski vacuum to the (left) Rindler wedge ($U>0, V<0$). The null Eddington–Finkelstein type coordinates, adapted to the new restriction, are then defined as $U = +\kappa^{-1} e^{-\kappa u}$, $V = -\kappa^{-1} e^{\kappa v}$. The resulting expression for the Minkowski vacuum is then similar to Eq. (4.122).

4.3.3.1 Unruh modes

One can combine the above two sets of positive frequency Rindler modes: $\overset{\rightarrow R}{X_w} \equiv (4\pi w)^{-1/2} e^{-iwu}$, $\overset{\leftarrow R}{X_w} \equiv (4\pi w)^{-1/2} e^{-iwv}$, for the right wedge, and $\overset{\rightarrow L}{X_w} \equiv (4\pi w)^{-1/2} e^{iwu}$, $\overset{\leftarrow L}{X_w} \equiv (4\pi w)^{-1/2} e^{iwv}$ for the left wedge, to form a complete set of modes for the full Minkowski spacetime. Obviously, the full set of Rindler modes is not equivalent to the standard Minkowskian modes. In fact, the Minkowski vacuum is not perceived as a true vacuum in each Rindler wedge. To show this one can follow the elegant method discussed in [Unruh (1976)]. The function

$$f(U) = \int_{-\infty}^{+\infty} dw\, e^{-iwU} \tilde{f}(w)\,, \qquad (4.123)$$

contains only positive frequency modes (i.e., $\tilde{f}(w)$ has support in $w > 0$) if it is analytic and bounded at infinity in the lower half complex U plane.

This analyticity property is not respected by the Rindler modes: they are not analytic as one passes from the right to the left Rindler wedge ($U = 0$ or $V = 0$). They indeed contain negative frequencies and this implies, as we already know, that the Rindler vacuum (either for the right or the left wedge) does not coincide with the Minkowski vacuum. However, there are simple combinations of Rindler modes that are analytic and bounded in the lower half complex U and V planes. They are called the *Unruh modes* and are defined, up to normalization, as

$$\begin{aligned}
\overrightarrow{X}_w^{U_{(1)}} &\equiv \overrightarrow{X}_w^{R} + e^{-\pi w \kappa^{-1}} \overrightarrow{X}_w^{L*} \propto (-\kappa U)^{iw\kappa^{-1}} \\
\overleftarrow{X}_w^{U_{(1)}} &\equiv \overleftarrow{X}_w^{R} + e^{-\pi w \kappa^{-1}} \overleftarrow{X}_w^{L*} \propto (\kappa V)^{-iw\kappa^{-1}} \\
\overleftarrow{X}_w^{U_{(2)}} &\equiv \overleftarrow{X}_w^{R*} + e^{\pi w \kappa^{-1}} \overleftarrow{X}_w^{L} \propto (-\kappa V)^{iw\kappa^{-1}} \\
\overrightarrow{X}_w^{U_{(2)}} &\equiv \overrightarrow{X}_w^{R*} + e^{\pi w \kappa^{-1}} \overrightarrow{X}_w^{L} \propto (\kappa U)^{-iw\kappa^{-1}}
\end{aligned} \qquad (4.124)$$

where, in the right column, we have given their expression in terms of the Minkowskian coordinates. It is easy to see that they are analytic and bounded in the lower half complex U and V planes if the branch cut of the complex powers is taken to lie in the upper half plane ($\ln(-1) = i\pi$). Therefore, the Unruh modes and the standard Minkowskian modes share the same vacuum state $|0_M\rangle$. A simple algebra then leads to the following relations

$$\begin{aligned}
(\overrightarrow{b}_w^{R} - e^{-\pi w \kappa^{-1}} \overrightarrow{b}_w^{L\dagger})|0_M\rangle &= 0\,, \\
(\overrightarrow{b}_w^{L} - e^{-\pi w \kappa^{-1}} \overrightarrow{b}_w^{R\dagger})|0_M\rangle &= 0\,,
\end{aligned} \qquad (4.125)$$

where $\overrightarrow{b}_w^{R}$ and $\overrightarrow{b}_w^{L}$ are annihilation operators for Rindler particles in the right and left wedges, respectively. Similar expressions for the ingoing sector can be equally obtained. Comparing these results to those obtained in Subsection 3.3.6 we immediately find the relation between the Minkowski and Rindler vacuum states

$$|0_M\rangle = \langle 0_R|0_M\rangle \exp(\sum_w e^{-\pi w \kappa^{-1}}(\overset{\to R\dagger}{b}_w \overset{\to L\dagger}{b}_w + \overset{\leftarrow R\dagger}{b}_w \overset{\leftarrow L\dagger}{b}_w))|0_R\rangle . \qquad (4.126)$$

We obtain again the same type of result already derived in the description of the "in" vacuum state in a gravitational collapse producing a black hole. The purity of the Minkowski vacuum is guaranteed by the existence of correlations between right and left Rindler particles, as explicitly shown in the formula (4.126).

A (right) Rindler observer, being confined to his own region, will then describe his local measurements in terms of a reduced thermal density matrix ρ obtained by tracing over all the external (left) Rindler states. Therefore, as far as he is concerned the initial vacuum state $|0_M\rangle$ has "evolved" into a mixed state. For the Rindler observer the correlations between the right and left regions are lost due to the tracing operation. Also the coherence of the expansion (4.126) is lost. Indeed, from the exact knowledge of the density matrix ρ one cannot reconstruct uniquely the pure Minkowski vacuum state. Another way of expressing this fact is saying that the transformation $|0_M\rangle \to \rho$ is not unitary and one necessarily looses information about the initial state. If the initial state is $|0_M\rangle$ plus left Rindler excitations the state measured by the right Rindler observer will still be described by the same ρ.

4.3.3.2 *Möbius invariance of the vacuum state*

The Unruh modes turn out to be particularly useful to discuss the question of the invariance of the Minkowski vacuum under the conformal Möbius transformations. A simple look at the expression (4.64) obtained in Subsection 4.2.3 for the expectation value of the particle number operator shows clearly the absence of spontaneous particle production from the vacuum $|0_x\rangle$, as measured by observers using coordinates $y^{\pm}$,[11]

$$\langle 0_x|\vec{N}_k|0_x\rangle = 0 = \langle 0_x|\overset{\leftarrow}{N}_k|0_x\rangle , \qquad (4.127)$$

if the two-point correlation functions for the fields $\partial_{y^{\pm}} f(y^{\pm})$ coincide with the corresponding values in the vacuum $|0_y\rangle$

$$\langle 0_x|\partial_{y^{\pm}} f(y^{\pm})\partial_{y^{\pm}} f(y'^{\pm})|0_x\rangle = \langle 0_y|\partial_{y^{\pm}} f(y^{\pm})\partial_{y^{\pm}} f(y'^{\pm})|0_y\rangle . \qquad (4.128)$$

---

[11] See [Aldaya et al. (1999)] for a different perspective.

This is equivalent to

$$\left(\frac{dx^\pm(y^\pm)}{dy^\pm}\right)\left(\frac{dx^\pm(y'^\pm)}{dy^\pm}\right)\frac{1}{(x^\pm(y^\pm)-x^\pm(y'^\pm))^2} = \frac{1}{(y^\pm - y'^\pm)^2} \ , \quad (4.129)$$

and the only transformations verifying this requirement are the Möbius ones. This should not be a surprise since we have already seen that the Möbius transformations are just those having a zero Schwarzian derivative and thus produce no flux at all

$$\langle 0_x | : T_{\pm\pm}(y^\pm) : |0_x\rangle = 0 \ . \quad (4.130)$$

The above results are trivial for translations, Lorentz boosts and dilatations: they do not modify the positivity of frequencies of the standard plane wave modes $(4\pi w)^{-1/2}e^{-iwU}$, $(4\pi w)^{-1/2}e^{-iwV}$. However it is more difficult to see why the special conformal transformations do not produce any mixing of positive and negative frequencies. Typical Möbius transformations that are not dilatations nor Poincaré are of the form

$$U \to -\frac{1}{a^2 U} \qquad V \to -\frac{1}{b^2 V} \quad (4.131)$$

where the parameters a^2, b^2 are positive constants. In terms of the Unruh modes it is easy to see why these transformations leave invariant the positive frequency solutions. Indeed, they interchange the two different types of positive frequency Unruh modes of each sector

$$\begin{aligned}(-\kappa U)^{iw\kappa^{-1}} &\leftrightarrow (\kappa U)^{-iw\kappa^{-1}} \\ (-\kappa V)^{iw\kappa^{-1}} &\leftrightarrow (\kappa V)^{-iw\kappa^{-1}} \ .\end{aligned} \quad (4.132)$$

Note that the form of the Unruh modes is especially adapted to the form of the above Möbius transformations to guarantee their invariance.

4.3.4 Three different vacuum states

All the discussion so far in this chapter has involved three different quantum states. We will see in the next chapter that there is a one to one correspondence between the states here defined and the usual states associated to black holes:

- $|0_R\rangle$ is the analog of the Boulware vacuum state [Boulware (1975)]. The corresponding modes are

$$\frac{1}{\sqrt{4\pi w}}e^{-iwu}\,,\qquad\qquad \frac{1}{\sqrt{4\pi w}}e^{-iwv}\,. \qquad (4.133)$$

- $|in\rangle$ corresponds to the Unruh state [Unruh (1976)]. The corresponding modes are

$$\frac{1}{\sqrt{4\pi w}}e^{-iwu_{in}} \sim \frac{1}{\sqrt{4\pi w}}e^{-iwU}\,,\qquad \frac{1}{\sqrt{4\pi w}}e^{-iwv}\,. \qquad (4.134)$$

- $|0_M\rangle$ is the analog of the Hartle–Hawking state [Hartle-Hawking (1976); Israel (1976)]. The corresponding modes are

$$\frac{1}{\sqrt{4\pi w}}e^{-iwU}\,,\qquad\qquad \frac{1}{\sqrt{4\pi w}}e^{-iwV}\,. \qquad (4.135)$$

A detailed study of these quantum states, as well as their physical interpretation, will be performed in the next chapter.

4.4 Anti-de Sitter Space as a Near-Horizon Geometry

We shall now jump to another physically interesting situation arising in the context of the near-horizon approximation. As we have seen in Chapter 2 the Schwarzschild geometry is not the only known black hole solution. By considering the Einstein–Maxwell theory we have the more general Reissner–Nordström solution described by the line element

$$ds^2_{(4)} = -(1 - \frac{2M}{r} + \frac{Q^2}{r^2})dt^2 + (1 - \frac{2M}{r} + \frac{Q^2}{r^2})^{-1}dr^2 + r^2 d\Omega^2\,, \qquad (4.136)$$

where Q is the electric (or magnetic) charge. The main difference with respect to Eq. (4.1) is that now we have two horizons $r_\pm = M \pm \sqrt{M^2 - Q^2}$ for $M > |Q|$ (non-extremal case) and in the limiting situation $M = |Q|$ (extreme case) one degenerate horizon ($r_+ = r_-$). Let us now study the near-horizon geometry in close analogy with Schwarzschild. The analysis is slightly more involved due the presence of the two horizons. A natural way to perform the near-horizon limit is to expand around the outer horizon:

$$r = r_+ + x \,, \tag{4.137}$$

for $x \ll r_+$. We find then that, for non-extremal black holes ($r_+ \neq r_-$)

$$ds^2_{(4)} \sim -\frac{x}{r_+}(1 - \frac{r_-}{r_+})dt^2 + \frac{r_+}{x(1 - \frac{r_-}{r_+})}dx^2 + r_+^2 d\Omega^2 \,. \tag{4.138}$$

A redefinition of the x coordinate

$$x \to \frac{(r_+ - r_-)}{4r_+^2} x^2 \tag{4.139}$$

puts the metric in the Rindler-like form

$$ds^2_{(4)} = -(\kappa_+ x)^2 dt^2 + dx^2 + r_+^2 d\Omega^2 \,, \tag{4.140}$$

where $\kappa_+ = \frac{(r_+ - r_-)}{2r_+^2}$ is the surface gravity at the horizon.

4.4.1 Extremal black holes

For extremal black holes ($r_+ = r_- \equiv r_0$) the approximated metric, replacing (4.138), when $r = r_0 + x$ for $x \ll r_0$ is

$$ds^2_{(4)} \sim -\frac{x^2}{r_0^2}dt^2 + \frac{r_0^2}{x^2}dx^2 + r_0^2 d\Omega^2 \,. \tag{4.141}$$

The above near-horizon geometry for the extremal case is radically different from the Rindler geometry encountered for non-extremal black holes. In fact, the two-dimensional radial ($t-r$) geometry in (4.141) is non-flat and it has negative constant curvature

$$R^{(2)} = -\frac{2}{r_0^2} \,. \tag{4.142}$$

More precisely, it is just the portion of two-dimensional Anti-de Sitter space AdS_2 covered by the (null) Poincaré coordinates ($x^\pm$)

$$ds^2 = -4r_0^2 \frac{dx^+ dx^-}{(x^+ - x^-)^2} \,. \tag{4.143}$$

The change of coordinates

$$x^\pm = t \pm \frac{r_0^2}{x} \tag{4.144}$$

brings the $(t-r)$ part of the metric (4.141) into (4.143).[12] The full four-dimensional geometry (4.141) is the so called Bertotti–Robinson spacetime [Bertotti (1959); Robinson (1959)], $AdS_2 \times S^2$, which turns out to be an exact solution of the Einstein–Maxwell equations in four dimensions.[13] The region described by the AdS_2 near-horizon geometry for the extremal black hole is depicted in Fig. 4.4(a). This turns out to be the full AdS_2 space,

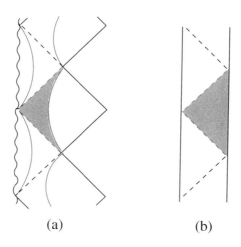

Fig. 4.4 (a) Near-horizon geometry of an extremal Reissner–Nordström black hole. The shaded region is mapped to the Poincaré wedge of full AdS_2 space in (b).

whose Penrose diagram is given in Fig. 4.4(b).[14] In particular, the Poincaré coordinates cover only the shaded wedge. They are the analog of the Rindler coordinates for the near-horizon analysis performed for Schwarzschild. As for Rindler space one can also construct global coordinates. They are given by the transformation

$$x^{\pm} = \tan \frac{1}{2}(\tau \pm \sigma \pm \frac{\pi}{2}) \ . \tag{4.145}$$

---

[12] In terms of the coordinates t and $y = r_0^2/x$ the metric takes the usual "Poincaré" form $r_0^{-2}ds^2 = y^{-2}(-dt^2 + dy^2)$.

[13] In three dimensions with a negative cosmological constant it is possible to have black hole configurations by global identifications of AdS_3 [Bañados et al. (1992)].

[14] For details on how to construct the Penrose diagram of AdS space see [Hawking and Ellis (1973)]. For the particular case considered here, see also [Spradlin and Strominger (1999)].

The $(t-r)$ part of the metric takes then the form

$$r_0^{-2} ds^2 = \frac{-d\tau^2 + d\sigma^2}{\cos^2 \sigma} , \qquad (4.146)$$

where $-\infty < \tau < +\infty$ and $-\pi/2 \leq \sigma \leq \pi/2$.

4.4.2 Near-extremal black holes

Anti-de Sitter space also arises from non-extremal configurations which are close to extremality. They are defined by those solutions for which $M = Q + \Delta m$ and $\Delta m << Q$, or, put in another way $r_\pm \sim Q \pm \sqrt{2Q\Delta m}$. The near-horizon approximation can be performed in a way similar to the extremal case: expanding around $r_0 = Q$, instead of r_+. One of the advantages of doing so is that, unlike the standard procedure leading to the Rindler geometry, we shall encounter the relic of both horizons also in the near-horizon geometry. Expanding

$$r = Q + x , \qquad (4.147)$$

with $x << Q$, we find

$$ds^2_{(4)} \sim -\frac{x^2 - 2Q\Delta m}{r_0^2} dt^2 + \frac{r_0^2}{x^2 - 2Q\Delta m} dx^2 + r_0^2 d\Omega^2 . \qquad (4.148)$$

As before, the two-dimensional part of the above metric has constant negative curvature $R^{(2)} = -\frac{2}{r_0^2}$, and therefore it is locally equivalent to the analog metric for the extremal case. Globally, however, they are different and this difference is parametrized by Δm. In fact, they describe different portions of the maximally extended AdS_2 spacetime. This can be easily seen from Fig. 4.5.

The shaded region is described, in null coordinates, by the metric (we write only the $(t-r)$ part)

$$ds^2 = -4r_0^2 (\frac{\kappa_+}{2})^2 \frac{du\, dv}{\sinh^2 \frac{\kappa_+}{2}(u-v)} , \qquad (4.149)$$

where, for near-extremal configurations,

$$\kappa_+ = \frac{r_+ - r_-}{2r_+^2} \approx \sqrt{\frac{2\Delta m}{Q^3}} . \qquad (4.150)$$

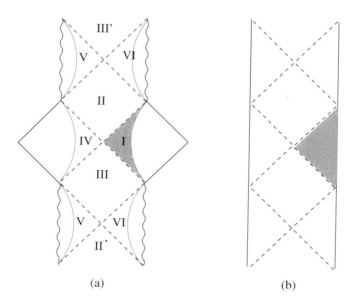

Fig. 4.5 (a) Near-horizon geometry of a near-extremal Reissner–Nordström black hole. The shaded wedge is now mapped to the corresponding one in (b), which is different from the Poincaré wedge of the full AdS_2 space.

This can be obtained from (4.148) by the change of coordinates

$$t = \frac{u+v}{2}$$
$$x = \sqrt{2Q\Delta m}\coth\frac{\sqrt{2Q\Delta m}(u-v)}{2r_0^2}\ . \qquad (4.151)$$

4.5 Radiation in Anti-de Sitter Space: the Hawking Effect

We shall work in parallel with Section 4.3. The analysis of the radiation for the non-extremal case can be performed in close analogy with the Schwarzschild black hole if the near-horizon geometry involved is the one obtained by expanding around r_+, leading to the metric (4.140). The formation of a non-extremal black hole can still be viewed, in the near-horizon limit and in the $(t-r)$ sector, as the formation of Rindler space starting from Minkowski spacetime. For this case all the discussion of Section 4.3 goes unchanged up to the replacement $\kappa \to \kappa_+$. However, the near-horizon approximation can also be defined, for near-extremal configurations, around the extremal horizon $r = r_0$. In this case the physical dynamical setting: Minkowsky + incoming matter → non-extremal black hole + Hawking radiation, which is described by portions of flat spaces, should be replaced by extremal black holes + incoming matter → near-extremal black hole + Hawking radiation, which can be described with spaces of constant negative curvature. This is so because, as we have explained in the previous section, both extremal and near-extremal black holes have near-horizon geometries described by portions of Anti-de Sitter space.[15]

The simplest dynamical process which makes a black hole to depart from extremality due to (low-energy) incoming neutral matter is given by the following Vaidya-type metric:

$$ds^2_{(4)} = -(1 - \frac{2M(v)}{r} + \frac{Q^2}{r^2})dv^2 + 2drdv + r^2 d\Omega^2 , \qquad (4.152)$$

where

$$M(v) = Q + \Delta m \Theta(v - v_0) . \qquad (4.153)$$

This corresponds to a classical stress tensor

$$T^{(4)}_{vv} = \frac{\Delta m}{4\pi r^2} \delta(v - v_0) . \qquad (4.154)$$

The corresponding Penrose diagram is given in Fig. 4.6. Expanding r around the extremal radius: $r = r_0 + x$ we obtain the near-horizon metric

$$ds^2_{(4)} \sim -\frac{x^2 - 2Q\Delta m \Theta(v - v_0)}{r_0^2} dv^2 + 2dxdv + r_0^2 d\Omega^2 . \qquad (4.155)$$

Before the shock wave $v < v_0$, the radial part of the above metric can be written in conformal coordinates

$$ds^2 = -\frac{4r_0^2}{(u_{in} - v)^2} du_{in} dv , \qquad (4.156)$$

where

$$x = \frac{2r_0^2}{u_{in} - v} . \qquad (4.157)$$

The null Poincaré coordinates (u_{in}, v) are the Eddington–Finkelstein type coordinates of the extremal black hole. We can immediately regard (4.157)

---
[15] For an account on the quantum properties of topological black holes in Anti-de Sitter space see [Klemm and Vanzo (1998)].

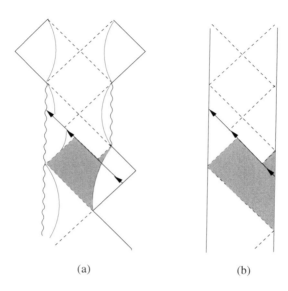

Fig. 4.6 (a) Full and (b) near-horizon Penrose diagrams corresponding to the creation of a near-extremal Reissner–Nordström black hole from the extremal one.

as the near-horizon limit of the tortoise coordinate for the extremal configuration

$$\frac{v - u_{in}}{2} = r^* = r + 2Q\left(\ln(r-Q) - \frac{Q}{2(r-Q)}\right) \approx -\frac{Q^2}{x} . \quad (4.158)$$

The coordinate u_{in} covers the exterior region of the extremal black hole, in the near-horizon approximation.

After the shock wave $v > v_0$ the two-dimensional metric in conformal form reads

$$ds^2 = -\left(\frac{x^2 - 2Q\Delta m}{r_0^2}\right)dudv , \quad (4.159)$$

where

$$x = \sqrt{2Q\Delta m}\coth\frac{\sqrt{2Q\Delta m}(u-v)}{2r_0^2} . \quad (4.160)$$

The coordinates (u, v) are the Eddington–Finkelstein type coordinates for the near-extremal charged black hole and therefore this requires $x^2 > 2Q\Delta m$ (i.e., they cover the black hole exterior horizon $r > r_+$). The explicit form of the metric in terms of these coordinates is given in Eq. (4.149).

Matching the metrics at $v = v_0$ implies the following relation between the coordinates u_{in} and u in the exterior region:

$$u_{in} = v_0 + \frac{2r_0^2}{\sqrt{2Q\Delta m}} \tanh \frac{\sqrt{2Q\Delta m}}{2r_0^2}(u - v_0) . \quad (4.161)$$

Note that for $\Delta m \to 0$ the above relation turns out to be the identity $u_{in} = u$, as one expects on physical grounds.

Proceeding as in Rindler space we should first construct the initial vacuum state $|in\rangle$ of the matter field. Here, however, we encounter a certain ambiguity in defining the natural initial state $|in\rangle$, due to the fact that the initial geometry is the extremal black hole. We can define the initial vacuum state as associated with the plane wave modes in the (u_{in}, v) coordinates or, alternatively, with respect to the plane wave modes in the Kruskal-type coordinates (U, V). In general these two sets of coordinates are related by (see Section 2.6)

$$u_{in} \sim -4Q\left(\ln \frac{|U|}{Q} + \frac{Q}{2U}\right) , \quad (4.162)$$

$$v \sim 4Q\left(\ln \frac{|V|}{Q} - \frac{Q}{2V}\right) . \quad (4.163)$$

In the near-horizon limit, the above relations become

$$u_{in} = -\frac{2Q^2}{U} , \quad (4.164)$$

$$v = -\frac{2Q^2}{V} . \quad (4.165)$$

The ambiguity is not relevant for our considerations here since, irrespective of the concrete definition of $|in\rangle$, we can always assume that

$$\langle in| : T_{u_{in}u_{in}} : |in\rangle = 0$$
$$\langle in| : T_{vv} : |in\rangle = 0 , \quad (4.166)$$

or, equivalently,

$$\langle in| : T_{UU} : |in\rangle = 0$$
$$\langle in| : T_{VV} : |in\rangle = 0 . \quad (4.167)$$

This is so because, in the near-horizon approximation, the relations $u_{in} = u_{in}(U)$ and $v = v(V)$ are given by pure Möbius transformations (see the considerations in Subsection 4.3.3).

After v_0, a near-extremal black hole is formed and the natural vacuum state is then constructed with respect to the coordinates (u,v). Since this state does not coincide with $|in\rangle$ we can naturally ask how the initial vacuum state $|in\rangle$ is described by the observer with (u,v) coordinates. The answer is obtained by direct evaluation of the normal ordered stress tensor

$$\langle in| : T_{uu}(u) : |in\rangle = -\frac{\hbar}{24\pi}\{u_{in}, u\}, \quad (4.168)$$

which turns out to be equal to

$$\langle in| : T_{uu}(u) : |in\rangle = \frac{\hbar \Delta m}{24\pi Q^3}. \quad (4.169)$$

This constant flux coincides with the thermal Hawking flux for near-extremal black holes

$$\langle in| : T_{uu}(u) : |in\rangle = \frac{\pi}{12\hbar}(k_B T_H)^2, \quad (4.170)$$

where T_H is Hawking's temperature around extremality:

$$k_B T_H = \frac{\hbar \kappa_+}{2\pi} \approx \frac{\hbar}{2\pi}\sqrt{\frac{2\Delta m}{Q^3}}. \quad (4.171)$$

There is a simple way to see that this flux is thermal. The relation between the u_{in} and u coordinates given by Eq. (4.161) can be rewritten as the usual exponential relation composed with a Möbius transformation:

$$u_{in} - v_0 = \frac{-\kappa_+^{-1} e^{-\kappa_+(u-v_0)} a + b}{-\kappa_+^{-1} e^{-\kappa_+(u-v_0)} c + d}, \quad (4.172)$$

where the coefficients of the Möbius transformation are: $a = 1, b = \kappa_+^{-1}, c = -\kappa_+/2, d = 1/2$. Since a Möbius transformation leaves the vacuum invariant, the properties of the radiation coincide exactly with those given by the usual thermal relation $u_{in} = -\kappa_+^{-1} e^{-\kappa_+ u}$.

To complete the analogy with the analysis of radiation in Rindler space, we can also investigate how the black hole radiation is described by the observer with coordinates $x^\pm$ such that the metric, for $v > v_0$,

$$ds^2 = -4r_0^2(\frac{\kappa_+}{2})^2 \frac{dudv}{\sinh^2 \frac{\kappa_+}{2}(u-v)}, \quad (4.173)$$

takes the same form as for $v < v_0$ in the coordinates (u_{in}, v)

$$ds^2 = -\frac{4r_0^2}{(x^+ - x^-)^2}dx^+dx^- . \tag{4.174}$$

Note that this can always be accomplished since the metric has constant curvature everywhere. These coordinates are unambiguously defined, up to Möbius transformations. Nevertheless, as we have stressed many times, this sort of ambiguity does not change the radiation emitted. We can transform (4.174) into (4.173) with the change of coordinates

$$\frac{\kappa_+ x^+}{2} = \tanh\frac{\kappa_+ v}{2} ,$$
$$\frac{\kappa_+ x^-}{2} = \tanh\frac{\kappa_+ u}{2} . \tag{4.175}$$

We can realize immediately that these coordinates are related with the (non-extremal) Kruskal coordinates (U, V)

$$U = -\frac{1}{\kappa_+}e^{-\kappa_+ u} ,$$
$$V = \frac{1}{\kappa_+}e^{\kappa_+ v} , \tag{4.176}$$

by the following Möbius transformations

$$x^+ = \frac{2}{\kappa_+}\frac{\kappa_+ V - 1}{\kappa_+ V + 1} ,$$
$$x^- = \frac{2}{\kappa_+}\frac{\kappa_+ U + 1}{-\kappa_+ U + 1} . \tag{4.177}$$

Taking into account the above result and that

$$\langle in| : T_{vv} : |in\rangle = 0 , \tag{4.178}$$

and

$$\langle in| : T_{uu}(u) : |in\rangle = \frac{\pi}{12\hbar}(k_B T_H)^2 , \tag{4.179}$$

it is easy to find that

$$\langle in| : T_{VV} : |in\rangle = -\frac{\hbar}{48\pi}\frac{1}{V^2} , \tag{4.180}$$

and

$$\langle in| : T_{UU}(u) : |in\rangle = 0 . \tag{4.181}$$

This represents again the negative influx of radiation entering the black hole horizon, as measured by a free falling observer with Kruskal-type coordinates (U, V).

4.5.1 Three vacuum states

Although the analysis so far has involved explicitly only one vacuum state, we have introduced three different coordinates systems which have associated three different vacuum states, in complete analogy with the Schwarszchild versus Rindler considerations of Section 4.3:

- The Boulware vacuum state is constructed using the plane wave modes with respect to the coordinates (u, v).

- $|in\rangle$ corresponds again to the Unruh state. The corresponding modes are plane waves with respect to the coordinates (u_{in}, v), or equivalently (U, v).

- The Hartle–Hawking state is constructed using the plane wave modes with respect to the Kruskal-type coordinates (U, V), or equivalently, with respect to the Poincaré coordinates $x^{\pm}$.

4.6 The Moving-Mirror Analogy for the Hawking Effect

This section is devoted to presenting another approximation scheme to model, in a flat spacetime, the basic features of Hawking radiation. In the near-horizon approximation considered in the previous sections the matter field f propagates freely and all the physics is contained in the relation between two different sets of coordinates that we generically refer to as $x^{\pm}$ and $y^{\pm}$. All the discussions can then be reinterpreted in terms of the so called moving-mirror analogy [Fulling and Davies (1976); Davies and Fulling (1977)].[16] The idea is the following. Instead of having a two-dimensional spacetime with two different and natural sets of modes $(u_i(x^-), v_i(x^+))$ and $(\tilde{u}_i(y^-), \tilde{v}_i(y^+))$ where the coordinates $y^{\pm}$ and $x^{\pm}$ are related by a

---

[16] A deeper analysis can be found in [Carlitz and Willey (1987); Wilczek (1993); Parentani (1996)].

conformal transformation:

$$y^- = y^-(x^-),$$
$$y^+ = y^+(x^+), \qquad (4.182)$$

one can introduce a *boundary* in the spacetime to produce the same physical consequences. The effect of the boundary is to disturb the modes in such a way that modes that at past null infinity behave as $(u_i(x^-), v_i(x^+))$, once evolved to future null infinity will take a form similar to $(\tilde{u}_i(y^-), \tilde{v}_i(y^+))$. This is the main property of a mirror model: it can nicely mimic the physics in a non-trivial background (i.e., Hawking radiation in a black hole geometry), or the effect of having two different sets of modes in a fixed background (as in the Unruh effect).

The basic ingredient to define a moving mirror model is the introduction of a (time-dependent) reflecting boundary in the space such that the field is assumed to satisfy the boundary condition $f = 0$ along its worldline. It is convenient to parametrize the trajectory of the mirror in terms of null coordinates

$$x^+ = p(x^-). \qquad (4.183)$$

Therefore the boundary condition is just

$$f(x^-, x^+ = p(x^-)) = 0. \qquad (4.184)$$

A null ray at fixed x^+ which reflects off the mirror becomes a null ray of fixed x^-. The concrete relation between the coordinates of this null ray is given by the mirror's trajectory $x^+ = p(x^-)$. In terms of mode functions it is easy to construct plane wave solutions of the equation $\Box f = 0$ vanishing on the worldline of the wall:

$$u_w^{in} = \frac{1}{\sqrt{4\pi w}} (e^{-iwx^+} - e^{-iwp(x^-)}). \qquad (4.185)$$

They represent a positive frequency wave e^{-iwx^+}, coming from I_R^-, that reflects on the curve $x^+ = p(x^-)$ and becomes an outgoing wave $e^{-iwp(x^-)}$ propagating to I_R^+, which in general is a superposition of positive and negative frequency parts. In addition we also have modes representing a pure outgoing positive frequency wave e^{-iwx^-} which is produced by the reflection of a wave $e^{-ip^{-1}(x^+)}$ from I_R^-

$$u_w^{out} = \frac{1}{\sqrt{4\pi w}} (e^{-iwx^-} - e^{-iwp^{-1}(x^+)}). \qquad (4.186)$$

The above two sets of modes are the natural mode basis for inertial observers at I_R^- and I_R^+, respectively, and for the dynamics of the field at the right hand side of the mirror. Similar basis can be constructed to describe the dynamics at the left of the mirror, but we shall restrict, as usual, to the right region. Moreover, we can construct wave packet basis from the plane wave modes and rederive the same results obtained in Section 4.2. The expectation value of the particle number in mode k is given by

$$\langle 0_{in}|N_k^{out}|0_{in}\rangle = -\frac{1}{\pi}\int_{I_R^+} dx^- dx'^- u_k^{out}(x^-) u_k^{out*}(x'^-) \times \qquad (4.187)$$
$$\left[\frac{dp}{dx^-}(x^-)\frac{dp}{dx^-}(x'^-)\frac{1}{(p(x^-)-p(x'^-))^2} - \frac{1}{(x^- - x'^-)^2}\right],$$

and the flux of energy radiated to the right is given by the Schwarzian derivative

$$\langle 0_{in}| :T_{--}(x^-): |0_{in}\rangle = -\frac{\hbar}{24\pi}\{p(x^-), x^-\}. \qquad (4.188)$$

We shall now consider with some detail those trajectories that play an important role from a physical point of view. The first one is the trajectory generating the thermal radiation that we get in a black hole formed by gravitational collapse at late times.

4.6.1 *Exponential trajectory: thermal radiation*

The results concerning thermal radiation obtained in Section 4.3 can be rederived in this context by considering the mirror trajectory $p(x^-) = -\kappa^{-1}e^{-\kappa x^-}$, as depicted in Fig. 4.7.

Using (4.187) and (4.188), straightforward calculations similar to those performed in Section 4.3 lead to

$$\langle 0_{in}|\vec{N}_{jn}|0_{in}\rangle = \frac{1}{e^{2\pi w_j \kappa^{-1}} - 1}, \qquad (4.189)$$

and

$$\langle 0_{in}| :T_{--}: |0_{in}\rangle = \frac{\hbar \kappa^2}{48\pi}. \qquad (4.190)$$

This is nothing else but the standard thermal radiation measured at I_R^+.

Due to the peculiar form of the trajectory, not all the "in" modes are reflected by the mirror. Indeed, those modes located at $x^+ > 0$ propagate

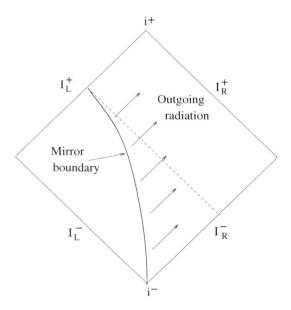

Fig. 4.7 Penrose diagram describing a moving mirror with exponential trajectory. The outgoing radiation produced by it at I_R^+ is exactly thermal.

without distortion to the left asymptotic region I_L^+. Therefore, for $x^+ > 0$ and at I_L^+ we have

$$\langle 0_{in} | : T_{++} : | 0_{in} \rangle = 0 . \qquad (4.191)$$

The multi-point correlators of the stress tensor are also identical to the vacuum ones. Moreover there exist correlations between I_L^+ and I_R^+, originated from the correlations in the "in" vacuum at I_R^- between the regions $x^+ > 0$ and $x^+ < 0$. For instance, we have[17]

$$\langle 0_{in} | \partial_+ f(x^+ > 0) \partial_- f(x^-) | 0_{in} \rangle = -\frac{\hbar}{4\pi} \frac{p'(x^-)}{(x^+ - p(x^-))^2} . \qquad (4.192)$$

The existence of the above correlations is crucial to ensure the purity of the "in" vacuum state. Only when one restricts measurements to I_R^+ and ignore the cross correlations one gets the mixed thermal state associated to the thermal radiation (4.189) and (4.190). However, one could be tempted

---

[17]This can be derived easily from the two-point function adapted to the boundary conditions imposed by the mirror $\langle 0_{in} | f(x) f(x') | 0_{in} \rangle = -\frac{\hbar}{4\pi} \ln \frac{(x^- - x'^-)(x^+ - x'^+)}{(x^- - x'^+)(x^+ - x'^-)}$.

to conclude that no real flux can be measured at I_L^+ and that the only role of this region is to allow for the existence of left-right correlations. A deeper analysis of the situation unravels that, due to the particular form of the trajectory, the restriction to the right of the mirror is such that only the portion $x^+ > 0$ of I_L^+ exists. Therefore, a proper quantization on this segment requires the introduction of a new null coordinate adapted to it. For instance one can introduce the Rindler-type coordinate y^+ defined by $y^+ = \kappa^{-1} \ln \kappa x^+$. With respect to y^+, we have a non-vanishing ingoing flux along I_L^+

$$\langle 0_{in}| : T_{y^+ y^+} : |0_{in}\rangle = \frac{\hbar \kappa^2}{48\pi} . \qquad (4.193)$$

The presence of a non-vanishing energy flux is not surprising. A careful analysis following [Fabbri et al. (2004)] shows that we have a non-trivial flux also using the x^+ coordinate. Indeed, Eq. (4.191) is valid everywhere for $x^+ > 0$, but not at $x^+ = 0$. Along this null line a "shock wave" of infinite energy is present. The reason for this is the absence of correlations for points before and after $x^+ = 0$ at I_L^+ in the state $|0_{in}\rangle$ (i.e., $\langle 0_{in} | \partial_+ f(x^+ < 0) \partial_+ f(x'^+ > 0) | 0_{in}\rangle_{I_L^+} = 0$ for $x^+ < 0$), which are instead present for the natural vacuum at I_L^+ covering all the real line $-\infty < x^+ < +\infty$. Therefore, in the limit $x^+ \to -\epsilon/2$, $x'^+ \to \epsilon/2$ with $\epsilon \to 0$ we have

$$\langle 0_{in}| : T_{++} : |0_{in}\rangle \sim \frac{\hbar}{4\pi \epsilon^2} . \qquad (4.194)$$

4.6.2 Radiationless trajectories

The natural timelike trajectories producing no radiation are the inertial ones $p(x^-) = \lambda x^- + constant$, with $\lambda > 0$. This example is trivial since the formulae (4.187) and (4.188) give immediately a vanishing result. However these do not exhaust all the possibilities. In fact, a generic trajectory of the form

$$p(x^-) = \frac{ax^- + b}{cx^- + d} , \qquad (4.195)$$

where $ad - bc = 1$ (i.e., $p(x^-)$ is a Möbius function), gives a vanishing Schwarzian derivative and then a vanishing flux. Moreover, since

$$\frac{dp}{dx^-}(x^-)\frac{dp}{dx^-}(x'^-)\frac{1}{(p(x^-) - p(x'^-))^2} - \frac{1}{(x^- - x'^-)^2} \qquad (4.196)$$

also vanishes for Möbius functions there is no particle production too.

4.6.2.1 Hyperbolic trajectories

The typical non-inertial trajectory defined by a Möbius function is

$$p(x^-) = -\frac{1}{a^2 x^-} \,. \qquad (4.197)$$

This consists of two different hyperbolic branches, one with $x^- < 0$ and the other with $x^- > 0$. They correspond to two uniformly accelerated mirrors, with proper acceleration given by $|a|$. See Fig. 4.8.

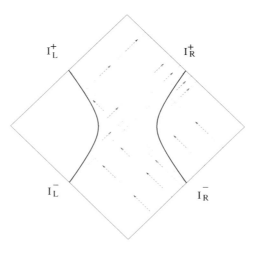

Fig. 4.8 A mirror with two hyperbolic branches.

Note that all the modes supported on I_R^- are reflected to I_R^+. The "in" modes supported on the interval $x^+ \in]-\infty, 0[$ reach I_R^+ on $x^- \in]0, \infty[$; the "in" modes supported on the interval $x^+ \in]0, \infty[$ reach I_R^+ on $x^- \in]-\infty, 0[$. In this way, the correlations existing between positive and negative x^+ are transferred without distortion to correlations between positive and negative x^-. Therefore, the "in" vacuum state is perceived also at I_R^+ as the natural vacuum state: $|0_{out}\rangle = |0_{in}\rangle$. This provides a trivial example where the "evolution" from I_R^- to I_R^+ preserves purity. A more involved example, probably more interesting from the physical point of view, is given in the next and final subsection.

4.6.3 Asymptotically inertial trajectories and unitarity

Following the example given in [Carlitz and Willey (1987)], let us consider a mirror which is initially at rest, i.e., $p(x^-) \approx x^-$ for $x^- \to -\infty$, then starts accelerating approaching, at intermediate times, the exponential trajectory, and finally decelerates and comes back to rest $p(x^-) \approx x^-$ for $x^- \to +\infty$. See Fig. 4.9.

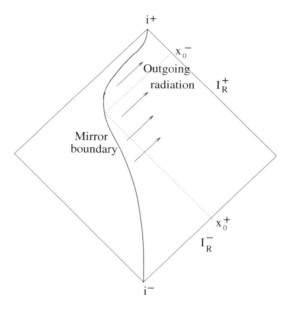

Fig. 4.9 Penrose diagram describing a moving mirror which is inertial in the asymptotic past and future and approaches the exponential trajectory at intermediate times.

The modification of the exponential trajectory at early and late times is designed to avoid that the mirror emits for all times.[18] This is similar to the analogous problem for the black hole in the fixed background approximation stressed in Section 3.5. In a sense, this modification of the mirror trajectory mimics the backreaction effects in a black hole spacetime. This scenario is unitary by construction, as all incident waves from I_R^- are reflected to I_R^+. Moreover, because the mirror is inertial both in the asymptotic past and in the future $\langle 0_{in}| : T_{--} : |0_{in}\rangle \to 0$ there. An interesting feature of this scenario is that the emitted flux is not always positive. Indeed, at the

---
[18] Note, however, that this is not the only possibility, i.e., the mirror can start inertial at i^- and end on I_L^+ following a hyperbolic trajectory.

saddle point x_0^-, where $p''(x_0^-) = 0$, it is

$$\langle 0_{in}| :T_{--}: |0_{in}\rangle|_{x_0^-} = -\frac{\hbar}{24\pi}\left(\frac{p'''(x_0^-)}{p'(x_0^-)} - \frac{3}{2}\frac{p''(x_0^-)^2}{p'(x_0^-)^2}\right)$$
$$= -\frac{\hbar}{24\pi}\frac{p'''(x_0^-)}{p'(x_0^-)}. \qquad (4.198)$$

Moreover, since $p' > 0$ and, as we will see in the following, $p'''(x_0^-) > 0$ we have[19]

$$\langle 0_{in}| :T_{--}: |0_{in}\rangle|_{x_0^-} < 0. \qquad (4.199)$$

One can think of the modes defined for $x^+ < x_0^+$ to be analogous of those with $x^+ < 0$ in the case of the exponential trajectory. Due to the modification of the trajectory at late times, the correlations between points $x_1^+ < x_0^+$ and $x_2^+ > x_0^+$ are recovered at future infinity. Moreover, the correlations at I_R^+, that in absolute value are always less than the vacuum ones and decay exponentially for the exponential trajectory[20] become bigger than the vacuum for some points $x_1^- < x_0^-$ and $x_2^- > x_0^-$. In particular, for the case of points infinitesimally close to x_0^- this can be seen from the expansion (4.65)

$$\frac{\langle 0_{in}|\partial_- f(x_1^-)\partial_- f(x_2^-)|0_{in}\rangle}{\langle 0_{out}|\partial_- f(x_1^-)\partial_- f(x_2^-)|0_{out}\rangle} = 1 - \frac{4\pi}{\hbar}\langle 0_{in}| :T_{--}: |0_{in}\rangle|_{x_0^-}(x_1^- - x_2^-)^2$$
$$- \frac{2\pi}{\hbar}\frac{d}{dx^-}\langle 0_{in}| :T_{--}: |0_{in}\rangle|_{x_0^-}(x_1^- - x_2^-)^3$$
$$+ \cdots. \qquad (4.201)$$

Also for distant points we can have correlations greater than those of the vacuum. To show this let us consider first the graph of the function $p'(x^-)$ for a mirror which is initially at rest and approaches the exponential trajectory at intermediate and late times, as depicted in Fig. 4.10.

The modification to this trajectory which accounts for the mirror to return to rest at late times is such that $p'(x^-)$ is generically of the form given in Fig. 4.11. From this figure it is clear that x_0^- is a minimum of the

---
[19] Negative energy can arise, as a quantum effect, even in flat space by considering suitable many particle states [Epstein et al. (1965)].

[20] In fact for the exponential trajectory we have

$$\frac{\langle 0_{in}|\partial_- f(x_1^-)\partial_- f(x_2^-)|0_{in}\rangle}{\langle 0_{out}|\partial_- f(x_1^-)\partial_- f(x_2^-)|0_{out}\rangle} = \frac{\kappa^2(x_1^- - x_2^-)^2}{4\sinh^2\kappa(x_1^- - x_2^-)/2} < 1.$$

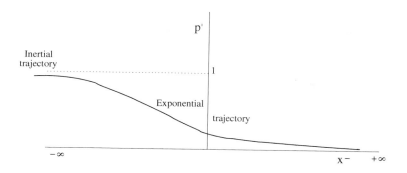

Fig. 4.10 Graphic representation of $p'(x^-)$ for a mirror starting at rest and approaching to the exponential trajectory.

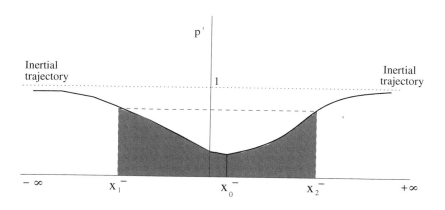

Fig. 4.11 Graphic representation of $p'(x^-)$ for asymptotically inertial trajectories.

function $p'(x^-)$, i.e., $p'''(x_0^-) > 0$, leading to the result (4.199). Moreover, take the two points $x_1^- < x_0^-$ and $x_2^- > x_0^-$, such that $p'(x_1^-) = p'(x_2^-)$. The ratio of correlations

$$\frac{\langle 0_{in}|\partial_- f(x_1^-)\partial_- f(x_2^-)|0_{in}\rangle}{\langle 0_{out}|\partial_- f(x_1^-)\partial_- f(x_2^-)|0_{out}\rangle} = \frac{p'(x_1^-)p'(x_2^-)(x_1^- - x_2^-)^2}{(p(x_1^-) - p(x_2^-))^2} \qquad (4.202)$$

is given by the square root of the quotient of the area of the rectangular region over the area defined by the function $p'(x^-)$ (i.e., the shaded region). It is clear that this quantity is greater than one. The emergence of strong correlations (i.e., bigger than the vacuum ones) in connection with negative fluxes and unitarity was also stressed in [Bose et al. (1996)].

Chapter 5

Stress Tensor, Anomalies and Effective Actions

One of the main problems in the analysis of black hole radiation is to improve the fixed background approximation used in the previous chapters. The Hawking radiation contributes to the stress tensor, modifying the classical one describing the collapsing matter, and therefore the classical gravity-matter equations should be modified accordingly. This is the motivation underlying the introduction of a modified set of equations, called the semiclassical Einstein equations

$$G_{\mu\nu}(g_{\mu\nu}) = 8\pi \langle T_{\mu\nu}(g_{\mu\nu}) \rangle , \qquad (5.1)$$

where the right hand side represents the expectation value of the stress tensor operator of the matter fields propagating on a spacetime with metric $g_{\mu\nu}$. Since the left hand side is a conserved tensor, $\nabla_\mu G^{\mu\nu} = 0$, consistency implies that $\langle T_{\mu\nu} \rangle$ must also be conserved

$$\nabla_\mu \langle T^{\mu\nu} \rangle = 0 . \qquad (5.2)$$

This requires that the quantization of the matter fields in the curved spacetime should be compatible with general covariance.

This means, for instance, that the normal ordered stress tensor operator used extensively in Chapter 4 is not a good candidate for the right hand side of Eqs. (5.1). In general it does not transform as a covariant tensor and therefore it is not compatible with the tensorial properties of $G_{\mu\nu}$. This should not be a surprise at all since normal ordering is just a removal of the (divergent) zero-point energy of the matter field. This can be justified in Minkowski space but not in general relativity, where the energy itself is a source of gravity. The absolute value of energy maintains a physical meaning. Therefore, more involved regularization schemes are required to deal with the underlying gravitational field [Birrell and Davies (1982)].

This makes the problem very difficult in general and, indeed, there is no generic expression for $\langle T_{\mu\nu}(g_{\mu\nu})\rangle$, except when the degree of symmetry of the particular matter fields and the specific spacetimes considered is sufficiently high. This happens, typically, in conformally flat spacetimes and for theories for which the classical action is invariant under conformal (Weyl) transformations

$$g_{\mu\nu} \to \Omega^2(x) g_{\mu\nu} \ . \tag{5.3}$$

In these situations one has all the ingredients to write down the semiclassical equations. This is not the case, however, for four-dimensional black hole spacetimes. Only in two dimensions, and for conformally invariant matter fields, there is an available expression for $\langle T_{\mu\nu}(g_{\mu\nu})\rangle$ for a generic metric. This is a consequence of the fact that every two-dimensional spacetime is (locally) conformally flat. Up to diffeomorphisms, every metric can be written as

$$ds^2 = e^{2\rho} ds^2_{(0)} \ , \tag{5.4}$$

where $ds^2_{(0)}$ represents a flat metric.

The classical Weyl invariance manifests itself in the existence of a traceless stress tensor $g^{\mu\nu} T_{\mu\nu} = 0$. This can be easily seen by considering infinitesimal Weyl transformations $\Omega(x) \simeq 1 + \omega(x)$ for which

$$\delta g_{\mu\nu} = 2\omega(x) g_{\mu\nu} \ . \tag{5.5}$$

The classical matter action S varies as

$$\delta S = \int d^n x \sqrt{-g} T_{\mu\nu} g^{\mu\nu} \omega(x) \ , \tag{5.6}$$

where n is the dimension of the spacetime. It is then clear that the local Weyl invariance of S implies that the classical stress tensor is traceless. However, at the quantum level things are different. The requirement of compatibility of the regularization procedure with general covariance forces to produce a (quantum) non-vanishing trace, first discovered by [Capper and Duff (1974)]. The quantum value of the trace is independent of the state in which the expectation value is taken. It is given in terms of pure geometrical objects and, in four dimensions, is a linear combination of $C_{\alpha\beta\gamma\delta} C^{\alpha\beta\gamma\delta}$ ($C_{\alpha\beta\gamma\delta}$ is the Weyl tensor), $R_{\alpha\beta} R^{\alpha\beta}$, R^2 and $\Box R$ [Deser et al. (1976)]. The dimensionless coefficients are known and depend on the spin of the particular conformally invariant field considered.

In two dimensions the trace can only be proportional to the two-dimensional Ricci scalar $R^{(2)}$ (hereafter R) and it is given by

$$\langle T \rangle = \frac{\hbar}{24\pi} R \, , \qquad (5.7)$$

where $T \equiv T_a^a$. Moreover, the conservation equations allow to determine completely the remaining components of $\langle T_{\mu\nu} \rangle$. This is very important from the physical point of view since it allows to write explicitly the semiclassical equations and to attack, in simplified scenarios, the backreaction problem. This will be the focus of Chapter 6.

The aim of the next section is to provide an alternative and more intuitive derivation of these results avoiding the technicalities of any covariant regularization scheme. We shall base our derivation on physical considerations. We shall exploit the results of Chapter 4 on the normal ordered stress tensor and the concept of locally inertial frames in connection with Einstein's equivalence principle. We will then complement this derivation by introducing the Polyakov effective action, followed by a detailed analysis of the quantum states relevant for black holes.

Moreover, the above strategy will be applied to a more sophisticated case: a spherically reduced four-dimensional scalar field in the s-wave approximation. We will then write down the backreaction equations in this approximation and we will discuss the delicate question of choosing the state-dependent functions.

5.1 Relating the Virasoro and Trace Anomalies via Locally Inertial Coordinates

In this section we shall show how to determine all the components of $\langle T_{ab} \rangle$, in two dimensions, by using the Virasoro anomaly of the normal ordered stress tensor. The fact that normal ordering works well in Minkowski spacetime suggests that in the presence of gravity we might use it in the locally inertial frames which are defined at any spacetime point. This route is compatible with general covariance and indeed it allows to determine all the components of the covariant quantum stress tensor.

Let us consider a two-dimensional massless scalar field f with classical action

$$S = -\frac{1}{2} \int d^2x \sqrt{-g} (\nabla f)^2 \, . \qquad (5.8)$$

The physical relevance of this theory has already been motivated in Chapter 4. In the near-horizon approximation the four-dimensional matter scalar field is governed, for each angular momentum mode, by the above action.

In a generic conformal coordinate system $x^\pm$, with metric given by $ds^2 = -e^{2\rho}dx^+dx^-$, we can define a quantization associated to the choice of the positive frequency modes

$$(4\pi w)^{-1/2}e^{-iwx^+}, \quad (4\pi w)^{-1/2}e^{-iwx^-}. \tag{5.9}$$

With this choice of modes one can immediately construct the normal ordered stress tensor $: T_{\pm\pm}(x^\pm) :$. This operator cannot be identified with the quantities (we indicate with $|\Psi\rangle$ the quantum state in the expectation values of the operators) $\langle\Psi|T_{\pm\pm}(x^\pm)|\Psi\rangle$ appearing in the right hand side of the semiclassical equations. The reason is clear, $: T_{\pm\pm}(x^\pm) :$ does not transform as a tensor under coordinate transformations. Indeed, under conformal transformations it picks up a non-tensorial term proportional to the Schwarzian derivative. This is a consequence of the fact that normal ordering is linked to the particular coordinate system $x^\pm$ used and therefore breaks general covariance. So $\langle\Psi| : T_{\pm\pm}(x^\pm) : |\Psi\rangle$ and $\langle\Psi|T_{\pm\pm}(x^\pm)|\Psi\rangle$ are, in general, different objects. However, there is a particular situation in which they coincide. It is well established that normal ordering works perfectly well in Minkowski space using inertial coordinates. Therefore, in Minkowski space and for (null) *Minkowskian coordinates* $\xi^\pm$

$$ds^2 = -d\xi^+ d\xi^-, \tag{5.10}$$

we have

$$\langle\Psi|T_{\pm\pm}(\xi^\pm)|\Psi\rangle \equiv \langle\Psi| : T_{\pm\pm}(\xi^\pm) : |\Psi\rangle . \tag{5.11}$$

In any other conformal coordinate system the stress tensor should be obtained by the usual tensorial law:

$$\langle\Psi|T_{\pm\pm}(x^\pm)|\Psi\rangle \equiv \left(\frac{d\xi^\pm}{dx^\pm}\right)^2 \langle\Psi|T_{\pm\pm}(\xi^\pm)|\Psi\rangle . \tag{5.12}$$

To clarify the physical meaning of the above relations we can consider the scenario of the Unruh effect explained in Chapter 4. The null Minkowskian coordinates were called (U, V) and according to our notation we have the identification $\xi^+ \equiv V$ and $\xi^- \equiv U$. As explained in Subsection 4.3.3, a uniformly accelerating observer measures a thermal bath of (Rindler) particles and, related to this, a nonvanishing normal ordered stress "tensor": $\langle 0_M | : T_{uu} : | 0_M \rangle = \frac{\hbar \kappa^2}{48\pi}$, $\langle 0_M | : T_{vv} : | 0_M \rangle = \frac{\hbar \kappa^2}{48\pi}$, where, we remember, $|0_M\rangle$ is the ordinary Minkowski vacuum. However, these nonvanishing and observer-dependent quantities can not be used as the right hand side of the semiclassical equations, otherwise they would violate general covariance which instead requires that

$$\langle 0_M | T_{uu}(u) | 0_M \rangle \equiv \left(\frac{dU}{du}\right)^2 \langle 0_M | T_{UU}(U) | 0_M \rangle = 0$$

$$\langle 0_M | T_{vv}(v) | 0_M \rangle \equiv \left(\frac{dV}{dv}\right)^2 \langle 0_M | T_{VV}(V) | 0_M \rangle = 0 . \quad (5.13)$$

The above discussion can be naturally extended to a curved background. In this case one should do it point by point and replace the (global) inertial frame $\xi^\pm$ by a "locally inertial frame" $\xi_X^\pm$, associated to the point X, according to Einstein's equivalence principle. In this frame the metric and the connection are similar to those of Minkowski space in the sense that

$$ds^2|_X = -d\xi_X^+ d\xi_X^-|_X$$
$$\Gamma^\alpha_{\mu\nu}(\xi_X^\pm)|_X = 0 . \quad (5.14)$$

The last condition is equivalent to

$$\partial_{\xi_X^\mu} \rho(\xi_X^\pm)|_X = 0 . \quad (5.15)$$

As already seen in Chapter 2, this implies that the metric in terms of the coordinates $\xi_X^\pm$ can be expanded around the point X up to second order as

$$ds^2 = -d\xi_X^+ d\xi_X^- + O\left((\xi_X^\pm - \xi_X^\pm(X))^2\right) d\xi_X^+ d\xi_X^- . \quad (5.16)$$

Varying X one generates a family of locally inertial frames that will be very useful to construct a covariant quantum stress tensor. As suggested by Eq. (5.11), we shall impose that $\langle \Psi | : T_{\pm\pm}(x^\pm) : | \Psi \rangle$ and $\langle \Psi | T_{\pm\pm}(x^\pm) | \Psi \rangle$ coincide at the point X in the locally inertial frame $\xi_X^\pm$

$$\langle \Psi | T_{\pm\pm}(\xi_X^\pm(X)) | \Psi \rangle \equiv \langle \Psi | : T_{\pm\pm}(\xi_X^\pm(X)) : | \Psi \rangle . \quad (5.17)$$

In any other frame they will no longer coincide since $\langle \Psi | T_{\pm\pm}(x^\pm) | \Psi \rangle$ transforms as a tensor and $\langle \Psi | : T_{\pm\pm}(x^\pm) : | \Psi \rangle$ does not. We will now show that the above equality (5.17) is strong enough to determine all the components of $\langle \Psi | T_{\pm\pm}(x^\pm) | \Psi \rangle$ everywhere in an arbitrary conformal frame. Since we are assuming that $\langle \Psi | T_{\pm\pm}(x^\pm) | \Psi \rangle$ transforms as a tensor, the explicit definition of $\langle\Psi|T_{\pm\pm}(x^\pm(X))|\Psi\rangle$, at the point X, is given by

$$\langle\Psi|T_{\pm\pm}(x^\pm(X))|\Psi\rangle \equiv \left(\frac{d\xi_X^\pm}{dx^\pm}\right)^2 (X)\langle\Psi|:T_{\pm\pm}(\xi_X^\pm(X)):|\Psi\rangle \ . \qquad (5.18)$$

It is easy to see compatibility with general covariance. For a different conformal frame $y^\pm$ we also have

$$\langle\Psi|T_{\pm\pm}(y^\pm(X))|\Psi\rangle \equiv \left(\frac{d\xi_X^\pm}{dy^\pm}\right)^2 (X)\langle\Psi|:T_{\pm\pm}(\xi_X^\pm(X)):|\Psi\rangle \ , \qquad (5.19)$$

and from the above two equations we immediately recover a tensorial transformation law

$$\langle\Psi|T_{\pm\pm}(y^\pm)|\Psi\rangle = \left(\frac{dx^\pm}{dy^\pm}\right)^2 \langle\Psi|T_{\pm\pm}(x^\pm)|\Psi\rangle \ , \qquad (5.20)$$

when passing from $x^\pm$ to $y^\pm$.

At this point one might be worried about the use of *local* coordinates in our discussion. The definition of a quantum state is *global* and it could potentially be in conflict with the introduction of normal ordering with respect to a set of positive frequency modes

$$(4\pi w)^{-1/2}e^{-iw\xi_X^+}, \quad (4\pi w)^{-1/2}e^{-iw\xi_X^-} \ , \qquad (5.21)$$

defined in terms of local coordinates. However, as far as one is concerned with a local observable like the stress tensor, and at the point X, which is always reached by the local inertial coordinates, one can legitimately introduce such a normal ordering. In fact, as we have already explained in the previous chapter, there is a well established expression relating the normal ordered stress tensor in two different conformal coordinate systems, irrespective of the nature of the coordinates.[1]

The normal ordered stress tensor with respect to the locally inertial coordinates $\xi_X^\pm$ can be reexpressed, via the Virasoro anomaly, in terms of the normal ordered operator with respect to the coordinates $x^\pm$:

$$:T_{\pm\pm}(x^\pm(X)): = \left(\frac{d\xi_X^\pm}{dx^\pm}\right)^2 (X):T_{\pm\pm}(\xi_X^\pm(X)): -\frac{\hbar}{24\pi}\{\xi_X^\pm,x^\pm\}|_X \ , \quad (5.22)$$

---

[1] For instance, local Rindler coordinates cannot cover the full Minkowski space, and despite this it makes perfect sense to talk about the normal ordered stress tensor in Rindler coordinates. Of course, this implies restricting measurements to the Rindler wedge.

where $\{\xi_X^\pm, x^\pm\}$ is the Schwarzian derivative. We can then rewrite Eq. (5.18) as

$$\langle\Psi|T_{\pm\pm}(x^\pm(X))|\Psi\rangle = \langle\Psi| : T_{\pm\pm}(x^\pm(X)) : |\Psi\rangle + \frac{\hbar}{24\pi}\{\xi_X^\pm, x^\pm\}|_X \ . \quad (5.23)$$

This shows clearly that the difference between the covariant stress tensor and the normal ordered one is given, at the point X, by the Schwarzian derivative between $\xi_X^\pm$ and $x^\pm$.

Note that if the spacetime is Minkowski, as in the previous discussion of the Unruh effect, and the coordinates $x^\pm$ are the Rindler coordinates, the additional terms coming from the Schwarzian derivative will cancel the contribution of the normal ordered stress tensor, recovering in this way the result of Eq. (5.13). In general, the evaluation of the Schwarzian derivative depends on the particular locally inertial frame chosen. This is so because it involves third order derivatives

$$\{\xi_X^\pm, x^\pm\} = \frac{d^3\xi_X^\pm}{dx^{\pm 3}} \bigg/ \frac{d\xi_X^\pm}{dx^\pm} - \frac{3}{2}\left(\frac{d^2\xi_X^\pm}{dx^{\pm 2}} \bigg/ \frac{d\xi_X^\pm}{dx^\pm}\right)^2 \ , \quad (5.24)$$

while the locally inertial frame is defined, up to two-dimensional Poincaré tranformations, to second order

$$\xi_X^\alpha = a^\alpha + b_\mu^\alpha(x^\mu - x^\mu(X)) + \frac{1}{2}b_\lambda^\alpha\Gamma_{\mu\nu}^\lambda(x(X))(x^\mu - x^\mu(X))(x^\nu - x^\nu(X)) + \ldots , \quad (5.25)$$

where a^α are two arbitrary constants and the coefficients b_μ^α are such that

$$b_\mu^\alpha b_\nu^\beta \eta_{\alpha\beta} = g_{\mu\nu}(X) \ . \quad (5.26)$$

As already stressed in Chapter 2, irrespective of the higher order terms, the coordinates ξ_X^α in Eq. (5.25) are locally inertial. However the third order is important since the Schwarzian derivative involves third derivatives as well.[2] So an additional input is necessary to define the components $\langle\Psi|T_{\pm\pm}(x^\pm(X))|\Psi\rangle$ of the quantum stress tensor. Among all the possible change of coordinates defined by Eq. (5.25) we can take them to be conformal by choosing

$$b_+^- = 0 = b_-^+ . \quad (5.27)$$

---

[2]Intuitively, one can think the reason for this is that although the acceleration of a classical point particle at a given point X can always be eliminated by going to any locally inertial frame there, a quanta is an "extended object" and therefore it appears to be sensitive to the particular choice of ξ_X^α (i.e., to the third order terms in the expansion given by Eq. (5.25)).

Equation (5.26) is then just

$$b_+^+ b_-^- = e^{2\rho(X)} . \tag{5.28}$$

For convenience, we can also choose the locally inertial coordinates such that $\xi_X^\pm(X) = 0$, so

$$a^+ = 0 = a^- . \tag{5.29}$$

Therefore Eq. (5.25) turns out to be, taking into account that $\Gamma_{\pm\pm}^\pm = 2\partial_\pm \rho$,

$$\xi_X^\pm = b_\pm^\pm [(x^\pm - x^\pm(X)) + \partial_\pm \rho(X)(x^\pm - x^\pm(X))^2 \\ + F_\pm (x^\pm - x^\pm(X))^3 + \cdots] . \tag{5.30}$$

There is a natural way to extend the definition of the locally inertial coordinates to third order. To see it, it is convenient to write down the explicit expansion of the metric (5.16) to include the second order terms

$$ds^2 = -d\xi_X^+ d\xi_X^- - e^{2\rho(X)} [\partial^2_{\xi_X^+} \rho(X)(\xi_X^+ - \xi_X^+(X))^2 \\ + \partial^2_{\xi_X^-} \rho(X)(\xi_X^- - \xi_X^-(X))^2 \\ + 2\partial_{\xi_X^+} \partial_{\xi_X^-} \rho(X)(\xi_X^+ - \xi_X^+(X))(\xi_X^- - \xi_X^-(X))] d\xi_X^+ d\xi_X^- + \cdots . \tag{5.31}$$

The term containing

$$\partial_{\xi_X^+} \partial_{\xi_X^-} \rho(X) \tag{5.32}$$

cannot be removed by fixing appropriately the third order terms in the expansion (5.30). The quantity (5.32) is directly related to the scalar curvature. In fact, in conformal gauge we have

$$R = 8e^{-2\rho} \partial_+ \partial_- \rho . \tag{5.33}$$

The possibility of removing this term would imply a meaningless restriction ($R = 0$) on the geometry of the spacetime at the point X. The other two terms, instead, can be removed without affecting the geometry.

For a flat metric the natural coordinates $\xi_X^\pm$ are the Minkowskian ones. For them all second order terms in (5.31) vanish and Eq. (5.11) holds. Therefore, for an arbitrary metric it is natural to impose the vanishing of the following second derivatives

$$\partial^2_{\xi_X^+} \rho(X) = 0 \tag{5.34}$$

$$\partial^2_{\xi_X^-} \rho(X) = 0 . \tag{5.35}$$

A straightforward and simple calculation allows to determine, from the above conditions, the coefficients $F_\pm$ in Eq. (5.30). We find that

$$F_\pm = \frac{1}{3}\partial_\pm^2 \rho(X) + \frac{2}{3}(\partial_\pm \rho(X))^2 \ . \tag{5.36}$$

Using the above expressions it is easy to get, for an arbitrary point,

$$\langle \Psi | T_{\pm\pm}(x^\pm) | \Psi \rangle = -\frac{\hbar}{12\pi}(\partial_\pm \rho \partial_\pm \rho - \partial_\pm^2 \rho) + \langle \Psi | : T_{\pm\pm}(x^\pm) : | \Psi \rangle \ . \tag{5.37}$$

In Minkowski space, and for inertial coordinates, the first term of the right hand side vanishes and the quantum stress tensor is given by the standard normal ordering prescription. For a general curved spacetime we have an additional contribution to be added to the one coming from normal ordering.

The expectation values of $: T_{\pm\pm}(x^\pm) :$ define a set of functions, that are usually denoted in the literature on two-dimensional gravity by $t_\pm(x^\pm)$ (up to a multiplicative constant)

$$\langle \Psi | : T_{\pm\pm}(x^\pm) : | \Psi \rangle \equiv -\frac{\hbar}{12\pi} t_\pm(x^\pm) \ . \tag{5.38}$$

These functions depend, obviously, on the quantum state $|\Psi\rangle$ and therefore characterize it. The remaining terms are independent of the quantum state. Taking into account the definition (5.38) we finally arrive at the expression

$$\langle \Psi | T_{\pm\pm}(x^\pm) | \Psi \rangle = -\frac{\hbar}{12\pi}(\partial_\pm \rho \partial_\pm \rho - \partial_\pm^2 \rho + t_\pm(x^\pm)) \ . \tag{5.39}$$

Notice that if $|\Psi\rangle$ is chosen to be the vacuum state with respect to the modes (5.9) (it is then usually denoted as $|x^\pm\rangle$) the functions $t_\pm$ vanish and the expressions (5.39) turn out to be[3]

$$\langle x^\pm | T_{\pm\pm}(x^\pm) | x^\pm \rangle = -\frac{\hbar}{12\pi}(\partial_\pm \rho \partial_\pm \rho - \partial_\pm^2 \rho) \ . \tag{5.40}$$

Notice also that a change of vacuum state $|x^\pm\rangle \to |\tilde{x}^\pm\rangle$ produces a change in the expectation values of the $T_{\pm\pm}(x^\pm)$ components of the stress tensor of the form

$$\langle \tilde{x}^\pm | T_{\pm\pm} | \tilde{x}^\pm \rangle = \langle x^\pm | T_{\pm\pm} | x^\pm \rangle - \frac{\hbar}{24\pi}\{\tilde{x}^\pm, x^\pm\} \ . \tag{5.41}$$

---

[3]This expression was first derived in [Davies et al. (1976)] by covariant point-splitting regularization.

We have recovered a generally covariant behavior for $\langle\Psi|T_{\pm\pm}(x^\pm)|\Psi\rangle$, but there is still a missing ingredient. We have to ensure the conservation law $\nabla^a\langle\Psi|T_{ab}|\Psi\rangle = 0$. In conformal gauge the analog of the classical conservation laws $\partial_\mp T_{\pm\pm} = 0$ take the form

$$\partial_\mp\langle\Psi|T_{\pm\pm}|\Psi\rangle = 0 \ . \tag{5.42}$$

However we find that

$$\partial_\mp\langle\Psi|T_{\pm\pm}|\Psi\rangle = \frac{\hbar}{12\pi}\{\partial_+\partial_-\partial_\pm\rho - 2\partial_\pm\rho\partial_-\partial_+\rho\}. \tag{5.43}$$

So Eq. (5.42) is not satisfied unless we allow for a nonvanishing component $\langle\Psi|T_{+-}|\Psi\rangle$. In this case the covariant conservation law $\nabla^a\langle\Psi|T_{ab}|\Psi\rangle = 0$ reads

$$\partial_\mp\langle\Psi|T_{\pm\pm}|\Psi\rangle + \partial_\pm\langle\Psi|T_{+-}|\Psi\rangle - \Gamma^\pm_{\pm\pm}\langle\Psi|T_{+-}|\Psi\rangle = 0 \ . \tag{5.44}$$

Taking into account that $\Gamma^\pm_{\pm\pm} = 2\partial_\pm\rho$ it is easy to see that Eq. (5.43) is now compatible with the conservation law (5.44) provided that

$$\langle\Psi|T_{+-}|\Psi\rangle = -\frac{\hbar}{12\pi}\partial_+\partial_-\rho - \frac{\lambda}{4}e^{2\rho} \ , \tag{5.45}$$

where λ is an arbitrary constant. Note that since $\partial_+t_- = 0 = \partial_-t_+$, this component has been automatically obtained to be independent of the quantum state. It is very easy to see that the above expression is equivalent to the usual trace anomaly (5.7) plus an arbitrary two-dimensional cosmological constant

$$\langle T\rangle = \frac{\hbar}{24\pi}R + \lambda \ . \tag{5.46}$$

The constant λ is usually neglected. We finally remark that this nonvanishing trace has emerged as a consequence of imposing general covariance to the quantum stress tensor. Classically the matter field does not feel the presence of the curved background and this is reflected by the vanishing of T_{+-}. The normal ordered prescription keeps zero trace, but breaks (partially) diffeomorphism invariance. Only when a generally covariant behavior is imposed to the quantum stress tensor a non vanishing trace emerges.

5.2 The Polyakov Effective Action

We now want to show how the expressions that we have obtained for the expectation values of the stress tensor can be derived by functional differentiation of an effective action S_{eff}:

$$-\frac{2}{\sqrt{-g}}\frac{\delta S_{eff}}{\delta g^{ab}} = \langle \Psi|T_{ab}|\Psi\rangle . \qquad (5.47)$$

We should remark that S_{eff} cannot be an ordinary (local) action since $\langle\Psi|T_{ab}|\Psi\rangle$ do not only depend only on the background geometry, but also on the quantum state $|\Psi\rangle$. Therefore, the Eqs. (5.47) are non-trivial. However, since the trace is state-independent the relation

$$-\frac{2}{\sqrt{-g}}g^{ab}\frac{\delta S_{eff}}{\delta g^{ab}} = \langle T\rangle \qquad (5.48)$$

allows to determine S_{eff} by functionally integrating the trace anomaly. In conformal gauge, and using the independent metric components $g^{+-} = g^{-+} = -2e^{2\rho}$, we have

$$-\frac{2}{g_{+-}}\frac{\delta S_{eff}}{\delta g^{+-}} = 2\Big[-\frac{\hbar}{12\pi}\partial_+\partial_-\rho\Big] . \qquad (5.49)$$

This gives

$$\frac{\delta S_{eff}}{\delta \rho} = \frac{\hbar}{6\pi}\partial_+\partial_-\rho , \qquad (5.50)$$

and by integration we get

$$S_{eff} = \frac{\hbar}{12\pi}\int d^2x \rho \partial_+\partial_-\rho . \qquad (5.51)$$

The full covariant expression is more involved, in fact it is *nonlocal* and it is given by the so called Polyakov effective action [Polyakov (1981)]:

$$S_P = -\frac{\hbar}{96\pi}\int d^2x\sqrt{-g}R\frac{1}{\Box}R . \qquad (5.52)$$

In conformal gauge $R = 8e^{-2\rho}\partial_+\partial_-\rho$ and $\Box = -4e^{-2\rho}\partial_+\partial_-$, so (5.52) recovers (5.51). Had we included the λ term as well the effective action (5.51) would have been modified by the addition of a Liouville-type potential term $\sim \lambda e^{2\rho}$ and (5.52) by a two-dimensional cosmological constant-type term.

A more concrete form of (5.52) is

$$S_P = -\frac{\hbar}{96\pi} \int d^2x d^2y \sqrt{-g(x)}\sqrt{-g(y)} R(x) G(x,y) R(y) \qquad (5.53)$$

where $G(x,y)$ is a Green function satisfying the equation

$$\Box_x G(x,y) = \frac{1}{\sqrt{-g(x)}} \delta^2(x-y) . \qquad (5.54)$$

Equation (5.53) makes manifest the nonlocality of the effective action. The boundary conditions for the Green function $G(x,y)$ correspond to the choice of the quantum state $|\Psi\rangle$. An alternative way to manage (5.52), and to explicitly check that it provides the expressions (5.39) for the null components of the quantum stress tensor, is to convert it into a local action by introducing an auxiliary scalar field φ constrained to obey the equation

$$\Box \varphi = R . \qquad (5.55)$$

The local action

$$S_P = -\frac{\hbar}{96\pi} \int d^2x \sqrt{-g}\, (-\varphi \Box \varphi + 2\varphi R) \qquad (5.56)$$

is equivalent to (5.52) if φ satisfies (5.55). To obtain the stress tensor we have to vary the above action with respect to the metric. We use for this the relation $\delta\sqrt{-g} = -1/2\sqrt{-g}\,g_{ab}\delta g^{ab}$ and the Palatini identity for the variation of the Ricci tensor (see, for instance, [Weinberg (1972)])

$$\delta R_{ab} = \frac{1}{2}\left(-\nabla^c\nabla_c \delta g_{ab} + \nabla^c\nabla_a \delta g_{cb} + \nabla^c\nabla_b \delta g_{ca} - \nabla_a \nabla_b g^{cd} \delta g_{cb}\right) . \qquad (5.57)$$

Since $\delta g_{ab} = -g_{ac}g_{bd}\delta g^{cd}$ we can also obtain the useful relation

$$g^{ab}\delta R_{ab} = g_{ab} \Box \delta g^{ab} - \nabla_a \nabla_b \delta g^{ab} . \qquad (5.58)$$

Neglecting total derivatives in (5.56) we then get

$$\delta S_P = -\frac{\hbar}{96\pi}\int d^2x \sqrt{-g}\Big[-\frac{1}{2}(\nabla\varphi)^2 g_{ab}\delta g^{ab} + \nabla_a\varphi \nabla_b\varphi \delta g^{ab}$$
$$- 2\varphi(\frac{1}{2}R g_{ab} - R_{ab})\delta g^{ab} + 2\varphi(g_{ab}\Box \delta g^{ab} - \nabla_a\nabla_b\delta g^{ab})\Big] . \qquad (5.59)$$

The term proportional to the (two-dimensional) Einstein tensor vanishes identically and, after integrating by parts, we are left with

$$\delta S_P = -\frac{\hbar}{96\pi} \int d^2x \sqrt{-g} \Big[-\frac{1}{2}(\nabla\varphi)^2 g_{ab}\delta g^{ab} + \nabla_a\varphi \nabla_b\varphi \delta g^{ab} \\ + 2g_{ab}\Box\varphi \delta g^{ab} - 2\nabla_a\nabla_b\varphi \delta g^{ab} \Big] . \tag{5.60}$$

From this we obtain the following stress tensor

$$-\frac{2}{\sqrt{-g}}\frac{\delta S_P}{\delta g^{ab}} = \frac{\hbar}{48\pi}\left[\nabla_a\varphi\nabla_b\varphi - \frac{1}{2}g_{ab}(\nabla\varphi)^2\right] \\ -\frac{\hbar}{24\pi}\left[\nabla_a\nabla_b\varphi - \frac{1}{2}g_{ab}\Box\varphi\right] + \frac{\hbar}{48\pi}g_{ab}R . \tag{5.61}$$

Note that we have also used the relation (5.55). It is easy to see from the above that we recover immediately the trace anomaly

$$-\frac{2}{\sqrt{-g}}g^{ab}\frac{\delta S_{eff}}{\delta g^{ab}} = \frac{\hbar}{24\pi}R . \tag{5.62}$$

Moreover we can also see how to recover the null components $\langle \Psi | T_{\pm\pm} | \Psi \rangle$. In conformal gauge the Eq. (5.55) reads

$$\partial_+\partial_-\varphi = -2\partial_+\partial_-\rho \tag{5.63}$$

and therefore

$$\varphi = -2\rho + 2(\varphi_+(x^+) + \varphi_-(x^-)), \tag{5.64}$$

where $\varphi_\pm(x^\pm)$ are arbitrary chiral functions generating the general solution of the corresponding homogeneous equation. Plugging (5.64) into (5.61) we get

$$-\frac{2}{\sqrt{-g}}\frac{\delta S_P}{\delta g^{\pm\pm}} = -\frac{\hbar}{12\pi}\left((\partial_\pm\rho)^2 - \partial_\pm^2\rho\right) + \frac{\hbar}{12\pi}\left((\partial_\pm\varphi_\pm)^2 - \partial_\pm^2\varphi\right) . \tag{5.65}$$

It is then clear that by denoting

$$-(\partial_\pm\varphi_\pm)^2 + \partial_\pm^2\varphi \equiv t_\pm(x^\pm) \tag{5.66}$$

we recover

$$-\frac{2}{\sqrt{-g}}\frac{\delta S_P}{\delta g^{\pm\pm}} = \langle T_{\pm\pm}\rangle = -\frac{\hbar}{12\pi}\left(\partial_\pm\rho\partial_\pm\rho - \partial_\pm^2\rho + t_\pm\right) . \tag{5.67}$$

We should remark that, from the defining equation (5.64), the functions $\varphi_\pm$ are not scalars. In order to make φ a scalar field, they must transform under conformal transformations as

$$\varphi_\pm(x^\pm) \to \varphi_\pm(y^\pm) = \varphi_\pm(y^\pm(x^\pm)) + \frac{1}{2}\ln\frac{dx^\pm}{dy^\pm} \qquad (5.68)$$

to compensate the transformation law of ρ

$$\rho(x^\pm) \to \rho(y^\pm) = \rho(x^\pm) + \frac{1}{2}\ln\frac{dx^+}{dy^+}\frac{dx^-}{dy^-}. \qquad (5.69)$$

All this means that $t_\pm(x^\pm)$ transform as

$$t_\pm(y^\pm) = (\frac{dx^\pm}{dy^\pm})^2 t_\pm(x^\pm) + \frac{1}{2}\{x^\pm, y^\pm\}, \qquad (5.70)$$

in agreement with the anomalous transformation law of the normal ordered stress tensor under the identification (5.38). So it should be clear now that the nonlocality of the Polyakov effective action (5.52) is associated to the auxiliary field in (5.56) and physically reflects the dependence on the state $|\Psi\rangle$ of $\langle\Psi|T_{ab}|\Psi\rangle$. In practice, the state dependence is contained in the functions $t_\pm$ and, therefore, in $\varphi_\pm$.

5.2.1 The role of the Weyl-invariant effective action

To get a clearer understanding of the expression for $\langle\Psi|T_{ab}|\Psi\rangle$ it is useful to reconsider the path integral interpretation of the effective action. It is produced by functional integration of the massless scalar field f

$$\int [df] e^{-\frac{i}{2\hbar}\int d^2x \sqrt{-g}(\nabla f)^2}. \qquad (5.71)$$

The functional integral requires the specification of the measure $[df]$. This, in turn, requires the definition of a scalar product for the deformations df [Polyakov (1981); Alvarez (1983)]. The choice

$$\|df\|^2 = \int d^2x\sqrt{-g}\, df\, df \qquad (5.72)$$

ensures the covariance of the effective action, but the price to pay is the loss of the classical Weyl invariance. Had we chosen a different measure the result would have been different. For instance, if we decide to maintain the

Weyl invariance of the original classical action we should choose

$$\|df\|^2 = \int d^2x\, df\, df \ . \tag{5.73}$$

This produces a Weyl invariant effective action [Jackiw, (1995); Karakhaniyan et al. (1994)] that we call S_W. The price to pay now is the (partial) loss of general covariance.[4] It turns out to be equal to the Polyakov effective action for the Weyl invariant combination $g^{ab}\sqrt{-g}$, i.e.,

$$S_W(g^{ab}) = S_P(g^{ab}\sqrt{-g}). \tag{5.74}$$

After a short calculation based on the form of the action (5.56), and taking into account that under the transformation $g^{ab} \to g^{ab}\sqrt{-g}$ we have

$$\sqrt{-g} \to 1\ , \quad \Box \to \sqrt{-g}\,\Box\ , \quad R \to \sqrt{-g}(R + \Box \ln \sqrt{-g})\ ,$$
$$\varphi \to \varphi + \ln \sqrt{-g}\ , \tag{5.75}$$

one can show that these two actions, S_P and S_W, are related in the following way [Navarro-Salas et al. (1995)]

$$S_P = S_W + S_{loc} \tag{5.76}$$

where

$$S_{loc} = \frac{\hbar}{96\pi} \int d^2x \sqrt{-g}\, [\ln \sqrt{-g}\, \Box \ln \sqrt{-g} + 2R \ln \sqrt{-g}]$$
$$- \frac{\hbar}{96\pi} \int d^2x \sqrt{-g}\, \nabla_a(\ln \sqrt{-g}\, \nabla^a \varphi - \varphi \nabla^a \ln \sqrt{-g})\ . \tag{5.77}$$

Note that the last term, the only one containing the auxiliary field φ, is a total derivative. The contribution of S_{loc} to the stress tensor is

$$-\frac{2}{\sqrt{-g}} \frac{\delta S_{loc}}{\delta g^{ab}} = \frac{\hbar}{48\pi} \left[\partial_a \ln \sqrt{-g}\, \partial_b \ln \sqrt{-g} - \frac{1}{2} g_{ab} \partial_\alpha \ln \sqrt{-g}\, \partial^\alpha \ln \sqrt{-g}\right]$$
$$+ \frac{\hbar}{24\pi} \left[\nabla_a \nabla_b \ln \sqrt{-g} - \frac{1}{2} g_{ab} \Box \ln \sqrt{-g}\right] + \frac{\hbar}{48\pi} g_{ab} R\ . \tag{5.78}$$

Note that the contribution of S_{loc} to the trace is the same as S_P

$$-\frac{2}{\sqrt{-g}} g^{ab} \frac{\delta S_{loc}}{\delta g^{ab}} = \frac{\hbar}{24\pi} R\ , \tag{5.79}$$

---

[4]The effective action remains invariant under those diffeomorphisms that possess a constant (unit) Jacobian [Jackiw, (1995)].

since, by construction, S_W does not contribute to the trace anomaly. However, S_{loc} is clearly non-covariant, and this is manifested in the evaluation of the covariant derivative

$$\nabla^a \left(-\frac{2}{\sqrt{-g}} \frac{\delta S_{loc}}{\delta g^{ab}} \right) = \frac{\hbar}{48\pi} \frac{1}{\sqrt{-g}} \partial_b R(\sqrt{-g} g^{cd}) \ . \tag{5.80}$$

Let us now restrict the expression (5.78) to conformal gauge. Taking into account that $\partial_\pm \ln \sqrt{-g} = 2\partial_\pm \rho$ and $\nabla_\pm (\partial_\pm \ln \sqrt{-g}) = 2\partial_\pm^2 \rho - 4(\partial_\pm \rho)^2$, we have

$$-\frac{2}{\sqrt{-g}} \frac{\delta S_{loc}}{\delta g^{\pm\pm}} = -\frac{\hbar}{12\pi} \left((\partial_\pm \rho)^2 - \partial_\pm^2 \rho \right) \ ,$$

$$-\frac{2}{\sqrt{-g}} \frac{\delta S_{loc}}{\delta g^{+-}} = -\frac{\hbar}{12\pi} \partial_+ \partial_- \rho \ . \tag{5.81}$$

Comparing with

$$\langle \Psi | T_{\pm\pm}(x^\pm) | \Psi \rangle = -\frac{\hbar}{12\pi} \left((\partial_\pm \rho)^2 - \partial_\pm^2 \rho \right) + \langle \Psi | : T_{\pm\pm}(x^\pm) : | \Psi \rangle \ , \tag{5.82}$$

it should then be clear that the quantization via pure normal ordering is related to S_W in the following way

$$-\frac{2}{\sqrt{-g}} \frac{\delta S_W}{\delta g^{\pm\pm}} = \langle \Psi | : T_{\pm\pm}(x^\pm) : | \Psi \rangle \ ,$$

$$-\frac{2}{\sqrt{-g}} \frac{\delta S_W}{\delta g^{+-}} = 0 \ . \tag{5.83}$$

We now have a clear understanding of the situation. The noncovariance of S_W is reflected in the noncovariance of the normal ordering quantization prescription. The Weyl invariance of S_W is reflected in the tracelessness of the associated stress tensor (as one can see from the second of Eqs. (5.83)). The noncovariance of S_{local} "cancels" against the noncovariance of S_W making their sum, namely S_P, generally covariant. In other words, the splitting of expression (5.82) between geometric and state-dependent terms is parallel to the splitting of the Polyakov effective action $S_P = S_{loc} + S_W$.

5.3 Choice of the Quantum State

In this section we shall study the natural choices for the quantum states $|\Psi\rangle$ in black hole physics and discuss the properties of the corresponding covariant stress energy tensor $\langle \Psi | T_{\pm\pm} | \Psi \rangle$. We have already mentioned the various possibilities in the previous chapter, in Subsections 4.3.4 and 4.5.1, in the context of the near-horizon approximation. Here we shall study them in detail and clarify their physical meaning by considering the full radial part of the Schwarzschild geometry.

5.3.1 Boulware state

The starting point of the analysis is the radial part of the Schwarzschild metric in the *static* double null form

$$ds^2 = -(1 - \frac{2M}{r})dudv ,\qquad (5.84)$$

where (u,v) are the retarded and advanced Eddington–Finkelstein coordinates. We can expand the field f in terms of plane wave modes associated to these coordinates

$$f = \int_0^\infty \frac{dw}{\sqrt{4\pi w}} \left(\vec{a}_w e^{-iwu} + \vec{a}^\dagger_w e^{iwu} + \overleftarrow{a}_w e^{-iwv} + \overleftarrow{a}^\dagger_w e^{iwv} \right). \qquad (5.85)$$

The Boulware state [Boulware (1975)] $|B\rangle$ is then defined as the vacuum associated to the above choice of modes. Therefore

$$\vec{a}_w|B\rangle = 0, \quad \overleftarrow{a}_w|B\rangle = 0 ,\qquad (5.86)$$

and the excited states can be obtained by the application of the creation operators $\vec{a}^\dagger_w, \overleftarrow{a}^\dagger_w$ out of this vacuum.

By definition of normal ordering, in this vacuum state we have the vanishing of the normal ordered expressions of the stress tensor components

$$\langle B|:T_{uu}(u):|B\rangle = 0 = \langle B|:T_{vv}(v):|B\rangle .\qquad (5.87)$$

Now straightforward application of Eqs. (5.37) and (5.45), with $\lambda = 0$, gives

$$\langle B|T_{uu}|B\rangle = \langle B|T_{vv}|B\rangle = \frac{\hbar}{24\pi}[-\frac{M}{r^3} + \frac{3}{2}\frac{M^2}{r^4}],$$
$$\langle B|T_{uv}|B\rangle = -\frac{\hbar}{24\pi}(1 - \frac{2M}{r})\frac{M}{r^3}. \qquad (5.88)$$

As one immediately sees, the modes in Eq. (5.85) reduce at infinity to the usual Minkowski ingoing and outgoing plane waves and there $\langle B|T_{ab}|B\rangle = 0$. So the state $|B\rangle$ reproduces at infinity the familiar notion of the "conventional" vacuum state as inferred from Minkowski field theory. The same property can be achieved by quantizing the field using a different set of modes but with the same asymptotic behavior. It is important to note, however, that only the above choice leads to a time-independent stress tensor. One can think of these features as the reasons for selecting $|B\rangle$ among the various candidates for a reasonable vacuum state of the theory. If the behavior of $|B\rangle$ at infinity seems quite reasonable, the same cannot be said for the horizon $r = 2M$. Although the expressions (5.88) are finite on the horizon $r = 2M$,

$$\langle B|T_{uu}|B\rangle = \langle B|T_{vv}|B\rangle \sim -\frac{\hbar}{768\pi M^2}$$
$$\langle B|T_{uv}|B\rangle \sim 0 \,, \tag{5.89}$$

on it the coordinates used are ill-defined. One should expect that, if the horizon belongs to the physical spacetime, then $\langle T_{ab}\rangle$ should be finite there with respect to a local regular frame such as the Kruskal frame (U, V). From the definition of the Kruskal coordinates it follows then that on the future horizon the change of coordinates between u and U is singular, namely $\frac{du}{dU} \sim \frac{1}{U} \sim \frac{1}{(r-2M)}$, while $\frac{dv}{dV}$ is regular (on the past horizon, instead, $\frac{du}{dU}$ is regular and $\frac{dv}{dV} \sim \frac{1}{r-2M}$). Therefore $\langle T_{UU}\rangle \sim \frac{\langle T_{uu}\rangle}{(r-2M)^2}$, $\langle T_{UV}\rangle \sim \frac{\langle T_{uv}\rangle}{r-2M}$, $\langle T_{VV}\rangle \sim \langle T_{vv}\rangle$ from which regularity on the future horizon is ensured whenever [Christensen and Fulling (1977)]

$$|\langle T_{vv}\rangle| < \infty,$$
$$(r - 2M)^{-1}|\langle T_{uv}\rangle| < \infty,$$
$$(r - 2M)^{-2}|\langle T_{uu}\rangle| < \infty \,. \tag{5.90}$$

The regularity on the past horizon is expressed by similar inequalities with u and v interchanged.

It is now clear that although $\langle B|T_{ab}|B\rangle$ is finite in the limit $r \to 2M$, the above conditions are clearly not fulfilled at the horizons. This behavior is connected to the fact that the state $|B\rangle$ is defined in terms of the (u, v) modes in Eq. (5.85) which oscillate infinitely on the horizon. From the physical point of view, the state $|B\rangle$ describes the *vacuum polarization* of the spacetime exterior to a static massive body whose radius is bigger than $2M$. In this way the physically relevant portion of the Schwarzschild spacetime does not contain horizons.

5.3.2 *Hartle–Hawking state*

Intuitively, if we want to construct a state with regular properties at the horizons we have to use modes that are regular there, for instance those associated to the Kruskal coordinates:

$$(4\pi w)^{-1/2} e^{-iwV}, \quad (4\pi w)^{-1/2} e^{-iwU} . \tag{5.91}$$

This defines a new vacuum state [Hartle-Hawking (1976); Israel (1976)] $|H\rangle$, usually called the Hartle–Hawking state. This state is clearly defined not only in the exterior asymptotically flat region, but in the maximally extended spacetime. The restriction to the external region implies tracing over the interior states and this produces a mixed state. Since the relations between the coordinates (U,V) and (u,v) are the same as the Minkowskian and Rindler coordinates analysed in Section 4.1, we can apply the results obtained of Section 4.3 and describe the Hartle–Hawking state as a thermal density matrix

$$|H\rangle|_{r>2M} \Leftrightarrow \rho = \overleftarrow{\rho} \otimes \overrightarrow{\rho}, \tag{5.92}$$

where $\overrightarrow{\rho}$ is given by

$$\overrightarrow{\rho} = \prod_w \left(1 - e^{-2\pi w \kappa^{-1}}\right) \sum_N e^{-2\pi N w \kappa^{-1}} |\overrightarrow{N}_w\rangle\langle\overrightarrow{N}_w|, \tag{5.93}$$

$|\overrightarrow{N}_w\rangle$ is the state in the Fock space with N outgoing particles of frequency w constructed from the Boulware vacuum $|\overrightarrow{N}_w\rangle = \hbar^{-N/2}(N!)^{-\frac{1}{2}}(a_w^\dagger)^N|B\rangle$, and $\overleftarrow{\rho}$ is the analogous thermal density matrix for the left-moving sector.

To evaluate the covariant stress tensor components we need to first evaluate the normal ordered expressions. As we have already explained in Subsection 4.3.3, the thermal character of the state gives[5]

$$\langle H| : T_{uu} : |H\rangle = \frac{\hbar}{768\pi M^2} = \langle H| : T_{vv} : |H\rangle . \tag{5.94}$$

Adding now the "vacuum polarization" terms (which are the only contributions in the Boulware state) we get[6]

$$\langle H|T_{uu}|H\rangle = \langle H|T_{vv}|H\rangle = \frac{\hbar}{768\pi M^2}(1 - \frac{2M}{r})^2 [1 + \frac{4M}{r} + \frac{12M^2}{r^2}],$$
$$\langle H|T_{uv}|H\rangle = -\frac{\hbar}{24\pi}(1 - \frac{2M}{r})\frac{M}{r^3}. \tag{5.95}$$

Notice that, on the horizon, the normal ordered quantities (5.94) cancel against the vacuum polarization contributions of Eq. (5.89). The above

---
[5]This is equivalent to the conditions $\langle H| : T_{UU}(U) : |H\rangle = 0 = \langle H| : T_{VV}(V) : |H\rangle$.
[6]This can also be obtained from the results for the Boulware state by using the formula (5.41) involving the Schwarzian derivatives between $U(V)$ and $u(v)$.

formulae have been written in such a way to explicitly show the regularity on the horizons, according to Eqs. (5.90).

The Kruskal modes, however, do not reduce asymptotically to standard Minkowski plane waves. As a consequence, $\langle H|T_{ab}|H\rangle$ does not vanish at infinity, instead we have the constant thermal flux

$$\langle H|T_{uu}|H\rangle = \langle H|T_{vv}|H\rangle \sim \frac{\hbar}{768\pi M^2},$$
$$\langle H|T_{uv}|H\rangle \sim 0. \qquad (5.96)$$

Therefore, $|H\rangle$ is a *thermal state* at the Hawking temperature

$$T_H = \frac{\hbar}{8\pi k_B M} \qquad (5.97)$$

and describes the thermal equilibrium of a black hole enclosed in a box with its own radiation. In principle, one can use other conformal coordinates, which behave like Kruskal at infinity and at the horizons, and the related vacuum states to reproduce the same properties there. However, only in the Hartle–Hawking state $\langle H|T_{uu}|H\rangle$ and $\langle H|T_{vv}|H\rangle$ will be time independent [Kay and Wald (1991)]. So only the state $|H\rangle$ properly describes a thermal equilibrium state.

5.3.3 The "in" and Unruh vacuum states: the Hawking flux

In this subsection we will consider the most interesting quantum states, i.e., those related to black hole evaporation.

5.3.3.1 "in" vacuum state

For this situation the physical configuration is the dynamical gravitational collapse. As usual let us consider a Schwarzschild black hole formed from Minkowski space by the collapse of an ingoing null shell located at $v = v_0$, as shown in Fig. 3.3.

In the "in" region $v < v_0$ the spacetime is flat

$$ds_{in}^2 = -du_{in}dv, \qquad (5.98)$$

where u_{in} and v are the usual retarded and advanced Minkowski null coordinates. For $v > v_0$ the "out" geometry describes a black hole of mass M

$$ds_{out}^2 = -(1 - 2M/r)dudv. \qquad (5.99)$$

Matching the two geometries at $v = v_0$ we have as usual[7]

$$u = u_{in} - 4M \ln\left(\frac{v_0 - u_{in} - 4M}{4M}\right). \tag{5.100}$$

The natural state describing this dynamical situation is the Minkowski vacuum in the "in" region. Therefore, the relevant modes for the quantization are[8]

$$(4\pi w)^{-1/2} e^{-iwv} \quad , \quad (4\pi w)^{-1/2} e^{-iwu_{in}}. \tag{5.101}$$

This state is called the "in" vacuum state $|in\rangle$. From this we can immediately obtain

$$\langle in| : T_{u_{in} u_{in}}(u_{in}) : |in\rangle = 0 = \langle in| : T_{vv}(v) : |in\rangle. \tag{5.102}$$

Notice that the coordinates (v, u_{in}) cover the full spacetime, while (u, v) do not cover the black hole region. Therefore the restriction of $|in\rangle$ to the exterior region produces a mixed state. The explicit expression of the corresponding density matrix is complicated, but we can evaluate the components of the normal ordered stress tensor

$$\langle in| : T_{uu}(u) : |in\rangle = -\frac{\hbar}{24\pi} \{u_{in}, u\}$$

$$= -\frac{\hbar}{24\pi} \left(\frac{8M}{(u_{in} - v_0)^3} + \frac{24M^2}{(u_{in} - v_0)^4}\right)$$

$$\langle in| : T_{vv}(v) : |in\rangle = 0. \tag{5.103}$$

We can then evaluate the covariant expression by adding the vacuum polarization "Boulware" terms. The result is [Hiscock (1981)]

$$\langle in|T_{uu}|in\rangle = \frac{\hbar}{24\pi}\left[-\frac{M}{r^3} + \frac{3}{2}\frac{M^2}{r^4} - \frac{8M}{(u_{in}-v_0)^3} - \frac{24M^2}{(u_{in}-v_0)^4}\right],$$

$$\langle in|T_{vv}|in\rangle = \frac{\hbar}{24\pi}\left[-\frac{M}{r^3} + \frac{3}{2}\frac{M^2}{r^4}\right] = \langle B|T_{vv}|B\rangle,$$

$$\langle in|T_{uv}|in\rangle = -\frac{\hbar}{24\pi}(1 - \frac{2M}{r})\frac{M}{r^3} = \langle B|T_{uv}|B\rangle. \tag{5.104}$$

There are two interesting limits in the analysis of the behavior of these quantities:

---

[7] For the details of the matching see Subsection 3.3.1.
[8] Remember that due to the regularity condition of the field f at $r = 0$ these modes are not independent. This fact does not change the results.

- At early times $u_{in} \sim u \to -\infty$ the normal ordered contribution vanishes and therefore we are left with pure vacuum polarization, which is zero at infinity.

- At late times $(u \to +\infty)$ u_{in} is related to u via the "Kruskal" relation

$$u_{in} \sim v_0 - 4M - 4Me^{-u/4M} , \qquad (5.105)$$

i.e., it behaves as the Kruskal coordinate U. In the above limit $u_{in} \to v_0 - 4M$ (i.e., the shell is close to crossing the horizon) we find, at future null infinity I^+, a *net flux*

$$\langle in|T_{uu}|in\rangle \to \frac{\hbar}{768\pi M^2} \qquad (5.106)$$

representing the *Hawking thermal outgoing flux* at the Hawking temperature T_H.

Notice also that at the future horizon $r \to 2M$ and $u_{in} \to v_0 - 4M$, where $\langle in|T_{ab}|in\rangle$ fulfills the regularity conditions (5.90), the only nonvanishing component is

$$\langle in|T_{vv}|in\rangle \to -\frac{\hbar}{768\pi M^2} \qquad (5.107)$$

representing an *influx of negative energy* radiation (the counterpart of the Hawking radiation at infinity) making the black hole shrink due to the evaporation.

5.3.3.2 *Unruh state*

The above discussion also serves to motivate the introduction of the so called Unruh state $|U\rangle$ [Unruh (1976)], which is constructed to reproduce the late time thermal radiation. It has been pointed out via Eq. (5.105) that at late times u_{in} behaves like the Kruskal coordinate U. This suggests the use of the following modes

$$(4\pi w)^{-1/2} e^{-iwv}, \quad (4\pi w)^{-1/2} e^{-iwU} , \qquad (5.108)$$

for the quantization of the field f in the Schwarzschild geometry. These modes and the corresponding vacuum state are defined both in the exterior and in the black hole region. Therefore, the restriction of the Unruh state to the asymptotically flat exterior region leads to a mixed state describing outgoing thermal radiation. Indeed it is given by[9]

$$|U\rangle|_{r>2M} \Leftrightarrow \vec{\rho} ,\qquad(5.109)$$

where $\vec{\rho}$ is again given by Eq. (5.93). The normal ordered components are then[10]

$$\langle U|:T_{uu}(u):|U\rangle = \frac{\hbar}{768\pi M^2} ,$$
$$\langle U|:T_{vv}(v):|U\rangle = 0 .\qquad(5.110)$$

The covariant components of the stress tensor then read

$$\langle U|T_{uu}|U\rangle = \frac{\hbar}{768\pi M^2}(1-\frac{2M}{r})^2\left[1+\frac{4M}{r}+\frac{12M^2}{r^2}\right] ,$$
$$\langle U|T_{vv}|U\rangle = \frac{\hbar}{24\pi}\left[-\frac{M}{r^3}+\frac{3}{2}\frac{M^2}{r^4}\right] = \langle B|T_{vv}|B\rangle ,$$
$$\langle U|T_{uv}|U\rangle = -\frac{\hbar}{24\pi}(1-\frac{2M}{r})\frac{M}{r^3} = \langle B|T_{uv}|B\rangle .\qquad(5.111)$$

It is easy to check that the regularity conditions (5.90) on the future event horizon are satisfied, but not at the past horizon. However, this singular behavior on the past horizon is completely spurious since for a black hole formed by gravitational collapse there is no past horizon. The correct physical interpretation of the Unruh state is that it represents the gravitational collapse in the late time near-horizon limit, where $|in\rangle \sim |U\rangle$ (this is the reason why in Chapter 4 we made no distinction between Unruh and "in" vacuum states). Note finally that in this limit the time dependence in Eq. (5.104) disappears.

The analysis presented so far concerning the form of the stress tensor and its properties in fixed background for two-dimensional conformal fields naturally extends and completes the near-horizon picture considered in Chapter 4. However, in parallel with the analysis of Chapter 3 of the Hawking effect in the full four-dimensional spacetime there remain to be addressed the effects due to the potential (the backscattering effects). In particular, this requires to evaluate separately the contribution of each angular momentum mode to the Hawking radiation. Therefore, it is natural to try to extend the formalism developed up to now to consider this issue. This implies to abandon the spacetime conformal symmetry and for this reason things become more complicated. The rest of the chapter is devoted

---
[9] The ingoing sector is omitted in this formula since it is trivial.
[10] This is equivalent to $\langle U|:T_{UU}(U):|U\rangle = 0 = \langle U|:T_{vv}(v):|U\rangle$.

to the simplest possible extension of these results by considering the s-wave component.

5.4 Including the Backscattering in the Stress Tensor: the s-wave

Let us start with the four-dimensional scalar field considered in Chapter 3. For convenience we use Gaussian units. The action is given by

$$S^{(4)} = -\frac{1}{8\pi}\int d^4x\sqrt{-g^{(4)}}(\nabla f)^2. \tag{5.112}$$

As usual we shall restrict our considerations to spherically symmetric spacetimes

$$ds^2_{(4)} = g_{ab}dx^a dx^b + l^2 e^{-2\phi(x^a)}d\Omega^2, \tag{5.113}$$

where we have parametrized the radius of the two-sphere $r = le^{-\phi}$ through a *dilaton* field ϕ.[11] Moreover we will also require the field f to be a function of the coordinates of the radial part of the metric, which we generically call x^a. Since in general the field in the spacetime (5.113) naturally admits an expansion in spherical harmonics, the condition $f = f(x^a)$ corresponds to picking out only the *s-wave* mode. Now, taking into account that $\sqrt{-g^{(4)}} = \sqrt{-g}e^{-2\phi}\sin\theta$ the integration over the angles θ and φ for the classical action (5.112) is easy and the result is

$$S = -\frac{1}{2}\int d^2x\sqrt{-g}e^{-2\phi}(\nabla f)^2. \tag{5.114}$$

In the above action the dilaton should be regarded as an external field.

We will now see that the above action reproduces the effect of the potential barrier for the propagation of the field in the Schwarzschild spacetime. The equation of motion derived by varying f is

$$\nabla^a(e^{-2\phi}\nabla_a f) = 0. \tag{5.115}$$

In null conformal coordinates $ds^2 = -e^{2\rho}dx^+dx^-$ it turns out to be

$$\partial_-(e^{-2\phi}\partial_+ f) + \partial_+(e^{-2\phi}\partial_- f) = 0. \tag{5.116}$$

---

[11] l is an arbitrary constant which will play no role in the discussion that follows and will be neglected in the rest of this chapter.

For the Schwarzschild metric the dilaton function is given implicitly by

$$e^{-\phi} = r(r^* = (x^+ - x^-)/2) , \qquad (5.117)$$

where $x^+ = v$, $x^- = u$ are the null Eddington–Finkelstein coordinates.[12] Taking into account the identification

$$f = \frac{f_{l=0}}{r} Y_{00} , \qquad (5.118)$$

Eq. (5.116) corresponds to the standard wave equation for the s-wave

$$\partial_u \partial_v f_{l=0} + (1 - \frac{2M}{r})\frac{2M}{r^3} f_{l=0} = 0 . \qquad (5.119)$$

It is worth emphasizing the following relation between the four-dimensional stress tensor, derived from the action (5.112), and the corresponding two-dimensional one obtained from (5.114). The classical stress tensor $T^{(4)}_{\mu\nu}$ is

$$T^{(4)}_{\mu\nu} = -\frac{2}{\sqrt{-g^{(4)}}} \frac{\delta S^{(4)}}{\delta g^{\mu\nu}_{(4)}} . \qquad (5.120)$$

Reduction under spherical symmetry translates the above equations into

$$T^{(4)}_{ab} = \frac{T_{ab}}{4\pi e^{-2\phi}} ,$$
$$T^{(4)}_{\theta\theta} = \frac{T^{(4)}_{\phi\phi}}{\sin^2\theta} = -\frac{1}{8\pi\sqrt{-g^{(2)}}} \frac{\delta S}{\delta \phi} , \qquad (5.121)$$

where four-dimensional quantities have superscripts. The four-dimensional stress tensor is conserved by construction

$$\nabla_\mu T^{(4)\mu}_\nu = 0 , \qquad (5.122)$$

and equivalently, in terms of two-dimensional quantities [Balbinot and Fabbri (1999); Kummer and Vassilevich (1999)]

$$\nabla_a T^a_b - \frac{1}{\sqrt{-g}} \frac{\delta S}{\delta \phi} \nabla_b \phi = 0 . \qquad (5.123)$$

---

[12]When the mass vanishes we recover the four-dimensional Minkowski space and the expression for the dilaton is simply $e^{-\phi} = \frac{x^+ - x^-}{2}$.

The above equations can be checked using the two-dimensional equation of motion of the field f and taking into account that

$$T_{ab} = e^{-2\phi}\left(\nabla_a f \nabla_b f - \frac{1}{2}g_{ab}(\nabla f)^2\right) , \qquad (5.124)$$

$$\frac{1}{\sqrt{-g}}\frac{\delta S}{\delta \phi} = e^{-2\phi}(\nabla f)^2 . \qquad (5.125)$$

5.4.1 Two-dimensional symmetries

It is easy to realize that the action (5.114) is invariant under Weyl transformations $(g_{ab} \to \Omega^2(x)g_{ab})$, as the theory without dilaton considered in the previous sections. However, unlike that theory, it is not invariant under spacetime conformal transformations. To see this explicitly we can consider infinitesimal transformations:

$$x^\pm \to x^\pm + \epsilon^\pm(x^\pm) . \qquad (5.126)$$

These induce the following infinitesimal transformations on the field f:

$$\delta f = \partial_+ f \epsilon^+ + \partial_- f \epsilon^- . \qquad (5.127)$$

The dilaton is an external field and therefore it is not transformed. The corresponding transformation for the Lagrangian density

$$L = e^{-2\phi}\partial_+ f \partial_- f \qquad (5.128)$$

is

$$\delta L = \partial_+(L\epsilon^+) + \partial_-(L\epsilon^-) + 2L(\partial_+\phi\epsilon^+ + \partial_-\phi\epsilon^-). \qquad (5.129)$$

In general δL is not a divergence, only for $\partial_+\phi = 0$ $(\partial_-\phi = 0)$ we have symmetry under ϵ^+ (ϵ^-). So, in some intuitive sense we could say that the conformal symmetry is recovered when $\partial_\pm \phi \to 0$. Therefore, due to the loss of conformal invariance the theory becomes more involved. An example of this is that the null components of the stress tensor

$$T_{\pm\pm}(x^+, x^-) = e^{-2\phi}(\partial_\pm f)^2 \qquad (5.130)$$

are no longer chiral functions as can be seen from the equation of motion for f given by Eq. (5.116). Nevertheless, we stress again that the theory is Weyl invariant and this implies that

$$T_{+-} = 0 . \qquad (5.131)$$

5.4.2 The normal ordered stress tensor

We shall now turn to the quantum theory and analyse the properties of the normal ordered stress tensor, in parallel to what we have done for the theory without dilaton. We can then define the normal ordered stress tensor operator via point-splitting regularization

$$: T_{\pm\pm}(x^+, x^-) := \lim_{x'^{\pm} \to x^{\pm}} e^{-(\phi(x)+\phi(x'))} \frac{\partial}{\partial x^{\pm}} \frac{\partial}{\partial x'^{\pm}} (f(x)f(x') - \langle f(x)f(x') \rangle), \quad (5.132)$$

where $\langle f(x)f(x') \rangle$ is the symmetrized two-point function.[13] The specific value of $\langle f(x)f(x') \rangle$ depends in a nontrivial way on the function ϕ, however its short-distance behavior is generically of the form[14]

$$\langle f(x)f(x') \rangle = -\frac{\hbar}{4\pi} e^{\phi(x)+\phi(x')} [\ln(x^+ - x'^+)(x^- - x'^-) + const.$$
$$+ O((x^+ - x'^+)(x^- - x'^-) \ln(x^+ - x'^+)(x^- - x'^-))] . \quad (5.133)$$

In order to derive this formula it is implicitly assumed that the modes are chosen to be of positive frequency with respect to the time $(x^+ + x^-)/2$. For instance:

• In four-dimensional Minkowski space the mode expansion of the field f (with the regularity condition $f(r = \frac{x^+ - x^-}{2} = 0) = 0$) is

$$f = \int_0^\infty \frac{dw}{\sqrt{4\pi w}} \frac{1}{r} \left[a_w(e^{-iwx^-} - e^{-iwx^+}) + a_w^\dagger(e^{iwx^-} - e^{iwx^+}) \right] \quad (5.134)$$

where, we remember, $r^{-1} = e^\phi = 2/(x^+ - x^-)$.

• In the Schwarzschild spacetime a complete set of modes is given by

$$\vec{u}_w = \frac{1}{\sqrt{4\pi w}} \frac{\vec{R}(r,w)}{r} e^{-iw\frac{x^+ + x^-}{2}}, \quad (5.135)$$

$$\overleftarrow{u}_w = \frac{1}{\sqrt{4\pi w}} \frac{\overleftarrow{R}(r,w)}{r} e^{-iw\frac{x^+ + x^-}{2}}, \quad (5.136)$$

---
[13] It is usually called the Hadamard function $\langle f(x)f(x') \rangle \equiv G^{(1)}(x,x')$.
[14] This can be shown using the De Witt-Schwinger expansion of $G^{(1)}(x,x')$, given in [Bunch et al. (1978)] adapted to the case under consideration [Balbinot et al. (2001)]. Moreover, it can also be checked explicitly in some exactly solvable cases [Fabbri et al. (2003)].

where the radial functions $R(r, w)$ obey an ordinary differential equation. If (x^+, x^-) are the Eddington–Finkelstein coordinates $(v = t + r^*, u = t - r^*)$ the differential equation for $R(r, w)$ becomes

$$-\frac{d^2 R}{dr^{*2}} + \frac{2M}{r^3}\left(1 - \frac{2M}{r}\right)R - w^2 R = 0 \ . \tag{5.137}$$

Clearly, for $M \neq 0$ it is no longer possible to impose the regularity condition $f(r = 0) = 0$, $r = 0$ being now a spacelike curvature singularity. Exact solutions of Eq. (5.137) are not known. In parallel with Section 3.4, the asymptotic form of the solutions at the horizon is

$$\vec{R} \sim e^{iwr^*} + r_{l=0}(w) e^{-iwr^*},$$
$$\overleftarrow{R} \sim T_{l=0}(w) e^{-iwr^*}, \tag{5.138}$$

and at infinity

$$\vec{R} \sim t_{l=0}(w) e^{iwr^*},$$
$$\overleftarrow{R} \sim e^{-iwr^*} + R_{l=0}(w) e^{iwr^*}, \tag{5.139}$$

where the reflection and transmission coefficients satisfy the conditions given in Subsection 3.4.1.

5.4.2.1 Transformation laws

If we perform a conformal transformation $x^\pm \to y^\pm(x^\pm)$ the conformal factor of the metric transforms as

$$e^{2\rho(x)} \to \frac{dx^+}{dy^+}\frac{dx^-}{dy^-} e^{2\rho(x)} \tag{5.140}$$

and the dilaton function $\phi(x^+, x^-) \to \phi(x^+(y^+), x^-(y^-))$. The new form of the metric is not relevant due to the Weyl invariance of the theory, but the new form of the dilaton enters in the short-distance behavior of the two-point function

$$\langle f(y) f(y') \rangle \sim -\frac{\hbar}{4\pi} e^{\phi(y) + \phi(y')} \ln |(y^+ - y'^+)(y^- - y'^-)| \ . \tag{5.141}$$

The new stress tensor operator, normal ordered with respect to the coordinates $y^\pm$, is then

$$: T_{\pm\pm}(y^+, y^-) := \lim_{y'^\pm \to y^\pm} e^{-(\phi(y)+\phi(y'))} \frac{\partial}{\partial y^\pm}\frac{\partial}{\partial y'^\pm}(f(y)f(y') - \langle f(y)f(y')\rangle). \tag{5.142}$$

As we have already stressed several times, normal ordering breaks covariance. This is evident from the fact that

$$\langle f(y(x))f(y'(x'))\rangle \neq \langle f(x)f(x')\rangle . \tag{5.143}$$

A calculation similar to that performed in Subsection 4.2.1 implies that the transformed normal ordered stress tensor picks out the following anomalous nontensorial contributions

$$: T_{\pm\pm}(y^+, y^-) := \left(\frac{dx^\pm}{dy^\pm}\right)^2 : T_{\pm\pm}(x^+, x^-) : -\frac{\hbar}{24\pi}\{x^\pm, y^\pm\}$$
$$-\frac{\hbar}{4\pi}\left[\frac{d^2x^\pm}{dy^{\pm 2}}\left(\frac{dx^\pm}{dy^\pm}\right)^{-1}\frac{\partial\phi}{\partial y^\pm} + \ln\left(\frac{dx^+}{dy^+}\frac{dx^-}{dy^-}\right)\left(\frac{\partial\phi}{\partial y^\pm}\right)^2\right] . \tag{5.144}$$

This expression generalizes the Virasoro-type transformation law by adding terms depending on the derivatives of ϕ. We can see again by direct inspection of Eq. (5.144) that the conformal symmetry can be reestablished in regions where $\partial_\pm \phi \to 0$. This happens typically when r approaches infinity and at the black hole horizons.

Let us go further in the analysis of the quantum properties of the theory. Due to the presence of ϕ the classical conservation laws $\partial_\mp T_{\pm\pm} = 0$ get modified to

$$\partial_\mp T_{\pm\pm} + \partial_\pm \phi \frac{\delta S}{\delta \phi} = 0 , \tag{5.145}$$

where

$$\frac{\delta S}{\delta \phi} = -2e^{-2\phi}\partial_+ f \partial_- f . \tag{5.146}$$

Let us analyse the quantum analog of these equations. The transformation law for the expectation value $\langle : T_{\pm\pm} : \rangle$ (we here omit writing explicitly the quantum state $|\Psi\rangle$) is given by Eq. (5.144) and the corresponding one for $\left\langle \frac{\delta S}{\delta \phi} \right\rangle$ should be, on general grounds, of the form

$$\left\langle \frac{\delta S}{\delta \phi}(y^\pm) \right\rangle = \frac{dx^+}{dy^+}\frac{dx^-}{dy^-}\left\langle \frac{\delta S}{\delta \phi}(x^\pm) \right\rangle + \Delta(\phi; x^\pm, y^\pm) . \tag{5.147}$$

Let us suppose that

$$\partial_\mp \langle : T_{\pm\pm} : \rangle + \partial_\pm \phi \left\langle \frac{\delta S}{\delta \phi} \right\rangle = 0 . \tag{5.148}$$

If we transform this relation according to Eqs. (5.144) and (5.147) we get, by consistency,

$$-\frac{\hbar}{4\pi}\frac{\frac{\partial^2 x^{\pm}}{\partial y^{\mp 2}}}{\frac{\partial x^{\pm}}{\partial y^{\pm}}}\frac{\partial}{\partial y^+}\frac{\partial}{\partial y^-}\phi - \frac{\hbar}{2\pi}\ln\frac{\partial x^+}{\partial y^+}\frac{\partial x^-}{\partial y^-}(\frac{\partial \phi}{\partial y^{\pm}})\frac{\partial}{\partial y^+}\frac{\partial}{\partial y^-}\phi$$

$$-\frac{\hbar}{4\pi}\left(\frac{\partial \phi}{\partial y^{\pm}}\right)^2 \frac{\frac{\partial^2 x^{\mp}}{\partial y^{\mp 2}}}{\frac{\partial x^{\mp}}{\partial y^{\mp}}} + \frac{\partial \phi}{\partial y^{\pm}}\Delta(\phi; x^{\pm}, y^{\pm}) = 0 \ . \tag{5.149}$$

These two equations are compatible with the uniqueness of $\Delta(\phi; x^{\pm}, y^{\pm})$ only if

$$\Box\phi = (\nabla\phi)^2 \ . \tag{5.150}$$

If ϕ does not obey Eq. (5.150) the quantum conservation law (5.148) must be modified to

$$\partial_{\mp}\langle: T_{\pm\pm}:\rangle + \partial_{\pm}\phi\left\langle\frac{\delta S}{\delta\phi}\right\rangle = \frac{\hbar}{4\pi}\partial_{\pm}\left(\partial_{+}\phi\partial_{-}\phi - \partial_{+}\partial_{-}\phi\right) . \tag{5.151}$$

The anomalous transformation law for $\left\langle\frac{\delta S}{\delta\phi}\right\rangle$ is therefore given by (5.147), where

$$\Delta = \frac{\hbar}{2\pi}\ln\left(\frac{dx^+}{dy^+}\frac{dx^-}{dy^-}\right)\frac{\partial^2 \phi}{\partial y^+ \partial y^-}$$

$$+ \frac{\hbar}{4\pi}\left[\frac{d^2 x^-}{dy^{-2}}\left(\frac{dx^-}{dy^-}\right)^{-1}\frac{\partial\phi}{\partial y^+} + \frac{d^2 x^+}{dy^{+2}}\left(\frac{dx^+}{dy^+}\right)^{-1}\frac{\partial\phi}{\partial y^-}\right] . \tag{5.152}$$

5.4.3 The covariant quantum stress tensor

In parallel to what we did in Section 5.1 for the conformal (minimally coupled) field, we use the above results and the notion of locally inertial frames to derive an expression for the covariant quantum stress tensor. The expectation values of the covariant stress tensor at a point X are defined, in the locally inertial frame $\xi_X^{\pm}$, by

$$\langle\Psi|T_{\pm\pm}(\xi^+(X),\xi^-(X))|\Psi\rangle = \langle\Psi|:T_{\pm\pm}(\xi_X^+(X),\xi_X^-(X)):|\Psi\rangle \ . \tag{5.153}$$

In terms of the coordinates $x^{\pm}$, and requiring covariance, we have

$$\langle\Psi|T_{\pm\pm}(x^+(X),x^-(X))|\Psi\rangle = \left(\frac{d\xi_X^{\pm}}{dx^{\pm}}(X)\right)^2 \langle\Psi|:T_{\pm\pm}(\xi_X^+(X),\xi_X^-(X)):|\Psi\rangle \ . \tag{5.154}$$

The relation between $: T_{\pm\pm}(x^+(X), x^-(X)) :$ and $: T_{\pm\pm}(\xi_X^+(X), \xi_X^-(X)) :$ is given by

$$: T_{\pm\pm}(x^+(X), x^-(X)) := \left(\frac{d\xi_X^\pm}{dx^\pm}(X)\right)^2 : T_{\pm\pm}(\xi_X^+(X), \xi_X^-(X)) :$$
$$-\frac{\hbar}{24\pi}\{\xi_X^\pm, x^\pm\}|_X - \frac{\hbar}{4\pi}[\frac{d^2\xi_X^\pm}{dx^{\pm 2}}(X)\left(\frac{d\xi_X^\pm}{dx^\pm}(X)\right)^{-1}\frac{\partial\phi}{\partial x^\pm}(X)$$
$$+ \ln\frac{d\xi_X^+}{dx^+}(X)\frac{d\xi_X^-}{dx^-}(X)\left(\frac{\partial\phi}{\partial x^\pm}(X)\right)^2]. \qquad (5.155)$$

We note again, in comparison with Eq. (5.22), the presence of dilaton dependent terms making the theory more involved. Inserting (5.154) into (5.155) we finally obtain

$$\langle\Psi|T_{\pm\pm}(x^+(X), x^-(X))|\Psi\rangle = \langle\Psi| : T_{\pm\pm}(x^+(X), x^-(X)) : |\Psi\rangle$$
$$+ \frac{\hbar}{24\pi}\{\xi_X^\pm, x^\pm\}|_X + \frac{\hbar}{4\pi}[\frac{d^2\xi_X^\pm}{dx^{\pm 2}}(X)\left(\frac{d\xi_X^\pm}{dx^\pm}(X)\right)^{-1}\frac{\partial\phi}{\partial x^\pm}(X)$$
$$+ \ln\frac{d\xi_X^+}{dx^+}(X)\frac{d\xi_X^-}{dx^-}(X)\left(\frac{\partial\phi}{\partial x^\pm}(X)\right)^2]. \qquad (5.156)$$

Now using the relation between $\xi_X^\pm$ and $x^\pm$ given in Section 5.1, up to third order, a straightforward computation leads to the following form for the stress tensor, for an arbitrary point X,

$$\langle\Psi|T_{\pm\pm}(x^+, x^-)|\Psi\rangle = \langle\Psi| : T_{\pm\pm}(x^+, x^-) : |\Psi\rangle - \frac{\hbar}{12\pi}(\partial_\pm\rho\partial_\pm\rho - \partial_\pm^2\rho)$$
$$+ \frac{\hbar}{2\pi}\left[\partial_\pm\rho\partial_\pm\phi + \rho(\partial_\pm\phi)^2\right]. \qquad (5.157)$$

Note that for a Minkowskian metric $\rho = 0$ we recover the equality $\langle\Psi|T_{\pm\pm}(x^+, x^-)|\Psi\rangle = \langle\Psi| : T_{\pm\pm}(x^+, x^-) : |\Psi\rangle$.

Our job is not finished yet. We still need an expression for $\langle\frac{\delta S}{\delta\phi}\rangle$. We still have to use the quantum covariant conservation laws

$$\nabla^a\langle T_{ab}\rangle = \nabla_b\phi\frac{1}{\sqrt{-g}}\langle\frac{\delta S}{\delta\phi}\rangle, \qquad (5.158)$$

which in the conformal frame are translated into

$$\partial_\mp\langle T_{\pm\pm}\rangle + \partial_\pm\langle T_{+-}\rangle - 2\partial_\pm\rho\langle T_{+-}\rangle + \partial_\pm\phi\left\langle\frac{\delta S}{\delta\phi}\right\rangle = 0. \qquad (5.159)$$

We then obtain the following expression for $\langle T_{+-}\rangle$

$$\langle T_{+-}\rangle = -\frac{\hbar}{12\pi}\left(\partial_+\partial_-\rho + 3\partial_+\phi\partial_-\phi - 3\partial_+\partial_-\phi\right), \qquad (5.160)$$

which generalizes the corresponding expression in the Polyakov theory with the addition of dilaton dependent terms. Note that these terms were already implicitly given in Eq. (5.151). In covariant form we get the following form for the trace anomaly [15]

$$\langle T\rangle = \frac{\hbar}{24\pi}\left(R - 6(\nabla\phi)^2 + 6\Box\phi\right). \qquad (5.161)$$

Finally, combining Eqs. (5.157), (5.159) and (5.160) we get[16]

$$\langle\Psi|\frac{\delta S}{\delta\phi}|\Psi\rangle = \langle\Psi|\frac{\delta S}{\delta\phi}|\Psi\rangle_{\rho=0} - \frac{\hbar}{2\pi}\left(\partial_+\partial_-\rho + \partial_+\rho\partial_-\phi + \partial_-\rho\partial_+\phi + 2\rho\partial_+\partial_-\phi\right). \qquad (5.162)$$

The last three terms can be obtained from the anomalous transformation law Eq. (5.152), while the term $\partial_+\partial_-\rho$ comes directly from the imposition of the conservation laws Eqs. (5.159). It is very important to point out that the state-dependent quantities in Eqs. (5.157) and (5.162) must satisfy the conservation laws

$$\partial_{\mp}\langle\Psi|:T_{\pm\pm}(x^+,x^-):|\Psi\rangle + \partial_{\pm}\langle T_{+-}\rangle|_{\rho=0} + \partial_{\pm}\phi\langle\Psi|\frac{\delta S}{\delta\phi}|\Psi\rangle_{\rho=0} = 0. \qquad (5.163)$$

This is a crucial ingredient to ensure the covariant conservation laws (5.159). To match with the notation used in Section 5.1 we define the following functions $t_\pm(x^+,x^-)$ and introduce an additional function $t(x^+,x^-)$

$$-\frac{\hbar}{12\pi}t_\pm(x^+,x^-) \equiv \langle\Psi|:T_{\pm\pm}(x^+,x^-):|\Psi\rangle,$$

$$-\frac{\hbar}{2\pi}t(x^+,x^-) \equiv \langle\Psi|\frac{\delta S}{\delta\phi}|\Psi\rangle_{\rho=0} \qquad (5.164)$$

characterizing the quantum state $|\Psi\rangle$. Notice that now, in contrast with the conformal case, the functions $t_\pm$ are no more chiral (the same is true for the new function t) and satisfy a more involved set of equations reflecting the nontriviality of the theory even in two-dimensional flat spacetime.

---

[15] For the original derivation see [Mukhanov et al. (1994)]. The problem was reconsidered a few years later by [Bousso and Hawking (1997)]. More details can be found in [Grumiller et al. (2002)].

[16] From now on, we indicate $\langle|\frac{\delta S}{\delta\phi}|\rangle$ obtained within the normal ordering prescription as $\langle|\frac{\delta S}{\delta\phi}|\rangle_{\rho=0}$.

5.4.4 Effective action

The natural way to construct the effective action S_{eff} for this theory is to perform the functional integral with respect to the matter field f. By consistency, this effective action should agree with the value of the trace anomaly

$$-\frac{2}{\sqrt{-g}} g^{ab} \frac{\delta S_{eff}}{\delta g^{ab}} = \langle T \rangle = \frac{\hbar}{24\pi} \left(R - 6(\nabla\phi)^2 + 6\Box\phi \right). \quad (5.165)$$

This time the classical action, due to the presence of the dilaton field, is involved and the computation of the functional integral is a rather complicated problem [Mukhanov et al. (1994)].[17]

5.4.4.1 Anomaly induced effective action

A way to partly bypass this problem and go further is to construct the so-called *anomaly induced effective action* S_{eff}^{an}. If we now regard the above expression (5.165) as a functional differential equation for the effective action, the anomaly induced effective action is a particular solution. In our case one can find the following solution

$$S_{eff}^{an} = -\frac{1}{2\pi} \int d^2x \sqrt{-g} \left(\frac{1}{48} R \frac{1}{\Box} R - \frac{1}{4}(\nabla\phi)^2 \frac{1}{\Box} R + \frac{1}{4} \phi R \right). \quad (5.166)$$

The first term reproduces the Polyakov effective action, related to the geometric part R in the trace anomaly, and the additional terms, of which the first is nonlocal, come from the dilaton terms. An important point to remark is that the anomaly induced effective action constructed in this way is, unlike the Polyakov case, not uniquely determined. Indeed we can always add a Weyl invariant term which does not affect the defining equation (5.165). Therefore we can always write

$$S_{eff} = S_{eff}^{an} + \text{Weyl} - \text{invariant terms}. \quad (5.167)$$

For the conformal case, instead, the Polyakov action is the exact one since, in some sense, the trace anomaly contains all the information. Here we have much more freedom and, consequently, some additional ambiguities can arise. The actual difference between S_{eff} and the anomaly induced effective action considered above is not trivial and can potentially lead to non-reliable results. For example, calculations made in the literature by considering the $\langle T_{ab} \rangle$ derived by functional differentiation of (5.166) leads

---
[17]See also [Nojiri and Odintsov (2001)].

to the disturbing result that, for the Unruh state, the emitted radiation at late times for the Schwarzschild black hole is drastically changed from (5.106) and becomes negative [Mukhanov et al. (1994); Balbinot and Fabbri (1999)]

$$\langle U|T_{uu}|U\rangle \to \frac{(1-6)\hbar}{768\pi M^2} \qquad (5.168)$$

and therefore is physically meaningless. The general belief is that this is due to the incomplete knowledge of the exact effective action.[18]

However, one of the lessons of our discussion on the Polyakov effective action is that the splitting of the mean values of the stress tensor into geometric terms and the corresponding normal-ordered expectation values holds irrespective of the fact that the Polyakov action is the exact effective action. The only relevant property of the Polyakov action is that it is indeed an anomaly induced effective action. This is so because the local action S_{loc} (Subsection 5.2.1) is unambiguously defined, irrespective of the addition of Weyl-invariant terms. Having this in mind we shall exploit this property of the anomaly induced effective action to follow the same path for the theory with the dilaton.

5.4.4.2 The Weyl-invariant effective action

As we did for the Polyakov action, the nonlocality of the anomaly induced effective action (5.166) can be dealt with by introducing two auxiliary fields ψ and χ satisfying particular equations of motion. One can see that S_{eff}^{an} can be rewritten as [Balbinot and Fabbri (2003)]

$$S_{eff}^{aux} = S_1(g_{ab}, \phi, \psi) + S_2(g_{ab}, \phi, \chi) + S_3(g_{ab}, \phi) , \qquad (5.169)$$

where

$$S_1(\psi) = \hbar \int d^2x \sqrt{-g} \left[\frac{1}{2}\psi\Box\psi + \frac{1}{\sqrt{48\pi}}\left(R - 6(\nabla\phi)^2\right)\psi \right] , \qquad (5.170)$$

$$S_2(\chi) = \hbar \int d^2x \sqrt{-g} \left[-\frac{1}{2}\chi\Box\chi + \sqrt{\frac{3}{4\pi}}(\nabla\phi)^2\chi \right] \qquad (5.171)$$

and finally

$$S_3 = \hbar \int d^2x \sqrt{-g} \left[-\frac{1}{8\pi}\phi R \right] . \qquad (5.172)$$

---
[18]See [Kummer and Vassilevich (1999)] for a different interpretation.

The auxiliary fields ψ and χ satisfy the equations of motion

$$\Box \psi = -\frac{1}{\sqrt{48\pi}} \left(R - 6(\nabla \phi)^2 \right) , \qquad (5.173)$$

$$\Box \chi = \sqrt{\frac{3}{4\pi}} (\nabla \phi)^2 \qquad (5.174)$$

and by inserting them into the action (5.169) the nonlocal form of (5.166) is recovered.[19] We can follow the same reasoning as in Subsection 5.2.1 to generate, by functional differentiation, an exact expression for the stress tensor in terms of geometric quantities and state-dependent terms. To this end one can split the covariant (anomaly induced) effective action in two terms, one associated to a Weyl invariant quantization S_W (as normal ordering prescription requires), and a local one associated to state independent quantities. Note again that the ambiguities in the particular form of the (anomaly induced) effective action (Weyl invariant terms) can be reabsorbed into S_W. Therefore the form of the splitting would have been the same had we started from the (unknown) exact effective action S_{eff} instead of S_{eff}^{an}. Based on S_{eff}^{aux} we can construct a Weyl-invariant effective action in the following way

$$S_W(g^{ab}) = S_{eff}^{aux}(g^{ab}\sqrt{-g}) . \qquad (5.175)$$

For the explicit calculation we have to take into account the transformation law, under $g^{ab} \to g^{ab}\sqrt{-g}$, of the auxiliary fields

$$\psi \to \psi - \frac{1}{\sqrt{48\pi}} \ln \sqrt{-g}$$

$$\chi \to \chi . \qquad (5.176)$$

In terms of the Weyl-invariant effective action we can split S_{eff}^{an} as

$$S_{eff}^{an} = S_W + S_{loc} , \qquad (5.177)$$

where

$$S_{loc} = \frac{\hbar}{96\pi} \int d^2x \sqrt{-g} \big[\ln \sqrt{-g} \Box \ln \sqrt{-g} + 2R \ln \sqrt{-g}$$
$$- 12(\nabla \phi)^2 \ln \sqrt{-g} - 12(\nabla \phi)(\nabla \ln \sqrt{-g}) \big]$$
$$- \frac{\hbar}{2\sqrt{48\pi}} \int d^2x \sqrt{-g} \nabla_a \left(\ln \sqrt{-g} \nabla^a \psi - \psi \nabla^a \ln \sqrt{-g} \right) . \quad (5.178)$$

---
[19]The first local form, slightly different from this one, was given in [Buric et al. (1999)].

Note that as for the Polyakov theory the auxiliary fields appear in S_{loc} only through a total derivative term. This means that we are essentially left with a local action, depending only on g_{ab} and ϕ. All the nonlocality of S_{eff}^{an} is concentrated in the Weyl-invariant piece S_W. Varying S_{loc} with respect to the metric we get the following contribution to the stress tensor

$$-\frac{2}{\sqrt{-g}}\frac{\delta S_{loc}}{\delta g^{ab}} = \frac{\hbar}{48\pi}\left[\partial_a \ln\sqrt{-g}\partial_b \ln\sqrt{-g} - \frac{1}{2}g_{ab}(\nabla \ln\sqrt{-g})^2\right]$$
$$+ \frac{\hbar}{24\pi}\left[\nabla_a\nabla_b \ln\sqrt{-g} - \frac{1}{2}g_{ab}\Box \ln\sqrt{-g}\right] + \frac{\hbar}{48\pi}g_{ab}R$$
$$+ \frac{\hbar}{4\pi}\left[\ln\sqrt{-g}\partial_a\phi\partial_b\phi - \frac{1}{2}g_{ab}\ln\sqrt{-g}(\nabla\phi)^2\right]$$
$$- \frac{\hbar}{4\pi}\left[\partial_a\phi\partial_b \ln\sqrt{-g} - \frac{1}{2}g_{ab}\nabla\phi\nabla \ln\sqrt{-g}\right]$$
$$- \frac{\hbar}{8\pi}\left[(\nabla\phi)^2 - \Box\phi)\right]g_{ab} \ . \tag{5.179}$$

Note that, as in the Polyakov theory, we generate the trace anomaly from S_{loc}

$$-\frac{2}{\sqrt{-g}}g^{ab}\frac{\delta S_{loc}}{\delta g^{ab}} = \frac{\hbar}{24\pi}(R - 6(\nabla\phi)^2 + 6\Box\phi) \ . \tag{5.180}$$

Moreover, varying S_{loc} with respect to ϕ we find

$$\frac{\delta S_{loc}}{\delta\phi} = \frac{\hbar}{8\pi}\sqrt{-g}\left[2\nabla_a(\ln\sqrt{-g}\nabla^a\phi) + \Box \ln\sqrt{-g}\right] \ . \tag{5.181}$$

The above expressions, when restricted to conformal gauge,[20] take a very familiar form

$$-\frac{2}{\sqrt{-g}}\frac{\delta S_{loc}}{\delta g^{\pm\pm}} = -\frac{\hbar}{12\pi}\left((\partial_\pm\rho)^2 - \partial_\pm^2\rho\right) + \frac{\hbar}{2\pi}\left[\partial_\pm\rho\partial_\pm\phi + \rho(\partial_\pm\phi)^2\right] \ , \tag{5.182}$$

---

[20] We have to be careful to define conformal gauge in this context. Since the Weyl-invariant effective action is only invariant under area-preserving diffeomorphisms we have to redefine the null coordinates $x^\pm \to x^\pm/\sqrt{2}$ to ensure that the change of coordinates $(t,r) \to (x^+, x^-)$ be of unit jacobian. In terms of these redefined null coordinates the metric takes the form $ds^2 = -2e^{2\rho}dx^+dx^-$. $\rho = 0$ still represents the two-dimensional Minkowskian metric $ds^2 = -dt^2 + dr^2$. In this situation we have $\ln\sqrt{-g}=0$ for the Minkowskian metric in null coordinates. In Polyakov theory this subtle point is irrelevant since the term $\ln\sqrt{-g}$ always appears inside derivatives. We have to remark, nevertheless, that this does not imply any physical restriction at all. Had we defined initially the conformal gauge as $ds^2 = -2e^{2\rho}dx^+dx^-$ no modification of previous formulae would be required.

$$-\frac{2}{\sqrt{-g}}\frac{\delta S_{loc}}{\delta g^{+-}} = -\frac{\hbar}{12\pi}(\partial_+\partial_-\rho + 3\partial_+\phi\partial_-\phi - 3\partial_+\partial_-\phi) \;, \qquad (5.183)$$

$$\frac{\delta S_{local}}{\delta \phi} = -\frac{\hbar}{2\pi}(\partial_+\partial_-\rho + \partial_+\rho\partial_-\phi + \partial_-\rho\partial_+\phi + 2\rho\partial_+\partial_-\phi) \;. \qquad (5.184)$$

This way we exactly reproduce the local terms of the quantum stress tensor $\langle\Psi|T_{\pm\pm}|\Psi\rangle$ and $\langle\Psi|\frac{\delta S}{\delta\phi}|\Psi\rangle$ given by (5.157) and (5.162). It should be clear, at this point, that the remaining (state-dependent) parts in Eqs. (5.157) and (5.162) should be identified as

$$-\frac{2}{\sqrt{-g}}\frac{\delta S_W}{\delta g^{\pm\pm}} = \langle\Psi| :T_{\pm\pm}(x^+,x^-): |\Psi\rangle \;,$$
$$-\frac{2}{\sqrt{-g}}\frac{\delta S_W}{\delta g^{+-}} = 0$$
$$\frac{\delta S_W}{\delta\phi} = \langle\Psi|\frac{\delta S}{\delta\phi}|\Psi\rangle_{\rho=0} \;. \qquad (5.185)$$

We should remark here that we have obtained the exact form of the quantum stress tensor starting just from the anomaly induced effective action (5.166). The Weyl-invariant ambiguities present in the anomaly induced effective action play no fundamental role since they enter, at the end of the day, into the expectation values of the normal ordered stress tensor $\langle\Psi| :T_{\pm\pm}: |\Psi\rangle$ and $\langle\Psi|\frac{\delta S}{\delta\phi}|\Psi\rangle_{\rho=0}$.

5.4.5 *Quantum states and Hawking flux*

As an application of the expressions obtained for the covariant quantum stress tensor with the presence of the dilaton field (which takes into account the effects due to backscattering), we now perform, in a way parallel to Section 5.3, an analysis of the different choices of quantum states. The complications that we will find is that now the normal ordered expressions $\langle\Psi| :T_{\pm\pm}: |\Psi\rangle$, unlike in the Polyakov theory, cannot in general be determined exactly, but only asymptotically. To this end let us consider, as usual, the Schwarzschild metric in the double null form in terms of the Eddington–Finkelstein coordinates u and v, and the dilaton field given by

$$e^{-\phi} = r(r^* = \frac{v-u}{2})) \;. \qquad (5.186)$$

5.4.5.1 Boulware state

In this case we can propose an analytic expression for the full stress tensor. We can naturally choose a state such that $(x^+ = v, \ x^- = u)$

$$-\frac{\hbar}{12\pi}t_u = \langle \Psi | : T_{uu}(u,v) : | \Psi \rangle = 0$$

$$-\frac{\hbar}{12\pi}t_v = \langle \Psi | : T_{vv}(u,v) : | \Psi \rangle = 0 \ . \quad (5.187)$$

This allows to determine, through Eq. (5.163), the third function characterizing the quantum state. We obtain

$$-\frac{\hbar}{2\pi}t(u,v) = \langle \Psi | \frac{\delta S}{\delta \phi} | \Psi \rangle_{\rho=0} = -\frac{3\hbar}{8\pi}\frac{M}{r^3} + \frac{\hbar}{\pi}\frac{M^2}{r^4} \ . \quad (5.188)$$

This state is analogous of the Boulware vacuum state introduced in Subsection 5.3.1. Now applying expressions (5.157) we get

$$\langle \Psi | T_{uu} | \Psi \rangle = \langle \Psi | T_{vv} | \Psi \rangle = \frac{\hbar}{24\pi}\left(-\frac{4M}{r^3} + \frac{15}{2}\frac{M^2}{r^4}\right)$$

$$+ \frac{\hbar}{16\pi r^2}(1-\frac{2M}{r})^2 \ln(1-\frac{2M}{r}) \ . \quad (5.189)$$

The $+-$ component is state independent and fixed by the trace anomaly

$$\langle T_{uv} \rangle = \frac{\hbar}{12\pi}(1-\frac{2M}{r})\frac{M}{r^3} \ . \quad (5.190)$$

Finally, we also have, from Eqs. (5.162) and (5.188),

$$\langle \frac{\delta S}{\delta \phi} \rangle = -\frac{7\hbar}{8\pi}\frac{M}{r^3} + \frac{2\hbar}{\pi}\frac{M^2}{r^4} + \frac{\hbar}{8\pi r^2}(1-\frac{4M}{r})(1-\frac{2M}{r})\ln(1-\frac{2M}{r}) \ . \quad (5.191)$$

We remember, from Eq. (5.121), that this is related to the $\theta\theta$ component of the corresponding four-dimensional stress tensor through the relation

$$\langle T_{\theta\theta} \rangle = -\frac{1}{4\pi(1-2M/r)}\langle \frac{\delta S}{\delta \phi} \rangle \ . \quad (5.192)$$

These expressions vanish in the limit $M \to 0$ as well as for $r \to \infty$. Moreover, in the horizon limit they behave as in the Polyakov theory[21]

$$\langle \Psi | T_{uu} | \Psi \rangle = \langle \Psi | T_{vv} | \Psi \rangle \sim -\frac{\hbar}{768\pi M^2}$$

$$\langle \Psi | T_{uv} | \Psi \rangle \sim 0 \ . \quad (5.193)$$

---

[21] This behavior is independent of the regularization scheme used, see [Balbinot et al. (2001)].

5.4.5.2 Hartle–Hawking state

We can also consider a state defined in the maximally extended Schwarzschild spacetime such that, when restricted to the asymptotically flat exterior region, describes a thermal bath of radiation at the Hawking temperature T_H. The corresponding density matrix is $\rho = \overleftarrow{\rho} \otimes \overrightarrow{\rho}$, where $\overrightarrow{\rho}$ is given by an expression similar to that given in the Polyakov theory

$$\overrightarrow{\rho} = \prod_w \left(1 - e^{-2\pi w \kappa^{-1}}\right) \sum_N e^{-2\pi N w \kappa^{-1}} |\overrightarrow{N}_w\rangle\langle \overrightarrow{N}_w| , \qquad (5.194)$$

$|\overrightarrow{N}_w\rangle$ is the state in the Fock space with N outgoing particles of frequency w constructed from the Boulware vacuum, and $\overleftarrow{\rho}$ is the analogous thermal density matrix for the left-moving sector. Despite the formal analogy with the theory without dilaton the physical content is different because the modes are different.

Let us study now the asymptotic behavior of the stress tensor. To this end we have to evaluate the corresponding values of the normal ordered terms and this depends explicitly and uniquely on the form of the modes. The expansion of the field is

$$f = \int_0^\infty \frac{dw}{\sqrt{4\pi w}} e^{-iwt} \left(\overrightarrow{a}_w \frac{\overrightarrow{R}(r,w)}{r} + \overleftarrow{a}_w \frac{\overleftarrow{R}(r,w)}{r} + h.c. \right) , \qquad (5.195)$$

where the approximate forms of $\overrightarrow{R}(r,w)$ and $\overleftarrow{R}(r,w)$ at the horizon and at infinity are given in Eqs. (5.138) and (5.139). For instance, at I^+ we have

$$f_{r\to\infty} \sim \int_0^\infty \frac{dw}{\sqrt{4\pi w}} e^{-iwu} \left(\overrightarrow{a}_w \frac{t_{l=0}(w)}{r} + \overleftarrow{a}_w \frac{R_{l=0}(w)}{r} + h.c. \right) . \qquad (5.196)$$

This implies that the formula for $\langle \overrightarrow{N}_w | : T_{uu} : |\overrightarrow{N}_w\rangle$ given in Subsection 4.3.1 for plane wave modes is modified to

$$< \overrightarrow{N}_w | : T_{uu} : |\overrightarrow{N}_w >_{r\to\infty} \sim \frac{\hbar N w}{2\pi} |t_{l=0}(w)|^2 , \qquad (5.197)$$

and

$$< \overleftarrow{N}_w | : T_{uu} : |\overleftarrow{N}_w >_{r\to\infty} \sim \frac{\hbar N w}{2\pi} |R_{l=0}(w)|^2 . \qquad (5.198)$$

Therefore, a simple calculation similar to that given in Section 4.3.1 gives

$$Tr[: T_{uu} : \vec{\rho}]_{r \to \infty} \sim \frac{\hbar}{2\pi} \int_0^\infty \frac{w dw}{e^{2\pi w \kappa^{-1}} - 1} |t_{l=0}(w)|^2$$
$$Tr[: T_{uu} : \overleftarrow{\rho}]_{r \to \infty} \sim \frac{\hbar}{2\pi} \int_0^\infty \frac{w dw}{e^{2\pi w \kappa^{-1}} - 1} |R_{l=0}(w)|^2 \ . \quad (5.199)$$

Combining these two results we get

$$Tr[: T_{uu} : \rho]_{r \to \infty} \sim \frac{\hbar}{2\pi} \int_0^\infty \frac{w dw}{e^{2\pi w \kappa^{-1}} - 1} (|t_{l=0}(w)|^2 + |R_{l=0}(w)|^2)$$
$$\sim \frac{\hbar}{768 \pi M^2} \ , \quad (5.200)$$

where, in the last step, we have made use of the relations $t_{l=0}(w) = T_{l=0}(w)$ and $|T_{l=0}(w)|^2 + |R_{l=0}(w)|^2 = 1$ (see Subsection 3.4.1). A similar calculation in the near-horizon limit gives

$$Tr[: T_{uu} : \rho]_{r \to 2M} \sim \frac{\hbar}{768 \pi M^2} \ . \quad (5.201)$$

We observe therefore that the properties of this thermal state for the state-dependent terms are similar, in the asymptotic region and at the horizon, to those already found in the Polyakov theory. Moreover, since in these regions $\partial_u \phi \sim 0 \sim \partial_v \phi$ also the contribution of the geometric terms is the same as in the theory without dilaton. Therefore we conclude that

$$Tr[T_{uu} \rho]_{r \to \infty} = Tr[T_{vv} \rho]_{r \to \infty} \sim \frac{\hbar}{768 \pi M^2}$$
$$Tr[T_{uu} \rho]_{r \to 2M} = Tr[T_{vv} \rho]_{r \to 2M} \sim 0 \ . \quad (5.202)$$

5.4.5.3 The "in" and Unruh vacuum states: Hawking flux with backscattering

Finally let us turn our attention to the states describing black hole evaporation. We can consider the "in" vacuum state in the scenario described in Subsection 5.3.3. In the Eddington–Finkelstein coordinates in the "out" region the normal ordered stress tensor at $v = v_0$ can be evaluated from the new anomalous transformation law given in Eq. (5.144)

$$\langle in|:T_{uu}(u,v):|in\rangle|_{v_0} = -\frac{\hbar}{24\pi}\{u_{in},u\} - \frac{\hbar}{4\pi}\left[\frac{d^2 u_{in}}{du^2}\frac{du}{du_{in}}\frac{\partial\phi}{\partial u}\right.$$

$$\left.+(\frac{\partial\phi}{\partial u})^2\ln(\frac{du_{in}}{du})\right] = -\frac{\hbar}{24\pi}\left(\frac{8M}{(u_{in}-v_0)^3} + \frac{24M^2}{(u_{in}-v_0)^4}\right)$$

$$-\frac{\hbar}{4\pi}\left[-\frac{2M(r-2M)}{(v_0-u_{in})^2 r^2} + \frac{(r-2M)^2}{4r^4}\ln\frac{v_0-4M-u_{in}}{v_0-u_{in}}\right],$$

$$\langle in|:T_{vv}(u,v):|in\rangle = 0 . \tag{5.203}$$

In contrast with the Polyakov theory, the evolution of $\langle in|:T_{uu}(u,v):|in\rangle|_{v_0}$ for $v > v_0$ is altered by the presence of the potential barrier. It is difficult to explicitly evaluate this evolution and also the final result at I^+, where the covariant stress tensor coincides with the normal ordered one, for intermediate retarded time u. At early times $u \to -\infty$, as in the Polyakov theory, there is no outgoing flux

$$\langle in|:T_{uu}:|in\rangle_{u\to-\infty} \sim 0 . \tag{5.204}$$

At late times there is still thermal radiation as in the Polyakov case, but with backreaction effects included. From Eq. (5.203) we see that the dilaton contributions vanish in this limit. In this region the "in" vacuum is well approximated by the Unruh vacuum, which is described by the outgoing thermal density matrix $\vec{\rho}$. The calculation for the outgoing flux is then similar to the one already considered for the Hartle–Hawking state. The only difference is that the density matrix only concerns the outgoing sector, while the incoming one is trivial. The relevant formula is the first one in Eq. (5.199). Therefore, at late times the luminosity $L_{l=0}$ is[22]

$$L_{l=0} \equiv \langle in|T_{uu}|in\rangle \sim Tr[T_{uu}\vec{\rho}] \sim \frac{\hbar}{2\pi}\int_0^\infty \frac{wdw}{e^{2\pi w\kappa^{-1}}-1}\Gamma_{w0} , \tag{5.206}$$

where $\Gamma_{w0} \equiv |t_{l=0}(w)|^2$ is the s-wave grey-body factor.

The evaluation of the expectation value of the stress tensor at the future horizon also provides an interesting result. For the normal ordered operator

---

[22]In terms of the corresponding four-dimensional stress tensor we have

$$\langle in|T_{uu}^{(4)}|in\rangle = \frac{L_{l=0}}{4\pi r^2} .$$

we have

$$Tr[:T_{vv}:\vec{\rho}]_{r\to 2M} \sim \frac{1}{2\pi}\int_0^\infty \frac{wdw}{e^{8\pi Mw}-1}|r_{l=0}(w)|^2 \ . \tag{5.207}$$

Now taking into account that $|r_{l=0}(w)|^2 = 1 - |t_{l=0}(w)|^2$ and the vacuum polarization contribution (5.193) we get

$$Tr[T_{vv}\vec{\rho}]_{r\to 2M} \sim -\frac{\hbar}{2\pi}\int_0^\infty \frac{wdw}{e^{2\pi w\kappa^{-1}}-1}\Gamma_{w0} \ . \tag{5.208}$$

We see that this is a negative flux entering the black hole horizon which compensates exactly the Hawking radiation at infinity. Finally, as for the Hartle–Hawking state, the normal ordered and vacuum polarization terms combine in such a way that at the future horizon

$$Tr[T_{uu}\vec{\rho}]_{r\to 2M} \sim 0 \ . \tag{5.209}$$

5.5 Beyond the s-wave Approximation

It is natural, at this point, to wonder what happens if one tries to include the modes with $l \neq 0$. Mathematically, one can derive a distinct two-dimensional theory for any l and by repeating the steps that led to the action (5.114) we obtain

$$S = -\frac{1}{2}\int d^2x\sqrt{-g}\left[e^{-2\phi}(\nabla f)^2 + l(l+1)f^2\right] \ . \tag{5.210}$$

Therefore one of the main features of the construction performed in this chapter, namely the two-dimensional Weyl invariance, is broken by the $l(l+1)$ term which acts as a mass term. This means that also the trace of the covariant stress tensor becomes state-dependent and, therefore, it becomes even more difficult to provide an expression for the covariant quantum stress tensor. Moreover, even if we were able to perform such analysis we would still have to face another "daunting" problem, namely how to sum over l to get the full covariant four-dimensional quantum stress tensor. Another complication is that quantization does not commute with dimensional reduction.[23] All these difficulties are perhaps not surprising since the theory we started with, described by the action (5.112), is not Weyl invariant in four dimensions. Fortunately, however, the bulk of the Hawking radiation is in the s-wave sector and therefore it is reasonable,

---
[23]This is called the dimensional reduction anomaly [Frolov et al. (2000)].

5.6 The Problem of Backreaction

The basic problem of backreaction emerges when one tries to properly write and solve the semiclassical equations

$$G_{\mu\nu}(g_{\mu\nu}) = 8\pi \langle T_{\mu\nu}(g_{\mu\nu}) \rangle \ . \tag{5.211}$$

This clearly requires knowing the expectation values $\langle \Psi | T_{\mu\nu}(g_{\mu\nu}) | \Psi \rangle$ for a large class of metrics and for the relevant physical state. The calculation of $\langle \Psi | T_{\mu\nu}(g_{\mu\nu}) | \Psi \rangle$ is, in general, a rather elusive problem also involving some ambiguities [Wald (1994)]. As already stressed at the beginning of this chapter, only in a few specific situations the computation can be carried out. This is the case of conformally flat spacetimes and for conformal matter fields. In the absence of a general solution to this problem one should then resort to approximate schemes. A natural way is to proceed à la Hartree–Fock. First one evaluates $\langle \Psi | T_{\mu\nu}(g^0_{\mu\nu}) | \Psi \rangle$ for the classical background metric $g^0_{\mu\nu}$ and then solves Einstein's equations with the new source $\langle \Psi | T_{\mu\nu}(g^0_{\mu\nu}) | \Psi \rangle$. This will produce a first-order corrected metric $g^1_{\mu\nu}$. One can repeat the analysis and compute $\langle \Psi | T_{\mu\nu}(g^1_{\mu\nu}) | \Psi \rangle$ and then get, from Einstein's equations, a second correction to the metric: $g^2_{\mu\nu}$. Eventually this will converge to a self-consistent solution $g_{\mu\nu}$. In practice this programme is very difficult to implement, since one encounters strong difficulties even in the first step. The evaluation of $\langle \Psi | T_{\mu\nu}(g^0_{\mu\nu}) | \Psi \rangle$ for the classical Schwarzschild metric has not been done in an exact and analytical way in four dimensions. Only approximate analytical and numerical results are available in the literature [Candelas (1980); Howard (1984); Brown et al. (1986); Frolov and Zelnikov (1987); Vaz (1989); Anderson et al. (1995)]. To improve the situation we need to perform the additional simplifications already considered in this chapter, while maintaining the main physical ingredients of the problem. This is common practice in many branches of physics.

5.6.1 *Backreaction equations from spherical reduction*

In Section 5.4 we have shown how, for spherically symmetric spacetimes and in the s-wave approximation, one can provide a self-consistent expression for $\langle\Psi|T_{ab}(g_{ab})|\Psi\rangle$. Therefore this allows us to go further in the study of the backreaction problem in an analytical setting. We shall keep in mind, however, that there is a fundamental point which remains to be addressed, namely the characterization of the relevant quantum state.

To illustrate and discuss this problem thoroughly we shall consider the four-dimensional Hilbert-Einstein action

$$S^{(4)}_{H-E} = \frac{1}{16\pi}\int d^4x\sqrt{-g^{(4)}}R^{(4)}. \tag{5.212}$$

Restricting to spherically symmetric metrics

$$ds^2_{(4)} = g_{ab}dx^a dx^b + e^{-2\phi(x^a)}(d\theta^2 + \sin^2\theta d\varphi^2), \tag{5.213}$$

the variation of the action is[24]

$$\delta S^{(4)}_{H-E} = \frac{1}{16\pi}\int d^2x\sqrt{-g}d\theta d\varphi e^{-2\phi}\sin\theta\left(G^{(4)}_{ab}\delta g^{ab} + G^{(4)}_{\theta\theta}\delta g^{\theta\theta} + G^{(4)}_{\varphi\varphi}\delta g^{\varphi\varphi}\right). \tag{5.214}$$

Taking into account that $G^{(4)}_{\varphi\varphi} = \sin^2\theta G^{(4)}_{\theta\theta}$, $g^{\theta\theta} = e^{2\phi}$ and $g^{\varphi\varphi} = e^{2\phi}\sin^{-2}\theta$, the integration over the angular variables leads to

$$\delta S_{H-E} = \frac{1}{4}\int d^2x\sqrt{-g}e^{-2\phi}\left(G^{(4)}_{ab}\delta g^{ab} + 2G^{(4)}_{\theta\theta}\delta g^{\theta\theta}\right), \tag{5.215}$$

where

$$G^{(4)}_{ab} = 2\nabla_a\nabla_b\phi - 2\nabla_a\phi\nabla_b\phi + g_{ab}\left(3|\nabla\phi|^2 - 2\Box\phi - e^{2\phi}\right) \tag{5.216}$$

and

$$G^{(4)}_{\theta\theta} = e^{-2\phi}\left(|\nabla\phi|^2 - \Box\phi - \frac{1}{2}R\right). \tag{5.217}$$

The classical Einstein equations for spherically symmetric configurations are then

$$G^{(4)}_{ab} = 8\pi T^{(4)}_{ab}, \tag{5.218}$$

$$G^{(4)}_{\theta\theta} = 8\pi T^{(4)}_{\theta\theta}. \tag{5.219}$$

---

[24] As usual four-dimensional quantities are indicated with $^{(4)}$ to distinguish them from those obtained from the radial part of the metric g_{ab}, which have no superscript.

Alternatively, these equations follow from an effective two-dimensional action obtained by dimensional reduction of (5.212) under spherical symmetry. Taking into account that

$$R^{(4)} = R - 6|\nabla\phi|^2 + 4\Box\phi + 2e^{2\phi} , \qquad (5.220)$$

the resulting two-dimensional action is

$$S_{H-E} = \frac{1}{4} \int d^2x \sqrt{-g} e^{-2\phi} \left(R + 2|\nabla\phi|^2 + 2e^{2\phi} \right) . \qquad (5.221)$$

For the matter sector we have the spherically reduced scalar considered in Section 5.4

$$S_m = -\frac{1}{2} \int d^2x \sqrt{-g} e^{-2\phi} |\nabla f|^2 . \qquad (5.222)$$

The classical equations of motion can also be written in the equivalent form

$$\begin{aligned} G_{ab} &= T_{ab} \\ -\frac{\delta S_{H-E}}{\delta\phi} &= \frac{\delta S_m}{\delta\phi} , \end{aligned} \qquad (5.223)$$

where G_{ab}, not to be confused with the two-dimensional Einstein tensor $R_{ab} - \frac{1}{2} R g_{ab}$ which is trivially zero, is given by functional differentiation of the two-dimensional gravity action S_{H-E}

$$G_{ab} = \frac{2}{\sqrt{-g}} \frac{\delta S_{H-E}}{\delta g^{ab}} = \frac{1}{2} e^{-2\phi} G^{(4)}_{ab} , \qquad (5.224)$$

and

$$T_{ab} = -\frac{2}{\sqrt{-g}} \frac{\delta S_m}{\delta g^{ab}} = 4\pi e^{-2\phi} T^{(4)}_{ab} . \qquad (5.225)$$

Moreover, we also have the relation

$$\frac{1}{\sqrt{-g}} \frac{\delta S_{H-E}}{\delta\phi} = G^{(4)}_{\theta\theta} . \qquad (5.226)$$

The semiclassical equations are then

$$\begin{aligned} G_{ab} &= \langle\Psi|T_{ab}|\Psi\rangle \\ -\frac{\delta S_{H-E}}{\delta\phi} &= \langle\Psi|\frac{\delta S_m}{\delta\phi}|\Psi\rangle . \end{aligned} \qquad (5.227)$$

Using the expressions for $\langle\Psi|T_{ab}|\Psi\rangle$ and $\langle\Psi|\frac{\delta S_m}{\delta\phi}|\Psi\rangle$ obtained, in conformal gauge, in Subsection 5.4.3 we finally have

$$G_{\pm\pm} = -\frac{\hbar}{12\pi}\left((\partial_\pm\rho)^2 - \partial_\pm^2\rho\right) + \frac{\hbar}{2\pi}\left[\partial_\pm\rho\partial_\pm\phi + \rho(\partial_\pm\phi)^2\right]$$
$$+ \langle\Psi|:T_{\pm\pm}(x^+,x^-):|\Psi\rangle,$$
$$G_{+-} = -\frac{\hbar}{12\pi}\left(\partial_+\partial_-\rho + 3\partial_+\phi\partial_-\phi - 3\partial_+\partial_-\phi\right), \quad (5.228)$$
$$-\frac{\delta S_{H-E}}{\delta\phi} = -\frac{\hbar}{2\pi}(\partial_+\partial_-\rho + \partial_+\rho\partial_-\phi + \partial_-\rho\partial_+\phi + 2\rho\partial_+\partial_-\phi)$$
$$+ \langle\Psi|\frac{\delta S_m}{\delta\phi}|\Psi\rangle_{\rho=0},$$

where

$$\mathcal{G}_{\pm\pm} = e^{-2\phi}\left(\partial_\pm^2\phi - 2\partial_\pm\rho\partial_\pm\phi - (\partial_\pm\phi)^2\right), \quad (5.229)$$
$$\mathcal{G}_{+-} = e^{-2\phi}\left(-\partial_+\partial_-\phi + 2\partial_+\phi\partial_-\phi + \frac{1}{4}e^{2(\rho+\phi)}\right), \quad (5.230)$$
$$-\frac{\delta\mathcal{S}_{H-E}}{\delta\phi} = 2e^{-2\phi}(\partial_+\partial_-\rho + \partial_+\phi\partial_-\phi - \partial_+\partial_-\phi). \quad (5.231)$$

These are the equations to be solved to take into account the backreaction of the quantum matter fields in the spacetime geometry.

5.6.2 The problem of determining the state-dependent functions

The crucial difficulty we encounter before attacking the above equations concerns the terms characterizing the quantum state. The reason is that these terms cannot be specified in general in an easy way. In fact they must obey a set of conservation equations

$$\partial_\mp\langle\Psi|:T_{\pm\pm}(x^+,x^-):|\Psi\rangle + \partial_\pm\phi\langle\Psi|\frac{\delta S_m}{\delta\phi}|\Psi\rangle_{\rho=0}$$
$$-\frac{\hbar}{4\pi}\partial_\pm\left(\partial_+\phi\partial_-\phi - \partial_+\partial_-\phi\right) = 0, \quad (5.232)$$

which, in turn, involve the dilaton field ϕ. But the values of this field are determined by the semiclassical equations (5.228) also involving the terms characterizing the quantum state. Therefore we face a rather involved set of equations mixing local geometry with global properties of the quantum state.

We shall explain these difficulties by analysing the various possibilities connected with the quantum states already studied in a fixed background.

- *Boulware state (quantum corrections outside static stars)*

For the Boulware-type construction it is possible, as we have already seen in fixed background in Subsection 5.4.5.1, to determine the state-dependent functions. Assuming that the quantum corrected geometry is static in the coordinates $x^\pm$ it is natural to impose that

$$\langle \Psi | : T_{\pm\pm}(x^+, x^-) : | \Psi \rangle = 0 . \tag{5.233}$$

The remaining state-dependent term $\langle \Psi | \frac{\delta S_m}{\delta \phi} | \Psi \rangle_{\rho=0}$ is then determined by the conservation equations (5.232). Staticity means that ρ and ϕ are functions of the spatial coordinate $x = (x^+ - x^-)/2$, i.e., $\rho = \rho(x)$ and $\phi = \phi(x)$. This implies that

$$-\frac{\hbar}{4\pi}(\partial_+\phi\partial_-\phi - \partial_+\partial_-\phi) = \frac{\hbar}{16\pi}(\phi'^2 - \phi'') , \tag{5.234}$$

where the prime means derivative with respect to x. Then plugging this relation into Eq. (5.232) we get

$$\frac{\hbar}{16\pi}(\phi'^2 - \phi'')' + \phi'\langle \Psi | \frac{\delta S_m}{\delta \phi} | \Psi \rangle_{\rho=0} = 0 , \tag{5.235}$$

which implies

$$\langle \Psi | \frac{\delta S_m}{\delta \phi} | \Psi \rangle_{\rho=0} = -\frac{\hbar}{16\pi} \frac{(\phi'^2 - \phi'')'}{\phi'} . \tag{5.236}$$

- *Hartle–Hawking state (Black hole in thermal equilibrium)*

We have given an intuitive definition of the Hartle–Hawking state in Subsection 5.4.5.2 in terms of density matrices. These allowed us to infer the asymptotic values of the stress tensor both at infinity and at the horizon. However, to solve the backreaction problem one needs to know the state-dependent functions everywhere. This is in principle a very difficult problem, but with the experience gained in the conformal case (Subsection 5.3.2) one can naturally make the following ansatz. We know that this state describes the thermal equilibrium of the black hole with its own radiation. The quantum corrected geometry is then static as in Boulware, but the analysis is more involved. Indeed, we cannot impose the condition (5.233) in the static coordinates $x^\pm$. However, unlike the previous situation the quantum corrections are compatible with the existence of the horizon. Therefore we introduce the related "Kruskal" coordinates $y^\pm$ by means of the relations [25]

$$y^\pm = \pm \kappa^{-1} e^{\pm \kappa x^\pm} . \tag{5.237}$$

Our state $|\Psi\rangle$ state is then naturally defined as that state which has vanishing normal ordered stress tensor in $y^\pm$ coordinates, that is

$$\langle \Psi | : T_{\pm\pm}(y^+, y^-) : |\Psi\rangle = 0 . \tag{5.238}$$

These coordinates, however, are not the static coordinates and therefore we need to translate this condition to the $x^\pm$ frame using the transformation law (5.144). In terms of derivatives of $x = (x^+ - x^-)/2$ we get

$$\langle \Psi | : T_{\pm\pm}(x^+, x^-) : |\Psi\rangle = \frac{\hbar \kappa^2}{48\pi} - \frac{\hbar \kappa}{8\pi} \left[\phi' + x\phi'^2 \right] . \tag{5.239}$$

The remaining state-dependent function $\langle \Psi | \frac{\delta S_m}{\delta \phi} |\Psi\rangle_{\rho=0}$ is then

$$\langle \Psi | \frac{\delta S}{\delta \phi} |\Psi\rangle_{\rho=0} = -\frac{\hbar \kappa}{8\pi} \left[\frac{\phi''}{\phi'} + \phi' + 2x\phi'' \right] - \frac{\hbar}{16\pi} \frac{(\phi'^2 - \phi'')'}{\phi'} . \tag{5.240}$$

We remark that if we set $\kappa = 0$ we return to the Boulware case.

From these examples we see that the state-dependent functions, and therefore the semiclassical equations, contain higher-order derivatives. The term $1/\phi'$ must be treated with care. These equations are too difficult to be solved analytically, only a numerical treatment is possible.

- "in" and Unruh states (Black hole evaporation)

The crucial ingredient for working out ansatz solutions for the problem of determining the state-dependent functions in the previous two situations was staticity. This is not the case for the states describing black hole evaporation. Following the construction made in Subsection 5.3.3.1 the natural definition of the "in" state is

$$\langle in | : T_{u_{in} u_{in}} : |in\rangle = 0 . \tag{5.241}$$

---

[25] In fixed background we would have $\kappa = 1/4M$; in the exact treatment κ is expected to acquire quantum corrections.

However, in the "out" region $v > v_0$ we cannot write down the corresponding normal ordered expression in terms of the u coordinate, since the relation $u_{in}(u)$ is *dynamical* and can only be determined by solving the backreaction equations. The same considerations apply for the Unruh state, which is defined from the Kruskal coordinate U, but for the same reason in the dynamical situation the precise form $U(u)$ is unknown. An additional complication is, as already mentioned in the fixed background approximation, to take into account the effects due to backscattering.

5.6.3 *Returning to the near-horizon approximation*

The discussion made so far based on the backreaction equations, and in particular all the difficulties of solving them even in the s-wave approximation considered, make it clear that if we want to have a chance to describe in detail the evaporation process we have to find ways to simplify the problem. We shall then return, in the next and final chapter, to the near-horizon approximation considered in Chapter 4. In this limit the task of determining the state-dependent functions is simplified because the ϕ dependence in Eqs. (5.232) drops out and we are left with two simple equations

$$\partial_{\mp}\langle\Psi| :T_{\pm\pm}(x^+,x^-): |\Psi\rangle = 0 . \tag{5.242}$$

We will see that in some cases of physical relevance the backreaction equations can be solved exactly providing a full analytic picture of the evaporation process. This will be the main focus of the next and final chapter.

Chapter 6

Models for Evaporating Black Holes

The full backreaction problem for the spherically reduced gravity-matter system in the s-wave approximation has been established in Section 5.6. There we found that in general it is very hard even to write down the full semiclassical equations consistently because the task of determining the state-dependent functions $\langle\Psi| : T_{\pm\pm}(x^+, x^-) : |\Psi\rangle$ and $\langle\Psi|\frac{\delta S}{\delta \phi}|\Psi\rangle_{\rho=0}$ is highly non trivial. In particular, for the evaporating black hole scenario these functions are in general dynamical and cannot be determined *a priori*. This makes the problem very difficult. The details of the evaporation of a Schwarzschild black hole as well as the possible end point configurations are unknown, as illustrated in Fig. 6.1.

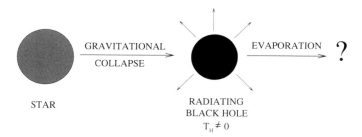

Fig. 6.1 Schematic representation of the evaporation process of a black hole formed by gravitational collapse.

The natural step is then to simplify the problem and come back to the near-horizon approximation described in Chapter 4. This will allow us to write down the corresponding backreaction equations by coupling the near-horizon gravity actions to Polyakov theory. In this chapter we will consider several physically relevant models where the semiclassical backreaction equations can be solved exactly.

The most famous model, which was also the responsible for the revival of interest in black hole evaporation, is the so called CGHS model [Callan et al. (1992)]. It is obtained as the near-horizon approximation of the near-extremal magnetically charged stringy black holes considered in Section 2.10. The study of black hole formation and evaporation in this model corresponds, in the higher dimensional theory, to the scattering of extremal black holes by neutral matter. In this context we have a natural candidate for the end point configuration, the extremal black hole itself, as depicted in Fig. 6.2. This is also the case for the evaporation of near-extremal Reissner–

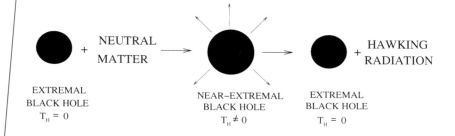

Fig. 6.2 By throwing low-energy neutral particles into an extremal black hole a nonextremal configuration is created and one expects it to decay via Hawking emission back to extremality.

Nordström black holes [Strominger and Trivedi (1993)].[1] The JT model [Jackiw (1984); Teitelboim (1984)] can serve to study these black holes, in the s-wave and near-horizon approximations, along similar lines. The resulting semiclassical theory is also solvable. In this scenario the inclusion of backreaction effects is especially important since at the late stages of the evaporation the fixed background approximation is not appropriate [Preskill et al. (1991)]. This is so because the thermal description of a black hole becomes problematic as the black hole approaches the extremal limit: fluctuations in temperature become large compared to the temperature itself.

The above two models are important because, despite the simplifications made to overcome the technical difficulties, the key conceptual issues of the evaporation process remain.

---

[1]This picture is more natural for magnetically charged black holes, since for the electrically charged ones the evaporation consists also of charged particles with the effect of reducing the hole charge [Hiscock and Weems (1990)].

6.1 The Near Horizon Approximation

We shall start by outlining the details concerning the near-horizon approximation in the Schwarzschild spacetime and improving the analysis of Section 4.1. The technique used will be then applied to the case of the near-extremal Reissner–Nordström black holes, in connection with the analysis of Section 4.4. Finally, we will derive the near-horizon limit for the near-extremal magnetically charged stringy black holes of Section 2.10, both in the Einstein and string frames.

6.1.1 *Schwarzschild*

Let us start from the dimensionally reduced Einstein-Hilbert action derived in Subsection 5.6.1

$$S_{H-E} = \frac{l^2}{4} \int d^2 x \sqrt{-g} e^{-2\phi} \left(R + 2|\nabla\phi|^2 + 2l^{-2} e^{2\phi} \right) , \tag{6.1}$$

where we have reinserted the parameter l defined by $r = le^{-\phi}$. Performing the near-horizon approximation around a given configuration of mass M_0 means to consider the following expansion[2] $r \sim r_0(1 + x)$, or, in terms of the dilaton field

$$e^{-\phi} \sim e^{-\phi_0}(1 + \tilde{\phi}) , \tag{6.2}$$

where $r_0 = le^{-\phi_0} = 2M_0$ is the radius of the horizon of the reference configuration. Expanding the action accordingly and keeping only the leading order terms in $\tilde{\phi}$ we get

$$S = \frac{l^2}{4} \int d^2 x \sqrt{-g} \left(e^{-2\phi_0}(1 + 2\tilde{\phi})R + \frac{2}{l^2} \right) . \tag{6.3}$$

The first term, proportional to $\sqrt{-g}R$, is a total derivative and therefore makes no contribution to the equations of motion. The relevant action that captures the near-horizon dynamics is then

$$S = \frac{r_0^2}{2} \int d^2 x \sqrt{-g} \left(\tilde{\phi} R + \frac{1}{r_0^2} \right) . \tag{6.4}$$

Let us now check that this model indeed reproduces the near-horizon geometry of the Schwarzschild black hole. The equations of motion derived

---
[2]Note that this expansion is a bit different from the one considered in Section 4.1. It follows more the philosophy of the one for near-extremal black holes, Subsection 4.4.2, where the reference configuration $r = r_0 = Q$ is more naturally selected.

from the action (6.4) obtained by functional differentiation with respect to g_{ab} are

$$\nabla_a \nabla_b \tilde{\phi} - g_{ab}\left(\Box\tilde{\phi} - \frac{1}{2r_0^2}\right) = 0 \tag{6.5}$$

and that derived from $\tilde{\phi}$ is simply

$$R = 0 \ . \tag{6.6}$$

This last equation is not surprising since we already know that the near-horizon region we are trying to reproduce is well described by Rindler space, which is a part of Minkowski space. Less trivial is the form of the dilaton $\tilde{\phi}$.

In conformal gauge $ds^2 = -e^{2\rho}dx^+dx^-$ the equations of motion are

$$\partial_+\partial_-\tilde{\phi} = -\frac{e^{2\rho}}{4r_0^2} \ , \tag{6.7}$$

$$\partial_+\partial_-\rho = 0 \ , \tag{6.8}$$

and

$$-\partial_\pm^2\tilde{\phi} + 2\partial_\pm\rho\partial_\pm\tilde{\phi} = 0 \ . \tag{6.9}$$

The last two equations, the $\pm\pm$ components of Eq. (6.5), are the so called constraint equations. One obvious solution for the metric is Minkowski spacetime

$$ds^2 = -dx^+dx^- \ , \tag{6.10}$$

i.e., $\rho = 0$, reducing the remaining equations to

$$\partial_+\partial_-\tilde{\phi} = -\frac{1}{4r_0^2} \ ,$$

$$\partial_\pm^2\tilde{\phi} = 0 \ . \tag{6.11}$$

The general solution for $\tilde{\phi}$ is then

$$\tilde{\phi} = -\frac{x^+x^-}{4r_0^2} + C \ , \tag{6.12}$$

where C is an arbitrary integration constant.[3] One apparently unpleasant feature of this solution is that it is not static. Indeed, in (t, r) coordinates

---
[3] A similar analysis, although in a different context, was given by [Cadoni and Mignemi (1995a)].

defined by $x^\pm = t \pm r$, where the metric becomes $ds^2 = -dt^2 + dr^2$, the dilaton reads

$$\tilde{\phi} = -\frac{(t^2 - r^2)}{4r_0^2} + C . \qquad (6.13)$$

Staticity can be achieved by choosing Rindler coordinates $x^\pm = \pm \kappa^{-1} e^{\pm \kappa \sigma^\pm}$

$$ds^2 = -e^{\kappa(\sigma^+ - \sigma^-)} d\sigma^+ d\sigma^- ,$$

$$\tilde{\phi} = \frac{e^{\kappa(\sigma^+ - \sigma^-)}}{4r_0^2 \kappa^2} + C . \qquad (6.14)$$

The identification with the near-horizon geometry of the Schwarzschild black hole considered in Section 4.1 can now be easily made. The expansion for r derived from the above solution is

$$r \equiv l e^{-\phi} \sim r_0 (1 + \tilde{\phi})$$
$$= r_0 (1 + C + \frac{e^{\kappa(\sigma^+ - \sigma^-)}}{4r_0^2 \kappa^2}) . \qquad (6.15)$$

On the other hand, from the near-horizon relation, in Eddington–Finkelstein coordinates

$$\frac{v - u}{2} \sim 2M \ln \frac{r - 2M}{2M} \qquad (6.16)$$

and considering small perturbations around the reference solution $M = M_0 + \Delta m$, with $\Delta m \ll M_0$, we get, at leading order in Δm

$$r \sim r_0 (1 + \frac{2\Delta m}{r_0} + e^{(v-u)/4M_0}) . \qquad (6.17)$$

This allows us to make the following identifications. The first is trivial, $\kappa = 1/4M_0$, and could have been derived only by looking at the radial part of the metric.[4] The second one clarifies the physical meaning of the constant C

$$C = 2\frac{\Delta m}{r_0} . \qquad (6.18)$$

---

[4]As already stressed in Section 4.1 the Rindler coordinates $\sigma^\pm$ are identified with the Eddington–Finkelstein coordinates (u, v) and the Minkowskian ones $x^\pm$ with the Kruskal coordinates. The non staticity of the above solutions in $x^\pm$ coordinates reflects the fact the Schwarzschild metric, when expressed in Kruskal coordinates, indeed looks time dependent.

One can also understand this result in the following alternative way. Performing the near-horizon approximation as in Section 4.1

$$r = 2M_0 + \frac{x^2}{8M_0} \qquad (6.19)$$

around the fixed radius $r = 2M_0$ we obtain, for a configuration of mass $M_0 + \Delta m$,

$$ds^2_{(4)} = -\left(\frac{x^2 - 16M_0\Delta m}{16M_0^2}\right)dt^2 + \frac{x^2\,dx^2}{x^2 - 16M_0\Delta m} + (4M_0^2 + \frac{x^2}{2})d\Omega^2 \ . \qquad (6.20)$$

The following change of coordinates

$$x^\pm = \pm 4M_0 e^{\pm(t\pm y)/4M_0} \ , \qquad (6.21)$$

where y is defined by

$$y = 2M_0 \ln \frac{|x^2 - 16M_0\Delta m|}{16M_0^2} \ , \qquad (6.22)$$

brings the metric and the dilaton to the form

$$ds^2 = -dx^+ dx^- \ ,$$
$$\tilde{\phi} = -\frac{x^+ x^-}{16M_0^2} + \frac{\Delta m}{M_0} \ , \qquad (6.23)$$

leading again to the identification (6.18).

6.1.2 Near-extremal Reissner–Nordström black holes: the JT model

As already stressed in Subsection 4.4.2, when the black hole is charged and we consider near-extremal configurations the near-horizon approximation can be naturally performed around the horizon of the extremal black hole. We shall now see how this approximation is realized in the action. In addition to the spherically reduced Einstein–Hilbert action (6.1) we need to know the corresponding one for the electromagnetic field. Starting from the four-dimensional action

$$S^{(4)}_{em} = -\frac{1}{16\pi}\int d^4x \sqrt{-g^{(4)}}\, F^{\mu\nu} F_{\mu\nu} \ , \qquad (6.24)$$

and assuming a radial electric field

$$A_\mu = (\frac{Q}{r}, 0, 0, 0) ,\qquad (6.25)$$

the action reduces to

$$S_{em} = -\frac{l^2}{4} \int d^2x \sqrt{-g} e^{-2\phi} F^{ab} F_{ab} . \qquad (6.26)$$

Since

$$F_{ab} = \frac{Q}{l^2} e^{2\phi} \epsilon_{ab} , \qquad (6.27)$$

where ϵ_{ab} is the two-dimensional antisymmetric tensor, the electromagnetic action turns out to be

$$S_{em} = -\frac{l^2}{4} \int d^2x \sqrt{-g} \frac{2Q^2}{l^4} e^{2\phi} . \qquad (6.28)$$

Had we started with a radial magnetic field, we would arrived at the same action (6.28) with Q being the magnetic charge. The two-dimensional Einstein–Maxwell action is then

$$S = \frac{l^2}{4} \int d^2x \sqrt{-g} e^{-2\phi} \left(R + 2|\nabla\phi|^2 + \frac{2}{l^2} e^{2\phi} - \frac{2Q^2}{l^4} e^{4\phi} \right) . \qquad (6.29)$$

Performing the near-horizon approximation around the extremal radius $r_0 = Q$

$$e^{-\phi} = e^{-\phi_0}(1 + \tilde\phi) , \qquad (6.30)$$

which mimics the expansion performed in Subsection 4.4.2 $r = r_0 + x$, with the identifications $x = Q\tilde\phi$, $le^{-\phi_0} = r_0 = Q$, we obtain

$$S = \frac{Q^2}{2} \int d^2x \sqrt{-g} \left(\tilde\phi R + \frac{2}{Q^2} \tilde\phi \right) . \qquad (6.31)$$

We note that the metric solution for this theory has negative constant curvature and therefore describes portions of AdS_2. The above theory was first proposed in a different context in [Jackiw (1984); Teitelboim (1984)].

6.1.3 Near-extremal dilaton black holes

We have seen in Section 2.10 that different black hole solutions emerge in the context of the low-energy approximation to string theory. Among the solutions analysed, the magnetically charged ones are the most interesting for our purposes. Also, since there is in principle an ambiguity in the choice of the physical metric, Einstein *versus* string frame, we will discuss both possibilities separately.

6.1.3.1 Einstein frame

Let us remind the four-dimensional action to start with

$$S^{(4)} = \frac{1}{16\pi} \int d^4x \sqrt{-g^{(4)}} \left(R - 2(\nabla\phi)^2 - \frac{1}{2}e^{-2\phi}F^2 \right) \tag{6.32}$$

and the spherically symmetric black hole solutions with non-zero magnetic charge Q

$$ds^2_{(4)} = -\left(1 - \frac{2M}{r}\right)dt^2 + \frac{dr^2}{(1 - \frac{2M}{r})} + r\left(r - \frac{Q^2}{2M}\right)d\Omega^2 ,$$

$$e^{-2\phi} = (1 - \frac{Q^2}{2Mr}) ,$$

$$F_{\theta\phi} = Q\sin\theta . \tag{6.33}$$

The extremal solution is given by the relation $M_0 = \frac{Q}{2}$, for which the singular "horizon" is at $r_0 = 2M_0 = Q$.

Let us now perform the near-horizon approximation of the solution (6.33) around the fixed surface $r = r_0$ by writing $r = r_0 + x$. As usual, we shall restrict to near-extremal configurations $M = M_0 + \Delta m$, where $\Delta m \ll M_0$. The resulting near-horizon solution is

$$ds^2_{(4)} = -\left(\frac{x - 2\Delta m}{r_0}\right)dt^2 + \frac{r_0\, dx^2}{x - 2\Delta m} + r_0(x + 2\Delta m)d\Omega^2 ,$$

$$e^{-2\phi} = \frac{1}{r_0}(x + 2\Delta m) . \tag{6.34}$$

We can immediately see the relevant fact that the radius of the two-sphere

$$r^2 \equiv l^2 e^{-2\hat\phi} = r_0(x + 2\Delta m) \tag{6.35}$$

is related to the dilaton field through the relation

$$e^{-2\hat\phi} = \frac{Q^2}{l^2}e^{-2\phi} . \tag{6.36}$$

We shall now derive the effective two-dimensional action which describes the above near-horizon solution. Spherical reduction of the action (6.32) can now be made under the splitting

$$ds^2_{(4)} = ds^2 + l^2 e^{-2\hat{\phi}} d\Omega^2 , \qquad (6.37)$$

where it must be emphasized that $\hat{\phi}$ is *a priori* different from the dilaton field ϕ. Moreover, in such a background F^2 reads

$$F^2 = \frac{2Q^2}{l^4} e^{4\hat{\phi}} . \qquad (6.38)$$

The resulting two-dimensional action depends on ϕ and $\hat{\phi}$

$$S = \frac{l^2}{4} \int d^2x \sqrt{-g} e^{-2\hat{\phi}} \left(R + 2(\nabla\hat{\phi})^2 + \frac{2e^{2\hat{\phi}}}{l^2} - 2(\nabla\phi)^2 - \frac{Q^2}{l^4} e^{-2\phi+4\hat{\phi}} \right) . \qquad (6.39)$$

The crucial point to note is that in the near-horizon limit ϕ and $\hat{\phi}$ are identical up to a constant shift, see Eq. (6.36). Therefore the final action will contain only one scalar field

$$S = \frac{Q^2}{4} \int d^2x \sqrt{-g} \left(e^{-2\phi} R + \frac{1}{Q^2} \right) . \qquad (6.40)$$

Notice that the structure of this action is the same as in the near-horizon approximation of the Schwarzschild solution (6.4). Comparing the two derivations we see that in the Schwarzschild case the near-horizon approximation was performed expanding around a fixed radius $le^{-\phi_0}$ (Eq. (6.2)), while here the physical requirement is (6.36) relating the dilaton field ϕ with the radius $\hat{\phi}$.[5]

One common feature of all the near-horizon approximations considered so far is that in all cases considered (Schwarzschild, Reissner–Nordström and dilaton black holes in the Einstein frame) the two-dimensional near-horizon geometry has constant curvature (zero for Schwarzschild and dilaton black holes and negative for Reissner–Nordström). The next example is qualitatively different since the near-horizon geometry is by itself a "genuine" black hole geometry, exhibiting curvature singularities as well as horizons.

---

[5] Actually, we could not impose (6.2) in this case because $e^{-2\hat{\phi}}$ is small, see Eq. (6.35), being both x and Δm small, and there's no analog of $e^{-2\phi_0}$ around which to expand.

6.1.3.2 String frame: the CGHS model

We now turn to the string frame, which, we recall, is defined by means of the conformal rescaling

$$g^S_{\mu\nu} = e^{2\phi} g^E_{\mu\nu} \,. \tag{6.41}$$

Here $g^E_{\mu\nu}$ is the metric in the Einstein frame and $g^S_{\mu\nu}$ is the metric in the string frame. The corresponding four-dimensional action is

$$S^{(4)} = \frac{1}{16\pi} \int d^4x \sqrt{-g^{(4)}}\, e^{-2\phi} \left(R + 4(\nabla\phi)^2 - \frac{1}{2}F^2 \right)\,. \tag{6.42}$$

The magnetically charged black hole solutions of this action are simply obtained starting from (6.33) and performing the transformation (6.41)

$$ds^2_{(4)} = -\frac{(1 - \frac{2M}{r})}{(1 - \frac{Q^2}{2Mr})} dt^2 + \frac{dr^2}{(1 - \frac{Q^2}{2Mr})(1 - \frac{2M}{r})} + r^2 d\Omega^2\,,$$

$$e^{-2\phi} = (1 - \frac{Q^2}{2Mr})\,,$$

$$F_{\theta\phi} = Q\sin\theta\,. \tag{6.43}$$

The details of the near horizon approximation can be worked out by considering $r = r_0 + x$ and expanding to first order both in x and the mass deviation from extremality $\Delta m = M - M_0 = M - r_0/2$. The resulting metric is

$$ds^2_{(4)} = -\frac{(x - 2\Delta m)}{(x + 2\Delta m)} dt^2 + \frac{r_0^2\, dx^2}{(x + 2\Delta m)(x - 2\Delta m)} + (r_0 + x)^2 d\Omega^2\,,$$

$$e^{-2\phi} = \frac{1}{r_0}(x + 2\Delta m)\,. \tag{6.44}$$

Note that this is a bit different from the solution that we get via conformal rescaling of (6.34):

$$ds^2_{(4)} = -\frac{(x - 2\Delta m)}{(x + 2\Delta m)} dt^2 + \frac{r_0^2\, dx^2}{(x + 2\Delta m)(x - 2\Delta m)} + r_0^2 d\Omega^2\,,$$

$$e^{-2\phi} = \frac{1}{r_0}(x + 2\Delta m)\,. \tag{6.45}$$

The difference is that here the radius of the two-sphere is "frozen" to r_0. In fact, in order to describe the relevant near-horizon dynamics, it is enough to restrict to the radial part of the metric, which exhibits a horizon at $x = 2\Delta m$ as well as a curvature singularity at $x = -2\Delta m$, and to the dilaton. The effective two-dimensional action reproducing this geometry can be obtained by performing the conformal rescaling (6.41) directly in the action (6.40). This produces the CGHS model [Callan et al. (1992)]

$$S = \frac{Q^2}{4} \int d^2x \sqrt{-g} \left[e^{-2\phi}(R + 4(\nabla\phi)^2 + \frac{1}{Q^2}) \right] . \quad (6.46)$$

Although the origin of this model is higher dimensional, it can be considered in its own right for the study of black hole formation and subsequent evaporation in a simplified setting.[6]

6.2 The CGHS Model

In this section we shall describe in detail the classical aspects of the CGHS model, first without matter and subsequently with the addition of matter fields. After having reviewed, in Subsections 6.2.3 and 6.2.4, the quantum aspects of the CGHS black hole in the fixed background approximation we will then turn, in Section 6.3, to the problem of backreaction in this model. The reader can complement these two sections with the following reviews [Harvey and Strominger (1992); Giddings (1994); Thorlacius (1994); Strominger (1995)].[7]

6.2.1 *The CGHS black hole*

We start our analysis by considering the pure gravitational CGHS action (6.46). In order to match with the notations used in the literature we shall write

$$Q^2 \equiv \frac{1}{4\lambda^2} \quad (6.47)$$

and fix the overall factor equal to 1/2. Thus we rewrite (6.46) as

$$S_{cghs} = \frac{1}{2} \int d^2x \sqrt{-g} \left[e^{-2\phi}(R + 4(\nabla\phi)^2 + 4\lambda^2) \right] . \quad (6.48)$$

Now we shall analyse in detail the properties of the CGHS black hole. First of all let us write down the equations of motion derived from the action (6.48). Variation with respect to the metric and the dilaton give,

---

[6] For earlier works on black holes in two dimensions see [Brown et al. (1986); Mann (1992)]. Systematic treatments of two-dimensional models can be found in [Frolov (1992); Klosch and Strobl (1996)] and also [de Alfaro et al. (1997)].

[7] See also [Benachenhou (1994)].

respectively,

$$e^{-2\phi}\left[-2\nabla_a\nabla_b\phi + \frac{1}{2}g_{ab}\left(-4(\nabla\phi)^2 + 4\nabla^2\phi + 4\lambda^2\right)\right] = 0 , \quad (6.49)$$

$$e^{-2\phi}\left[-2R - 8\lambda^2 + 8(\nabla\phi)^2 - 8\nabla^2\phi\right] = 0 . \quad (6.50)$$

Choosing conformal gauge $ds^2 = -e^{2\rho}dx^+dx^-$ the action (6.48) becomes

$$S_{cghs} = \int dx^+dx^- e^{-2\phi}\left[2\partial_+\partial_-\rho - 4\partial_+\phi\partial_-\phi + \lambda^2 e^{2\rho}\right] . \quad (6.51)$$

Variation with respect to ρ is equivalent to the $+-$ component of (6.49), i.e.,

$$e^{-2\phi}\left(2\partial_+\partial_-\phi - 4\partial_+\phi\partial_-\phi - \lambda^2 e^{2\rho}\right) = 0 , \quad (6.52)$$

which can be conveniently rewritten as

$$\partial_+\partial_- e^{-2\phi} + \lambda^2 e^{2(\rho-\phi)} = 0 . \quad (6.53)$$

Similarly, the ϕ equation

$$e^{-2\phi}\left(-16e^{-2\rho}\partial_+\partial_-\rho - 8\lambda^2 - 32e^{-2\rho}\partial_+\phi\partial_-\phi + 32e^{-2\rho}\partial_+\partial_-\phi\right) = 0 \quad (6.54)$$

can be rewritten as

$$2e^{-2\phi}\partial_+\partial_-(\rho - \phi) + \partial_+\partial_- e^{-2\phi} + \lambda^2 e^{2(\rho-\phi)} = 0 . \quad (6.55)$$

Finally, the constraint equations (i.e., the $\pm\pm$ components of Eqs. (6.49))

$$e^{-2\phi}\left(-2\partial_\pm^2\phi + 4\partial_\pm\rho\partial_\pm\phi\right) = 0 \quad (6.56)$$

can be cast in the form

$$\partial_\pm^2 e^{-2\phi} - 2\partial_\pm(\rho - \phi)\partial_\pm e^{-2\phi} = 0 . \quad (6.57)$$

Equations (6.53), (6.55) and (6.57) display a simple structure in terms of the fields $e^{-2\phi}$ and $2(\rho - \phi)$. Now, taking the difference between (6.55) and (6.53) one obtains the free field equation

$$2\partial_+\partial_-(\rho - \phi) = 0 . \quad (6.58)$$

This is equivalent to the requirement of zero curvature for the metric in the Einstein frame coming from (6.40).[8]

---

[8] Actually, up to the replacement $\rho \to \rho - \phi$ the field equations derived from (6.40) and (6.46) are the same.

6.2.1.1 Free field

The advantage of the existence of a free field is that it allows to fix completely the conformal coordinate system to work with. The conformal gauge $ds^2 = -e^{2\rho} dx^+ dx^-$ is not uniquely defined, as the transformations

$$x^\pm \to y^\pm = y^\pm(x^\pm) \tag{6.59}$$

preserve the conformal character of the metric. It is precisely this ambiguity which arises in the general solution of the free field equation (6.58)

$$2(\rho - \phi) = w_+(x^+) + w_-(x^-) , \tag{6.60}$$

where $w_+(x^+)$ and $w_-(x^-)$ are two arbitrary functions depending on x^+ and x^-, respectively. Substituting this into Eqs. (6.53) and (6.57) we get

$$\partial_+ \partial_- e^{-2\phi} = -\lambda^2 e^{w_+ + w_-} , \tag{6.61}$$

and

$$\partial_\pm^2 e^{-2\phi} + 2\partial_\pm \phi \partial_\pm w_\pm e^{-2\phi} = 0 . \tag{6.62}$$

The general solution to these equations is of the form

$$e^{-2\phi} = C - h_+ h_- , \tag{6.63}$$

where C is a constant and $h_+(x^+)$ and $h_-(x^-)$ are again chiral functions given by

$$h_+ = \lambda \int dx^+ e^{w_+}$$

$$h_- = \lambda \int dx^- e^{w_-} . \tag{6.64}$$

For the metric we have

$$ds^2 = -\frac{e^{(w_+ + w_-)} dx^+ dx^-}{(C - h_+ h_-)} . \tag{6.65}$$

The meaning of the emergence of the chiral functions w_+ and w_- is now clearly associated to the conformal transformations (6.59). The above generic solution can be generated from the simplest solution

$$ds^2 = -\frac{dy^+ dy^-}{(C - \lambda^2 y^+ y^-)}$$

$$e^{-2\phi} = C - \lambda^2 y^+ y^- \tag{6.66}$$

acting with the transformation (6.59) and identifying

$$\frac{dy^\pm}{dx^\pm} = e^{\omega_\pm} . \tag{6.67}$$

The simplest gauge choice $\omega_\pm = 0$ is equivalent to require

$$\rho = \phi \tag{6.68}$$

and leads to the solution

$$ds^2 = -\frac{dx^+ dx^-}{(C - \lambda^2(x^+ - x_0^+)(x^- - x_0^-))}$$
$$e^{-2\phi} = C - \lambda^2(x^+ - x_0^+)(x^- - x_0^-) , \tag{6.69}$$

where $x_0^\pm$ are constants that can always be eliminated by shifts of the coordinates. For simplicity we shall take them to be zero and replace C by m/λ.

$$ds^2 = -\frac{dx^+ dx^-}{(\frac{m}{\lambda} - \lambda^2 x^+ x^-)}$$
$$e^{-2\phi} = \frac{m}{\lambda} - \lambda^2 x^+ x^- . \tag{6.70}$$

The constant m cannot be removed by coordinate transformations and will have a clear physical meaning: it corresponds to the mass of the black hole. This can be directly justified from the original higher dimensional theory, where, as we shall see later, m is proportional to the mass deviation from extremality Δm. These (black hole) solutions were first discovered in [Witten (1991); Mandal et al. (1991)] in the context of two-dimensional non-critical string theory.

6.2.1.2 Linear dilaton vacuum

Let us now analyse in detail the properties of the above solutions. When $m = 0$ the metric

$$ds^2 = -\frac{dx^+ dx^-}{-\lambda^2 x^+ x^-} \tag{6.71}$$

is flat and one can introduce Minkowskian coordinates $\sigma^\pm$, defined by

$$\lambda x^\pm = \pm e^{\pm \lambda \sigma^\pm} , \tag{6.72}$$

such that the metric becomes

$$ds^2 = -d\sigma^+ d\sigma^- = -dt^2 + d\sigma^2 , \qquad (6.73)$$

where $\sigma^\pm = t \pm \sigma$. Furthermore, the form of the dilaton is

$$e^{-2\phi} = e^{\lambda(\sigma^+ - \sigma^-)} = e^{2\lambda\sigma} , \qquad (6.74)$$

or

$$\phi = -\lambda\sigma . \qquad (6.75)$$

This is the so called *linear dilaton vacuum*, already encountered at the end of Section 2.10.[9] The corresponding Penrose diagram is given in Fig. 2.21.

6.2.1.3 Black hole solutions

In the four-dimensional theory, the linear dilaton vacuum corresponds to the extremal dilaton black hole in the string frame, which is a completely regular solution free of singularities and horizons (see the discussion in Section 2.10). When $m \neq 0$, the higher dimensional solutions (6.43) exhibit both an event horizon and a singularity and these features are captured by the corresponding two-dimensional near-horizon configurations. The scalar curvature is

$$R = 8e^{-2\rho}\partial_+\partial_-\rho = \frac{4m\lambda}{\frac{m}{\lambda} - \lambda^2 x^+ x^-} . \qquad (6.76)$$

Therefore, we have a curvature singularity at

$$\lambda^3 x^+ x^- = m . \qquad (6.77)$$

The corresponding (x^+, x^-) diagram, very similar to the Kruskal diagram for the Schwarzschild black holes, is displayed in Fig. 6.3. In particular, the null lines $x^\pm = 0$ correspond to the future and past horizons.

Since it will be important in the following, let us see how we can regard the future event horizon also as an *apparent horizon*. For this we need to resort to the corresponding higher dimensional interpretation. Following the definitions given in Section 2.8 the trapped surfaces are given by the condition

$$\partial_+ A \leq 0 , \qquad (6.78)$$

---

[9]To match with the formulae of Section 2.10, we would like to reiterate that $\lambda \equiv 1/2Q$ and that σ is the same as the tortoise coordinate r^* along the throat.

Modeling Black Hole Evaporation

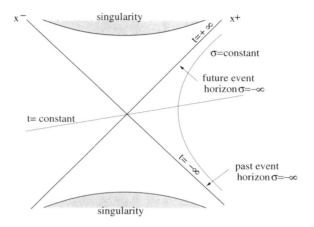

Fig. 6.3 Kruskal diagram of the CGHS black hole.

where A is the the area of the transverse two-sphere $A = 4\pi r^2$. From the four-dimensional solution (6.44) it turns out that

$$\partial_+ A = 8\pi r_0^2 \partial_+ e^{-2\phi} . \tag{6.79}$$

Therefore, the boundary of the trapped surfaces is given by the equation

$$\partial_+ e^{-2\phi} = 0 \tag{6.80}$$

leading to $x^- = 0$.

Since the metric in (6.70) is regular at the horizon, the coordinates $x^\pm$ are Kruskal-type coordinates for the CGHS-black hole. In fact, they provide the maximal analytical extension of the solution. For this reason the gauge choice $\rho = \phi$ is called the *Kruskal gauge*. *Asymptotically flat null coordinates* $\sigma^\pm$ are defined as in Eq. (6.72), that is $\lambda x^\pm = \pm e^{\pm \lambda \sigma^\pm}$, and they correspond to the gauge choice

$$w^\pm = \pm \lambda \sigma^\pm . \tag{6.81}$$

In this frame, which is of the null Eddington–Finkelstein type, the solution looks like

$$ds^2 = -\frac{d\sigma^+ d\sigma^-}{(1 + \frac{m}{\lambda} e^{-\lambda(\sigma^+ - \sigma^-)})} , \tag{6.82}$$

$$e^{-2\phi} = \frac{m}{\lambda} + e^{\lambda(\sigma^+ - \sigma^-)} . \tag{6.83}$$

This coordinate system covers only the exterior region of the black hole. In fact, $\sigma^+ - \sigma^- \to -\infty$ corresponds to the horizon (the surface $\sigma^+ = -\infty$ is the past and $\sigma^- = +\infty$ the future event horizon). Similarly, introducing (t, σ) coordinates by $\sigma^\pm = t \pm \sigma$ this is rewritten as

$$ds^2 = -\frac{dt^2 - d\sigma^2}{(1 + \frac{m}{\lambda}e^{-2\lambda\sigma})}, \qquad (6.84)$$

$$e^{-2\phi} = \frac{m}{\lambda} + e^{2\lambda\sigma}. \qquad (6.85)$$

One can also define the analog of the Schwarzschild gauge where the spacetime metric is written in the form

$$ds^2 = -f(r)dt^2 + \frac{dr^2}{f(r)}. \qquad (6.86)$$

The coordinate transformation leading from Eq. (6.84) to Eq. (6.86) involves only the spatial coordinate and is the following

$$r = \frac{1}{2\lambda} \ln(e^{2\lambda\sigma} + \frac{m}{\lambda}). \qquad (6.87)$$

The resulting metric is

$$ds^2 = -(1 - \frac{m}{\lambda}e^{-2\lambda r})dt^2 + \frac{dr^2}{(1 - \frac{m}{\lambda}e^{-2\lambda r})},$$
$$e^{-2\phi} = e^{2\lambda r}. \qquad (6.88)$$

This coordinate system is valid too outside the horizon, which is located at

$$r_h = \frac{1}{2\lambda} \ln \frac{m}{\lambda}. \qquad (6.89)$$

6.2.1.4 *CGHS and four-dimensional dilaton black holes*

To complete the picture we shall explicitly identify the CGHS solution (6.70) with the near-horizon geometry of the near-extremal dilaton black hole in the string frame (6.45). This can be done first by introducing the coordinate y defined by

$$y = r_0 \ln \frac{|x - 2\Delta m|}{r_0}, \qquad (6.90)$$

where $r_0 = Q$. This casts the radial part of the metric (6.45) and the dilaton in the form

$$ds^2 = -\frac{1}{1 + \frac{4\Delta m}{r_0} e^{-y/r_0}} (dt^2 - dy^2) ,$$

$$e^{-2\phi} = e^{y/r_0} + \frac{4\Delta m}{r_0} . \tag{6.91}$$

Definition of Kruskal coordinates

$$x^\pm = \pm 2 r_0 e^{\pm(t\pm y)/2r_0} \tag{6.92}$$

leads to

$$ds^2 = -\frac{dx^+ dx^-}{\frac{4\Delta m}{r_0} - \frac{x^+ x^-}{4r_0^2}} ,$$

$$e^{-2\phi} = \frac{4\Delta m}{r_0} - \frac{x^+ x^-}{4r_0^2} . \tag{6.93}$$

Identification with (6.70) can now be easily made by using the relation $\frac{1}{r_0} = 2\lambda$. Moreover the two-dimensional mass parameter m is related to the deviation of the mass from extremality Δm through the identification

$$m = \frac{2\Delta m}{r_0^2} . \tag{6.94}$$

6.2.1.5 CGHS and Schwarzschild black holes

To conclude the section we shall remark that, although the CGHS action (6.48) and the corresponding black hole solutions were originally obtained within the context of string theory, they can also be derived within the framework of pure general relativity. Starting from the Schwarzschild near-horizon action (6.4), and due to the similarity with the action (6.40), a conformal transformation similar to (6.41) but with conformal factor $\tilde{\phi}^{-1}$ brings (6.4) into the form of the CGHS action (6.48). To exactly match with (6.48), up to the overall coefficient, we would then need the identification

$$\frac{1}{r_0^2} = 4\lambda^2 . \tag{6.95}$$

These relations are even more clear at the level of the classical solutions. Performing this conformal transformation to the two-dimensional part of the near-horizon Schwarzschild metric (6.20)

$$ds^2 = \tilde{\phi}^{-1}\left[-\left(\frac{x^2 - 16M_0\Delta m}{16M_0^2}\right)dt^2 + \frac{x^2\,dx^2}{x^2 - 16M_0\Delta m}\right], \quad (6.96)$$

where

$$r_0^2\tilde{\phi} = \frac{x^2}{4}, \quad (6.97)$$

we obtain

$$ds^2 = -\frac{(x^2 - 16M_0\Delta m)r_0^2}{4M_0^2 x^2}dt^2 + \frac{4r_0^2\,dx^2}{(x^2 - 16M_0\Delta m)}. \quad (6.98)$$

Since $r_0^2 = 4M_0^2$, introducing the coordinate σ defined by

$$16M_0^2 e^{\sigma/2M_0} = x^2 - 16M_0\Delta m \quad (6.99)$$

we get

$$ds^2 = \frac{-dt^2 + d\sigma^2}{1 + \frac{\Delta m}{M_0}e^{-\sigma/2M_0}}, \quad (6.100)$$

and

$$\tilde{\phi} = e^{\sigma/2M_0} + \frac{\Delta m}{M_0}. \quad (6.101)$$

By comparison with (6.84) we find $\lambda = \frac{1}{4M_0}$, which is equivalent to (6.95), and also

$$m = \frac{\Delta m}{4M_0^2}. \quad (6.102)$$

6.2.2 Including matter fields

After having analysed the solutions of the relevant classical gravitational action which properly describes the near horizon region, let us now come to the matter sector. The near-horizon expansion of Eq. (6.45), with fixed radius of the sphere $r_0 = Q$, leads directly to

$$S_m = -\frac{Q^2}{2}\int d^2x\sqrt{-g}|\nabla f|^2. \quad (6.103)$$

The full two-dimensional action we shall analyse is

$$S = \frac{1}{2}\int d^2x\sqrt{-g}\left[e^{-2\phi}(R + 4(\nabla\phi)^2 + 4\lambda^2) - |\nabla f|^2\right], \quad (6.104)$$

where the overall coefficient and the normalization of the field f have been fixed to match the notation used in the literature. In this approximation we have a two-dimensional conformal scalar field which propagates freely in the background geometry. In conformal gauge the equation of motion for the field f is simply

$$\partial_+\partial_- f = 0 \ . \tag{6.105}$$

Since the matter part of the action is Weyl invariant, in conformal gauge it is given by

$$S = \int dx^+ dx^- \partial_+ f \partial_- f \ . \tag{6.106}$$

Therefore the ρ and ϕ equations of motion are not modified and they remain of the form:

$$\partial_+\partial_- e^{-2\phi} + \lambda^2 e^{2(\rho-\phi)} = 0 \ , \tag{6.107}$$

$$2e^{-2\phi}\partial_+\partial_-(\rho-\phi) + \partial_+\partial_- e^{-2\phi} + \lambda^2 e^{2(\rho-\phi)} = 0 \ . \tag{6.108}$$

The equations that are modified are the constraints

$$\partial_\pm^2 e^{-2\phi} + 4\partial_\pm\phi\partial_\pm(\rho-\phi)e^{-2\phi} + T_{\pm\pm} = 0 \ , \tag{6.109}$$

where the null components of the stress tensor are given by

$$T_{\pm\pm} = (\partial_\pm f)^2 \ . \tag{6.110}$$

The above equations remain solvable since the key ingredient, namely the existence of the free field $\rho - \phi$, is preserved by the matter field. In fact, the free field equation

$$2\partial_+\partial_-(\rho-\phi) = 0 \tag{6.111}$$

is a consequence of the ρ and ϕ equations of motion, which are not altered by the matter. This means, for instance, that the gauge fixing condition $\rho = \phi$ is consistent with the full set of equations of motion. In this Kruskal gauge the general solution to the above equations can be given in terms of both outgoing T_{--} and incoming T_{++} energy fluxes.

6.2.2.1 Dynamical black hole solutions

At the classical level we are mainly interested in those solutions describing the formation of a black hole by incoming matter

$$f = f(x^+) \tag{6.112}$$

from the linear dilaton configuration. Assuming that

$$T_{--} = 0 \tag{6.113}$$

the classical solution is given by

$$e^{-2\phi} = e^{-2\rho} = \frac{m(x^+)}{\lambda} - \lambda^2 x^+ \left(x^- + \frac{P(x^+)}{\lambda^2}\right), \tag{6.114}$$

where

$$m(x^+) = \lambda \int_{x_i^+}^{x^+} dy^+ y^+ T_{++}(y^+) \tag{6.115}$$

and

$$P(x^+) = \int_{x_i^+}^{x^+} dy^+ T_{++}(y^+) . \tag{6.116}$$

In the above expressions we have assumed that the ingoing flux starts at some time x_i^+, so at early advanced times $x^+ < x_i^+$ the solution reduces to the linear dilaton vacuum. Moreover, we also assume that the incoming matter flux vanishes after the time x_f^+ ($x_f^+ > x_i^+$). For $x^+ > x_f^+$ the solution takes the form of an eternal black hole

$$ds^2 = -\frac{dx^+ dx^-}{\frac{m(x_f^+)}{\lambda} - \lambda^2 x^+ (x^- + \frac{P(x_f^+)}{\lambda^2})} ,$$

$$e^{-2\phi} = \frac{m(x_f^+)}{\lambda} - \lambda^2 x^+ \left(x^- + \frac{P(x_f^+)}{\lambda^2}\right) . \tag{6.117}$$

Comparing with Eq. (6.70) we see that the mass parameter is $m = m(x_f^+)$ and that the x^- coordinate is shifted by the amount $P(x_f^+)/\lambda^2$. The physical meaning of $m = m(x_f^+)$ is clear when one considers the solution in asymptotically flat coordinates $\sigma^\pm$ defined by

$$\lambda x^+ = e^{\lambda \sigma^+} ,$$
$$-\lambda \left(x^- + \frac{P(x_f^+)}{\lambda^2} \right) = e^{-\lambda \sigma^-} , \qquad (6.118)$$

which convert the solution to the standard form (6.82). From (6.115) we have

$$m(x_f^+) = \int_{\sigma_i^+}^{\sigma_f^+} d\sigma^+ T_{\sigma^+ \sigma^+} , \qquad (6.119)$$

which is nothing but the total mass. This in turn justifies, as anticipated in Subsection 6.2.1, the identification of the integration constant C with m/λ in Eq. (6.70) where m is the mass of the two-dimensional black hole.[10]

6.2.2.2 *Singularities, event and apparent horizons*

Let us analyse the spacetime properties of the dynamical solution, as they are represented in Fig. 6.4.

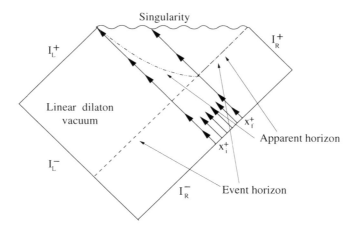

Fig. 6.4 Penrose diagram of a CGHS black hole formed by incoming matter.

---

[10] A more formal proof involves the calculation of the ADM mass, giving the same result [Witten (1991)].

The scalar curvature

$$R = \frac{4m(x^+)\lambda}{\frac{m(x^+)}{\lambda} - \lambda^2 x^+ (x^- + \frac{P(x^+)}{\lambda^2})} \qquad (6.120)$$

is singular along the spacelike line

$$x^- = -\frac{P(x^+)}{\lambda^2} + \frac{m(x^+)}{\lambda^3 x^+} . \qquad (6.121)$$

In the limit $x^+ \to +\infty$ this curve is asymptotic to the null line

$$x^- = -\frac{P(x_f^+)}{\lambda^2} , \qquad (6.122)$$

which, in turn, defines the event horizon of the black hole. The global nature of the event horizon is here quite explicit since its location depends only on the final value of the function $P(x^+)$. The apparent horizon, in contrast, is local and it is sensitive to the details of the dynamics. The natural definition, following from the discussion of Subsection 6.2.1.3, is[11]

$$\partial_+ e^{-2\phi} = 0 . \qquad (6.123)$$

In our case we find

$$x^- = -\frac{P(x^+)}{\lambda^2} . \qquad (6.124)$$

This curve is spacelike or null and coincides, after x_f^+, i.e., when the geometry has settled down to the final static black hole geometry, with the event horizon (6.122).

6.2.3 *Bogolubov coefficients and Hawking radiation*

Our aim here is to study the Hawking radiation for the matter field f in the fixed background geometry described previously, corresponding to the formation of a CGHS black hole from the linear dilaton vacuum. In this analysis we shall mainly use the standard approach of Bogolubov transformations for the natural mode basis defined in the initial and final spacetime regions. We shall also follow the line of [Giddings and Nelson (1992)]. To have a clear understanding of the physics it is convenient to keep in mind the corresponding higher dimensional picture. The formation of a CGHS black hole from the initial linear dilaton vacuum is a near-horizon description of

---

[11] This definition is more straightforward in the Einstein frame. See also [Russo et al. (1992a)].

the scattering process of a low-energy (neutral) particle off the extremal black hole. The linear dilaton vacuum represents the extremal charged black hole. The incoming flux of matter makes it non-extremal and allows to produce Hawking radiation. We shall analyse this radiation using the near-horizon CGHS model.

We have two relevant regions involved in the problem. One is the flat initial linear dilaton vacuum, described by the metric

$$ds^2 = -\frac{dx^+ dx^-}{-\lambda^2 x^+ x^-} , \quad (6.125)$$

or, equivalently, in Minkowskian coordinates

$$ds^2 = -d\sigma_{in}^+ d\sigma_{in}^- , \quad (6.126)$$

where

$$\pm \lambda x^\pm = e^{\pm \lambda \sigma_{in}^\pm} . \quad (6.127)$$

The other region is the final static black hole with metric

$$ds^2 = -\frac{dx^+ dx^-}{\frac{m(x_f^+)}{\lambda} - \lambda^2 x^+ \left(x^- + \frac{P(x_f^+)}{\lambda^2}\right)} . \quad (6.128)$$

The asymptotically flat coordinates are now

$$\lambda x^+ = e^{\lambda \sigma_{out}^+}$$
$$-\lambda \left(x^- + \frac{P(x_f^+)}{\lambda^2}\right) = e^{-\lambda \sigma_{out}^-} , \quad (6.129)$$

in terms of which the metric reads

$$ds^2 = -\frac{d\sigma_{out}^+ d\sigma_{out}^-}{(1 + \frac{m(x_f^+)}{\lambda} e^{-\lambda(\sigma_{out}^+ - \sigma_{out}^-)})} . \quad (6.130)$$

We can now relate the "in" and "out" coordinates through the Kruskal $x^\pm$ coordinates. We get

$$\sigma_{in}^+ = \sigma_{out}^+ \equiv \sigma^+$$
$$\sigma_{in}^- = -\frac{1}{\lambda} \ln \left(e^{-\lambda \sigma_{out}^-} + \frac{P(x_f^+)}{\lambda}\right) . \quad (6.131)$$

The relation $\sigma_{in}^+ = \sigma_{out}^+$ implies the triviality of the Bogolubov transformations for the ingoing sector. Therefore we shall concentrate only on the outgoing sector, which is the responsible for the Hawking radiation. Note that here there is no classical analog of the "reflecting surface" $r = 0$ of four-dimensional Minkowski space. As a consequence left and right movers are independent. See Fig. 6.5.

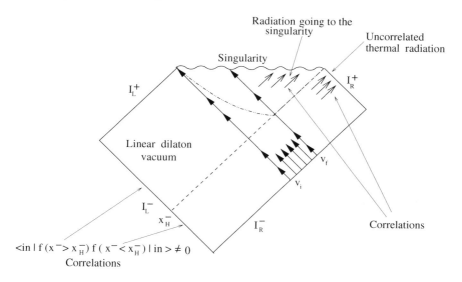

Fig. 6.5 Penrose diagram for the formation of a CGHS black hole. In the fixed background approximation left and right movers are independent. The correlations shown in the linear dilaton vacuum are transformed into correlations between the outgoing radiation and radiation going to the singularity.

The natural basis for the "in" region is given by the plane wave modes with respect to the null Minkowskian coordinate σ_{in}^-

$$u_w^{in} = \frac{1}{\sqrt{4\pi w}} e^{-iw\sigma_{in}^-} . \qquad (6.132)$$

The mode expansion for the outgoing sector of the field f is then

$$f = \int_0^\infty dw (a_w^{in} u_w^{in} + a_w^{in\dagger} u_w^{in*}) . \qquad (6.133)$$

For the "out" region we can construct the plane wave modes with respect to the asymptotically flat null coordinate σ_{out}^-

$$u_w^{out} = \frac{1}{\sqrt{4\pi w}} e^{-iw\sigma_{out}^-} . \qquad (6.134)$$

However, as we have already seen several times in this book, the basis defined by these modes is not complete, since they have support, by construction, in the exterior region of the black hole horizon. Modes with support beyond the black hole horizon can be constructed without difficulty, but for them there is no natural choice. We can leave them unspecified and call them generically u_w^{int}. This ambiguity in fact is irrelevant for the physical measurements of the asymptotic external observer. Therefore we can also expand the field f as follows

$$f = \int_0^\infty dw (a_w^{out} u_w^{out} + a_w^{out\dagger} u_w^{out*} + a_w^{int} u_w^{int} + a_w^{int\dagger} u_w^{int*}) \ . \tag{6.135}$$

The above two basis have been chosen orthonormal under the scalar product

$$(f_1, f_2) = -i \int_\Sigma d\Sigma^\mu (f_1 \partial_\mu f_2^* - \partial_\mu f_1 f_2^*) \ , \tag{6.136}$$

where Σ is an appropriate Cauchy hypersurface and the corresponding creation and annihilation operators satisfy the usual commutation relations.

The "in" vacuum state is defined by

$$a_w^{in} |0_{in}\rangle = 0 \ . \tag{6.137}$$

The problem now is to determine how this state is described within the Fock space of the observer in the "out" asymptotically flat region. As explained in Chapter 3 the main tool are the Bogolubov transformations. We now proceed to calculate the Bogolubov coefficients using the standard formulae derived in Chapter 3. In our particular case we have

$$\alpha_{ww'} = (u_w^{out}, u_{w'}^{in}) = -2i \int_{-\infty}^{+\infty} d\sigma_{in}^- u_w^{out} \frac{\partial u_{w'}^{in*}}{\partial \sigma_{in}^-} \ ,$$

$$\beta_{ww'} = (u_w^{out}, u_w^{in*}) = 2i \int_{-\infty}^{+\infty} d\sigma_{in}^- u_w^{out} \frac{\partial u_{w'}^{in}}{\partial \sigma_{in}^-} \ . \tag{6.138}$$

Taking into account Eqs. (6.131) the above integrals turn out to be

$$\alpha_{ww'} = \frac{1}{2\pi} \sqrt{\frac{w'}{w}} \int_{-\infty}^{\sigma_{in}^-(H)} d\sigma_{in}^- \exp\left(\frac{iw}{\lambda} \ln(e^{-\lambda \sigma_{in}^-} - \frac{P(x_f^+)}{\lambda}) + iw' \sigma_{in}^-\right)$$

$$\beta_{ww'} = \frac{1}{2\pi} \sqrt{\frac{w'}{w}} \int_{-\infty}^{\sigma_{in}^-(H)} d\sigma_{in}^- \exp\left(\frac{iw}{\lambda} \ln(e^{-\lambda \sigma_{in}^-} - \frac{P(x_f^+)}{\lambda}) - iw' \sigma_{in}^-\right) \tag{6.139}$$

where the upper integration point $\sigma^-_{in}(H)$, i.e., the location of the event horizon, is given by

$$\sigma^-_{in}(H) = -\frac{1}{\lambda} \ln \frac{P(x_f^+)}{\lambda} . \qquad (6.140)$$

Making the substitution

$$x = \frac{P(x_f^+)}{\lambda} e^{\lambda \sigma^-_{in}} , \qquad (6.141)$$

we get

$$\alpha_{ww'} = \frac{1}{2\pi\lambda} \sqrt{\frac{w'}{w}} \left(\frac{P(x_f^+)}{\lambda} \right)^{\frac{i(w-w')}{\lambda}} \int_0^1 dx (1-x)^{\frac{iw}{\lambda}} x^{-1-\frac{i(w-w')}{\lambda}}$$

$$\beta_{ww'} = \frac{1}{2\pi\lambda} \sqrt{\frac{w'}{w}} \left(\frac{P(x_f^+)}{\lambda} \right)^{\frac{i(w+w')}{\lambda}} \int_0^1 dx (1-x)^{\frac{iw}{\lambda}} x^{-1-\frac{i(w+w')}{\lambda}} . \qquad (6.142)$$

The integral in x is a β function, so we have[12]

$$\alpha_{ww'} = \frac{1}{2\pi\lambda} \sqrt{\frac{w'}{w}} \left(\frac{P(x_f^+)}{\lambda} \right)^{\frac{i(w-w')}{\lambda}} B(-\frac{i(w-w')}{\lambda}, 1 + \frac{iw}{\lambda})$$

$$\beta_{ww'} = \frac{1}{2\pi\lambda} \sqrt{\frac{w'}{w}} \left(\frac{P(x_f^+)}{\lambda} \right)^{\frac{i(w+w')}{\lambda}} B(-\frac{i(w+w')}{\lambda}, 1 + \frac{iw}{\lambda}) . \qquad (6.143)$$

We should note the important fact that for this model the Bogolubov coefficients are exactly calculable at all times, and not just at late times as in Chapter 3 for the Schwarzschild black hole. If we want to follow here the usual argument to work out the late time behavior of the Bogolubov coefficients we should replace the expression of the "out" modes u_w^{out} by their approximate value near the horizon $\sigma^-_{in} \approx \sigma^-_{in}(H)$

$$u_w^{out}(\sigma^-_{in}) = \frac{1}{\sqrt{4\pi w}} \exp\left(\frac{iw}{\lambda} \ln(e^{-\lambda \sigma^-_{in}} - \frac{P(x_f^+)}{\lambda}) \right)$$

$$\approx \frac{1}{\sqrt{4\pi w}} \exp\left(\frac{iw}{\lambda} \ln \frac{|\sigma^-_{in} - \sigma^-_{in}(H)|}{\lambda} \right) . \qquad (6.144)$$

---

[12] We have omitted the ϵ terms to properly carry out the integrals. See the discussion in Subsection 3.3.3.

More rigorously, since we have the exact form of the Bogolubov coefficients we have a direct way to estimate the relative quotient $|\alpha_{ww'}|/|\beta_{ww'}|$ at late times [Piqueras (1999)]. This can be easily obtained estimating the *large* w' behavior[13] of the exact relation

$$\frac{|\alpha_{ww'}|^2}{|\beta_{ww'}|^2} = \left(\frac{w'+w}{w'-w}\right)\frac{\sinh[\pi(w'+w)/\lambda]}{\sinh[\pi(w'-w)/\lambda]} \cdot \quad (6.145)$$

In the limit $w' \to +\infty$ we get the thermal relation

$$|\beta_{ww'}| = e^{-\pi w/\lambda}|\alpha_{ww'}|, \quad (6.146)$$

corresponding to a Hawking temperature

$$k_B T_H = \frac{\hbar \lambda}{2\pi} \cdot \quad (6.147)$$

We observe that this temperature is a *constant*, independent of the mass of the black hole. This fact has a higher dimensional counterpart since, as we have already seen in Section 2.10, the corresponding stringy dilatonic black holes have a surface gravity which approximates, near extremality, to the constant $\kappa \approx 1/2Q$, which agrees with the above result due to the identification $\lambda = 1/2Q$. In parallel with the analysis performed in Chapter 3, we see from Fig. 6.5 that there are correlations of the "in" vacuum which are transferred into correlations between the outgoing thermal radiation and radiation going to the singularity.

6.2.4 *Quantum states for the CGHS black hole*

We shall now derive the expectation values of the stress tensor for the quantum states of interest Boulware, Hartle–Hawking, Unruh for an eternal CGHS black hole. Moreover we shall also consider the "in" vacuum state for the dynamical formation of a CGHS black hole. This analysis is in parallel to that presented in Section 5.3 for the Schwarzschild black hole. First of all let us recall the general expression of $\langle\Psi|T_{ab}|\Psi\rangle$ in conformal gauge for conformally coupled scalar fields, as derived in Section 5.1:

$$\langle\Psi|T_{\pm\pm}|\Psi\rangle = -\frac{\hbar}{12\pi}\left((\partial_\pm \rho)^2 - \partial_\pm^2 \rho\right) + \langle\Psi|:T_{\pm\pm}(x^\pm):|\Psi\rangle,$$

$$\langle\Psi|T_{+-}|\Psi\rangle = -\frac{\hbar}{12\pi}\partial_+\partial_-\rho. \quad (6.148)$$

---

[13] Since w' is the frequency of the incoming wave, this is equivalent to considering the late-time limit.

We will work in Eddington–Finkelstein type coordinates $\sigma^\pm$ for which the spacetime metric of an eternal CGHS black hole reads

$$ds^2 = -\frac{d\sigma^+ d\sigma^-}{(1 + \frac{m}{\lambda} e^{-\lambda(\sigma^+ - \sigma^-)})} , \tag{6.149}$$

and therefore

$$\rho = -\frac{1}{2} \ln(1 + \frac{m}{\lambda} e^{-\lambda(\sigma^+ - \sigma^-)}) . \tag{6.150}$$

Substituting this expression into Eqs. (6.148) we obtain

$$\langle \Psi | T_{\pm\pm} | \Psi \rangle = \frac{\hbar \lambda^2}{48\pi} \left[\frac{1}{(1 + \frac{m}{\lambda} e^{-\lambda(\sigma^+ - \sigma^-)})^2} - 1 \right] + \langle \Psi | :T_{\pm\pm}(\sigma^\pm): | \Psi \rangle ,$$

$$\langle \Psi | T_{+-} | \Psi \rangle = -\frac{\hbar m \lambda}{24\pi} \frac{e^{-\lambda(\sigma^+ - \sigma^-)}}{(1 + \frac{m}{\lambda} e^{-\lambda(\sigma^+ - \sigma^-)})^2} . \tag{6.151}$$

Since $\langle \Psi | T_{+-} | \Psi \rangle$ is state independent, in the following we will focus only on $\langle \Psi | T_{\pm\pm} | \Psi \rangle$.

6.2.4.1 Boulware state

The Boulware state $|B\rangle$ is defined in $\sigma^\pm$ coordinates by the condition

$$\langle B | :T_{\pm\pm}(\sigma^\pm): | B \rangle = 0 . \tag{6.152}$$

From Eqs. (6.151) we get

$$\langle B | T_{\pm\pm} | B \rangle = \frac{\hbar \lambda^2}{48\pi} \left[\frac{1}{(1 + \frac{m}{\lambda} e^{-\lambda(\sigma^+ - \sigma^-)})^2} - 1 \right] . \tag{6.153}$$

Two interesting limits are infinity $\sigma^+ - \sigma^- = 2\sigma \to +\infty$, where

$$\langle B | T_{\pm\pm} | B \rangle \to 0 + O(e^{-2\lambda\sigma}) , \tag{6.154}$$

and the horizon $\sigma^+ - \sigma^- = 2\sigma \to -\infty$, for which

$$\langle B | T_{\pm\pm} | B \rangle \to -\frac{\hbar \lambda^2}{48\pi} . \tag{6.155}$$

The nonzero value of the stress tensor at the horizon in the Eddington–Finkelstein gauge is such that when evaluated in the Kruskal frame $x^\pm = \pm \lambda^{-1} e^{\pm \lambda x^\pm}$

$$\langle B | T_{\pm\pm}(x^\pm) | B \rangle \to -\frac{\hbar}{48\pi x^{\pm 2}} \tag{6.156}$$

it is divergent both at the past ($x^+ = 0$) and at the future ($x^- = 0$) horizons.

6.2.4.2 Hartle–Hawking state

The definition of the state-dependent functions $\langle \Psi | : T_{\pm\pm} : |\Psi\rangle$ for the Hartle–Hawking state $|H\rangle$ is natural in Kruskal coordinates $x^\pm$, where they are simply given by

$$\langle H | : T_{\pm\pm}(x^\pm) : |H\rangle = 0 \,. \tag{6.157}$$

Transformation to coordinates $\sigma^\pm$ (via the Schwarzian derivative) gives

$$\langle H | : T_{\pm\pm}(\sigma^\pm) : |H\rangle = \frac{\hbar \lambda^2}{48\pi} \,. \tag{6.158}$$

Insertion of this result into Eqs. (6.151) leads to

$$\langle H | T_{\pm\pm}(\sigma^\pm) | H \rangle = \frac{\hbar \lambda^2}{48\pi} \frac{1}{(1 + \frac{m}{\lambda} e^{-\lambda(\sigma^+ - \sigma^-)})^2} \,. \tag{6.159}$$

In the asymptotic region we have

$$\langle H | T_{\pm\pm}(\sigma^\pm) | H \rangle \to \frac{\hbar \lambda^2}{48\pi} = \frac{\pi (k_B T_H)^2}{12 \hbar} \,, \tag{6.160}$$

describing a thermal bath at the constant Hawking temperature $T_H = \lambda \hbar / 2\pi k_B$. At the horizon it is

$$\langle H | T_{\pm\pm}(\sigma^\pm) | H \rangle \to 0 \,. \tag{6.161}$$

To check that the regularity conditions are indeed satisfied at the horizon we transform the expressions (6.159) to Kruskal coordinates

$$\langle H | T_{\pm\pm}(x^\pm) | H \rangle = \frac{\hbar \lambda^2 (-\lambda^2 x^+ x^-)^2}{48\pi (-\lambda^2 x^+ x^- + \frac{m}{\lambda})^2} \frac{1}{(\pm \lambda x^\pm)^2} \tag{6.162}$$

and in the horizon's limit $x^\pm \to 0$ we obtain the finite result

$$\langle H | T_{\pm\pm}(x^\pm) | H \rangle \to \frac{\hbar \lambda^2}{48\pi} \frac{(\pm \lambda x^\mp)^2}{(m/\lambda)^2} \,. \tag{6.163}$$

6.2.4.3 Unruh state

The boundary functions for the Unruh state are given separately for the right-mover and the left-mover sectors. Using the Kruskal coordinate x^- it is

$$\langle U| :T_{--}(x^-): |U\rangle = 0 \,. \tag{6.164}$$

This leads to

$$\langle U| :T_{--}(\sigma^-): |U\rangle = \frac{\hbar\lambda^2}{48\pi} \,. \tag{6.165}$$

For the left-mover sector the state-dependent function is instead simply given by

$$\langle U| :T_{++}(\sigma^+): |U\rangle = 0 \,. \tag{6.166}$$

The results for the stress tensor in the Unruh state are then

$$\langle U|T_{--}|U\rangle = \frac{\hbar\lambda^2}{48\pi} \frac{1}{(1+\frac{m}{\lambda}e^{-\lambda(\sigma^+-\sigma^-)})^2} \,,$$

$$\langle U|T_{++}|U\rangle = \frac{\hbar\lambda^2}{48\pi} \left[\frac{1}{(1+\frac{m}{\lambda}e^{-\lambda(\sigma^+-\sigma^-)})^2} - 1\right] \,. \tag{6.167}$$

It is easy to check, along the lines of the previous two cases, that $\langle U|T_{--}|U\rangle$ is regular at the future horizon ($x^- = 0$ or $\sigma^- \to +\infty$) and at future infinity ($\sigma^+ \to +\infty$) it displays an outflux of thermal radiation

$$\langle U|T_{--}|U\rangle \to \frac{\hbar\lambda^2}{48\pi} \,, \tag{6.168}$$

i.e., the Hawking radiation. Finally, $\langle U|T_{++}|U\rangle$ shows a Boulware-type (nonphysical) divergence at the past horizon ($x^+ = 0$ or $\sigma^+ \to -\infty$).

6.2.4.4 "in" vacuum state

Finally, we comment on the "in" vacuum state $|in\rangle$ for the dynamical process of black hole formation, of which the Unruh state represents the late time limit. The only modification with respect to the last case considered is that the state function $\langle in| :T_{--}: |in\rangle$ is not defined with respect to the Kruskal coordinate x^-, but in terms of the "in" outgoing null coordinate σ^-_{in}

$$\sigma^-_{in} = -\frac{1}{\lambda}\ln(e^{-\lambda\sigma^-} + \frac{P(x_f^+)}{\lambda}) \,, \tag{6.169}$$

i.e.,

$$\langle in| : T_{--}(\sigma_{in}^-) : |in\rangle = 0 \ . \tag{6.170}$$

For the ++ component it is simply

$$\langle in| : T_{++}(\sigma^+) : |in\rangle = 0 \ . \tag{6.171}$$

Transforming Eq. (6.170) to the coordinate σ^- we get

$$\langle in| : T_{--}(\sigma^-) : |in\rangle = \frac{\hbar\lambda^2}{48\pi}\left[1 - \frac{1}{(1 + \frac{P(x_f^+)}{\lambda}e^{\lambda\sigma^-})^2}\right] \ . \tag{6.172}$$

This gives, for $x^+ > x_f^+$,

$$\langle in|T_{--}|in\rangle = \frac{\hbar\lambda^2}{48\pi}\left[\frac{1}{(1 + \frac{m(x_f^+)}{\lambda}e^{-\lambda(\sigma^+ - \sigma^-)})^2} - \frac{1}{(1 + \frac{P(x_f^+)}{\lambda}e^{\lambda\sigma^-})^2}\right] \ ,$$

$$\langle in|T_{++}|in\rangle = \frac{\hbar\lambda^2}{48\pi}\left[\frac{1}{(1 + \frac{m(x_f^+)}{\lambda}e^{-\lambda(\sigma^+ - \sigma^-)})^2} - 1\right] \ . \tag{6.173}$$

We see that at future null infinity ($\sigma^+ \to +\infty$) we have

$$\langle in|T_{--}|in\rangle \to \langle in| : T_{--}(\sigma_{in}^-) : |in\rangle = \frac{\hbar\lambda^2}{48\pi}\left[1 - \frac{1}{(1 + \frac{P(x_f^+)}{\lambda}e^{\lambda\sigma^-})^2}\right] \ , \tag{6.174}$$

and that only in the late time limit $\sigma^- \to +\infty$ we recover the Unruh state results. Moreover, we remark that the Hawking radiation (6.174), which asymptotically goes to a constant $\frac{\hbar\lambda^2}{48\pi}$, depends only on the parameter $P(x_f^+)$.

6.3 The Problem of Backreaction in the CGHS Model

We shall start by writing down the backreaction equations for the CGHS model. Since the matter field is conformally coupled to the two-dimensional metric, and there is no coupling with the dilaton field, the expectation values of the quantum stress tensor are given by the expressions (6.148). They have been studied in the previous section for the fixed black hole geometry. Our aim is to use its generic validity for an arbitrary background to write down the semiclassical equations:[14]

$$-\partial_+\partial_- e^{-2\phi} - \lambda^2 e^{2(\rho-\phi)} - \frac{N\hbar}{12\pi}\partial_+\partial_-\rho = 0, \quad (6.175)$$

$$2e^{-2\phi}\partial_+\partial_-(\rho - \phi) + \partial_+\partial_- e^{-2\phi} + \lambda^2 e^{2(\rho-\phi)} = 0, \quad (6.176)$$

$$\partial_\pm^2 e^{-2\phi} + 4\partial_\pm\phi\partial_\pm(\rho-\phi)e^{-2\phi} - \frac{N\hbar}{12\pi}\left((\partial_\pm\rho)^2 - \partial_\pm^2\rho\right) + \langle\Psi|:T_{\pm\pm}:|\Psi\rangle = 0. \quad (6.177)$$

These equations can be derived formally from the following semiclassical action

$$S = \frac{1}{2}\int d^2x\sqrt{-g}\left[e^{-2\phi}(R + 4(\nabla\phi)^2 + 4\lambda^2)\right] + NS_P, \quad (6.178)$$

where S_P is the Polyakov effective action already presented in Section 5.2

$$S_P = -\frac{\hbar}{96\pi}\int d^2x\sqrt{-g}R\Box^{-1}R \quad (6.179)$$

and N is the number of scalar fields. To make full sense of the semiclassical equations from a quantum gravity point of view one usually considers the large N limit, where the quantum gravitational corrections can be neglected.

As we have explained in Section 5.6, an important problem to be solved before attacking the semiclassical equations is to properly choose the state-dependent functions. This time, however, the problem is simpler since, in the assumed near-horizon approximation, the minimal coupling of the four-dimensional geometry with massless scalar fields produces a two-dimensional conformal coupling in the string frame. This implies that the functions $\langle\Psi|:T_{\pm\pm}:|\Psi\rangle$ are chiral and, therefore, the possible choices for them is quite restricted. Only for static configurations, corresponding to Boulware and Hartle–Hawking states, there is a clear way to select the adequate state-dependent functions. In static coordinates they are zero for Boulware and constant, proportional to the square of the Hawking temperature, for the thermal equilibrium state. Attempts to study numerically these configurations were given in [Birnir et al. (1992); Hawking (1992); Susskind and Thorlacius (1992); Lowe (1993)]. For the dynamical evaporating scenario the problem is more involved. It was speculated in the original

---

[14]The backreaction problem for two-dimensional gravity has been addressed for the first time in [Balbinot and Floreanini (1985)].

paper [Callan et al. (1992)], without explicitly solving the equations, that collapsing matter radiates away, in the form of Hawking radiation, all its energy before an event horizon forms. Therefore, they suggested that the process of black hole formation and evaporation preserves purity and there is no information loss. However, it was shown in [Banks et al. (1992); Russo et al. (1992a)] that this conclusion should be revised because a curvature singularity always forms at the critical value $e^{-2\phi_{cr}} = N\hbar/12$, separating the so called strong-coupling (or Liouville) region ($e^{-2\phi} \approx 0$), where the quantum terms dominate, and the classical weak coupling region ($e^{-2\phi} \to +\infty$). Moreover, this singularity is hidden inside an apparent horizon.

In general, the absence of analytical solutions makes it very difficult to follow the evolution of the evaporating black holes and to infer anything about the end point of the process. Also the selection of the state-dependent functions is not at all a trivial matter.

6.3.1 State-dependent functions for evaporating black holes

If the physical scenario is given by the initial linear dilaton vacuum that is later converted into a black hole due to the incoming matter, an intuitive way to find out $\langle\Psi| : T_{\pm\pm} : |\Psi\rangle$ is the following. From a quantum point of view, we should think of the state $|\Psi\rangle$ as the product of a coherent state for the incoming sector and the "in" vacuum for the outgoing one

$$|\Psi\rangle = |f^c\rangle \otimes |0_{\sigma_{in}^-}\rangle . \qquad (6.180)$$

The coherent state $|f^c\rangle$ is constructed, as usual [Klauder and Skagerstam (1985)], in such a way that the expectation value $\langle f^c| : T_{++}(\sigma_{in}^+) : |f^c\rangle$ reproduces the classical value of the incoming stress tensor T_{++}^{cl} given by the classical field f^c

$$\langle f^c| : T_{++}(\sigma_{in}^+) : |f^c\rangle = (\partial_+ f^c(\sigma_{in}^+))^2 \equiv T_{++}^{cl}(\sigma_{in}^+) . \qquad (6.181)$$

In addition, if $\sigma_{in}^\pm$ are the initial null Minkowskian coordinates it is natural to impose the absence of radiation in the "in" region

$$\langle\Psi| : T_{\pm\pm}(\sigma_{in}^\pm) : |\Psi\rangle\big|_{in} = 0 . \qquad (6.182)$$

The problem is now how to extend these conditions to the whole spacetime, including the "out" region where the asymptotically inertial null coordinates are $\sigma^\pm$. At past null infinity the only source is the classical collapsing matter

$$\langle\Psi|:T_{++}(\sigma^+):|\Psi\rangle|_{I_R^-} = T_{++}^{cl}(\sigma^+) \tag{6.183}$$

and at future null infinity the Hawking radiation flux is given by

$$\langle\Psi|:T_{--}(\sigma^-):|\Psi\rangle = -\frac{N\hbar}{24\pi}\{\sigma_{in}^-,\sigma^-\} . \tag{6.184}$$

Note that, in giving the above two expressions, we have used the fact that the vacuum polarization terms vanish at past and future null infinities, since there $\partial\rho \to 0$, $\partial^2\rho \to 0$. These conditions are not trivial to implement. The difficulties are the following:

- We do not have, *a priori*, a definition of the asymptotically flat coordinates $\sigma^\pm$. In fact these coordinates can be determined only on the basis of the solution of the semiclassical equations, which in turn depend crucially on $\langle\Psi|:T_{\pm\pm}:|\Psi\rangle$. In the fixed background approximation employed in the previous section to compute $\langle\Psi|T_{\pm\pm}|\Psi\rangle$, this problem disappears because there we can construct explicitly the asymptotically flat coordinates of the classical solution. For an evaporating solution, we do not know *a priori*, before solving the equations, how to identify the coordinates $\sigma^\pm$. A way to approach this problem is to resort to the Hartree-Fock method. We can use the classical solution to find a first order approximation for $\langle\Psi|:T_{\pm\pm}:|\Psi\rangle$. So,

$$\langle\Psi|:T_{++}(\sigma_{(0)}^+):|\Psi\rangle \approx T_{++}^{cl}(\sigma_{(0)}^+) , \tag{6.185}$$

and

$$\langle\Psi|:T_{--}(\sigma_{(0)}^-):|\Psi\rangle \approx -\frac{N\hbar}{24\pi}\{\sigma_{in}^-,\sigma_{(0)}^-\} , \tag{6.186}$$

where $\sigma_{(0)}^\pm$ are the asymptotically flat coordinates for the classical solution. From them by solving the semiclassical equations one derives the quantum corrected geometry, at first order, and this serves as the starting point to evaluate the new asymptotically inertial coordinates $\sigma_{(1)}^\pm$ and so on.

- The second difficulty concerns the choice of the gauge (i.e., the particular conformal frame) in which to solve the above equations. At the classical level we have seen that because $\rho - \phi$ is a free field, it allows to clearly identify the coordinate system in which one is working. In the above semi-classical equations $\rho - \phi$ is no more a free field (this is because Eq. (6.53) is modified to Eq. (6.175) due to the trace anomaly, while Eq. (6.176) is unchanged). Therefore the classical solvability of the CGHS model is broken by the Polyakov term. It is clear that, in the absence of exact solutions, the problem of getting the exact relation between $\sigma_{in}^{\pm}$ and $\sigma^{\pm}$ is more involved and it makes more difficult to bypass the first problem.

6.3.2 The RST model

Russo, Susskind and Thorlacius [Russo et al. (1992b)] addressed the difficulties presented in the previous subsection and, in particular, found a way to maintain the free field equation for $\rho - \phi$ also at the semiclassical level. The idea is to add to the Polaykov effective action a local term of the form

$$S_{rst} = -\frac{\hbar}{48\pi} \int d^2x \sqrt{-g} \phi R . \tag{6.187}$$

The new semiclassical action becomes

$$S = S_{cghs} + N(S_P + S_{rst}) . \tag{6.188}$$

In conformal gauge S_{cghs} is given in (6.51) and

$$S_P + S_{rst} = -\frac{\hbar}{12\pi} \int dx^+ dx^- [\partial_+ \rho \partial_- \rho + \phi \partial_+ \partial_- \rho] . \tag{6.189}$$

The equations obtained by variation with respect to ρ and ϕ are then modified to

$$-\partial_+\partial_- e^{-2\phi} - \lambda^2 e^{2(\rho-\phi)} - \frac{N\hbar}{12\pi}\partial_+\partial_-(\rho - \frac{\phi}{2}) = 0, \tag{6.190}$$

$$2e^{-2\phi}\partial_+\partial_-(\rho - \phi) + \partial_+\partial_- e^{-2\phi} + \lambda^2 e^{2(\rho-\phi)} + \frac{N\hbar}{24\pi}\partial_+\partial_-\rho = 0 . \tag{6.191}$$

Adding the previous two equations it is easy to see that we again obtain the free field equation

$$\partial_+\partial_-(\rho - \phi) = 0 . \tag{6.192}$$

It is this relation which, as we shall see, will ensure exact solvability also at the semiclassical level. In addition to the previous two equations, obtained by variation of S in conformal gauge, we have the modified constraint equations which read

$$\left(e^{-2\phi} - \frac{\hbar N}{48\pi}\right)(4\partial_\pm\rho\partial_\pm\phi - 2\partial_\pm^2\phi)$$
$$- \frac{N\hbar}{12\pi}\left((\partial_\pm\rho)^2 - \partial_\pm^2\rho\right) + \langle\Psi|:T_{\pm\pm}:|\Psi\rangle = 0 \ . \quad (6.193)$$

It should be noted that the term (6.187), being local, does not affect the rate of Hawking radiation. Therefore with respect to the semiclassical theory presented at the beginning of Section 6.3 only the local dynamics should be modified, not the overall picture. The main feature of the RST model is its *exact solvability* and this has allowed, as we shall now see, to find for the first time an *analytical solution describing an evaporating black hole*.

6.3.2.1 Liouville fields

An elegant reformulation of the RST model can be given, following [Bilal and Callan (1992); de Alwis (1993)], by introducing the new Liouville-type fields Ω and χ

$$\Omega = \frac{\sqrt{\kappa}}{2}\phi + \frac{e^{-2\phi}}{\sqrt{\kappa}} \ ,$$
$$\chi = \sqrt{\kappa}\rho - \frac{\sqrt{\kappa}}{2}\phi + \frac{e^{-2\phi}}{\sqrt{\kappa}} \ , \quad (6.194)$$

where $\kappa = \frac{N\hbar}{12\pi}$.[15] In terms of Ω and χ the semiclassical action in conformal gauge takes the form

$$S = \frac{1}{\pi}\int dx^+ dx^- \left[-\partial_+\chi\partial_-\chi + \partial_+\Omega\partial_-\Omega + \lambda^2 e^{\frac{2}{\sqrt{\kappa}}(\chi-\Omega)}\right] \ . \quad (6.195)$$

Variation of the above action with respect to χ and Ω gives the equations

$$\partial_+\partial_-\chi = -\frac{\lambda^2}{\sqrt{\kappa}}e^{\frac{2}{\sqrt{\kappa}}(\chi-\Omega)} \ ,$$
$$\partial_+\partial_-\Omega = -\frac{\lambda^2}{\sqrt{\kappa}}e^{\frac{2}{\sqrt{\kappa}}(\chi-\Omega)} \ , \quad (6.196)$$

which correspond to Eqs. (6.190) and (6.191). In particular, the free field equation $\partial_+\partial_-(\rho-\phi) = 0$ becomes

$$\partial_+\partial_-(\chi-\Omega) = 0 \ . \quad (6.197)$$

---

[15] We maintain the notation used in the literature. Of course, this κ should not be confused with the surface gravity.

Finally, the constraint equations (6.193) can be rewritten as

$$-\partial_\pm \chi \partial_\pm \chi + \sqrt{\kappa} \partial_\pm^2 \chi + \partial_\pm \Omega \partial_\pm \Omega + \langle \Psi | : T_{\pm\pm} : | \Psi \rangle = 0 \,. \tag{6.198}$$

6.3.2.2 General solution

Let us now discuss the general solution to Eqs. (6.196). Because $\chi - \Omega$ is a free field, following Eq. (6.60) we write

$$\chi - \Omega = \frac{\sqrt{\kappa}}{2}(w_+(x^+) + w_-(x^-)) \,. \tag{6.199}$$

Equations (6.196) then become

$$\partial_+ \partial_- \chi = -\frac{\lambda^2}{\sqrt{\kappa}} e^{w_+ + w_-} \,,$$

$$\partial_+ \partial_- \Omega = -\frac{\lambda^2}{\sqrt{\kappa}} e^{w_+ + w_-} \,. \tag{6.200}$$

We write the general solution to the above equations and Eqs. (6.198) in the form

$$\Omega = \tilde{C} - \frac{1}{\sqrt{\kappa}} h_+(x^+) h_-(x^-) - \frac{F(x^+) + G(x^-)}{\sqrt{\kappa}} \,,$$

$$\chi = \tilde{C} + \frac{\sqrt{\kappa}}{2}(w_+ + w_-) - \frac{1}{\sqrt{\kappa}} h_+(x^+) h_-(x^-) - \frac{F(x^+) + G(x^-)}{\sqrt{\kappa}} \tag{6.201}$$

where $\tilde{C}$ is an arbitrary integration constant,

$$h_+ = \lambda \int dx^+ e^{w_+} \,,$$

$$h_- = \lambda \int dx^- e^{w_-} \tag{6.202}$$

and

$$F(x^+) = \int^{x^+} dx'^+ \int^{x'^+} dx''^+ \langle \Psi | : T_{++}(x''^+) : | \Psi \rangle \,,$$

$$G(x^-) = \int^{x^-} dx'^- \int^{x'^-} dx''^- \langle \Psi | : T_{--}(x''^-) : | \Psi \rangle \,. \tag{6.203}$$

The functions w_+ and w_- have the same meaning as in the classical theory. They parameterize the "gauge" choice of conformal coordinates. The parameter $\tilde{C} = \frac{m}{\lambda\sqrt{\kappa}}$ gives the mass of the static solutions, that we are now going to analyse.

6.3.2.3 Semiclassical static solutions

The problem now is to properly choose the state-dependent functions $\langle \Psi | : T_{\pm\pm}(x^\pm) : |\Psi\rangle$ in the coordinates $x^\pm$. As we have already explained in Section 5.6, this problem is simpler for the static cases and we shall first focus on them. We shall assume that, in static coordinates $\sigma^\pm$, the state-dependent functions are trivial

$$\langle \Psi | : T_{\pm\pm}(\sigma^\pm) : |\Psi\rangle = 0 . \tag{6.204}$$

This corresponds to the *Boulware state*. Staticity for the free field $\rho - \phi$ implies that $w_+ = \lambda \sigma^+$ and $w_- = -\lambda \sigma^-$, up to irrelevant constants. It is then easy to see that, in the "Kruskal" coordinates $x^\pm$ for which $\rho - \phi = 0$, the state-dependent functions are

$$\langle \Psi | : T_{\pm\pm}(x^\pm) : |\Psi\rangle = -\frac{\kappa}{4(x^\pm)^2} . \tag{6.205}$$

The solution then turns out to be

$$\Omega = \chi = -\frac{\lambda^2 x^+ x^-}{\sqrt{\kappa}} - \frac{1}{4}\sqrt{\kappa}\ln(-\lambda^2 x^+ x^-) + \frac{m}{\lambda\sqrt{\kappa}} . \tag{6.206}$$

Going back to the $\sigma^\pm$ coordinates it is easy to see that the solution is static and that it approaches, for $\sigma \to +\infty$, the asymptotic form of a CGHS black hole of mass m. For $m = 0$ we have the remarkable fact that the solution is *exactly the linear dilaton vacuum*. However, for a non vanishing m we have a singular behavior when one approaches to the location of the classical horizons $x^+ = 0, x^- = 0$. In the semiclassical theory these horizons are converted into curvature singularities. This can be seen by computing the scalar curvature $R = 8e^{-2\rho}\partial_+\partial_-\rho$. It can be shown that

$$R = \frac{8e^{-2\rho}}{\Omega'}\left(\partial_+\partial_-\chi - \frac{\Omega''}{\Omega'^2}\partial_+\Omega\partial_-\Omega\right) , \tag{6.207}$$

where

$$\Omega' \equiv \frac{d\Omega}{d\phi} = \frac{\sqrt{\kappa}}{2} - \frac{2}{\sqrt{\kappa}}e^{-2\phi} . \tag{6.208}$$

Indeed, close to $x^+ x^- = 0$, occurring in the weak-coupling region $e^{-2\phi} \to +\infty$, we have

$$\partial_+\Omega\partial_-\Omega \sim \frac{1}{x^+ x^-} , \tag{6.209}$$

making these surfaces singular. The above enforces the idea, in a solvable semiclassical model, that the Boulware state can describe only static configurations (stars) for which there are no horizons.

We can also study the static solutions in thermal equilibrium by giving boundary conditions corresponding to the *Hartle–Hawking state*, namely

$$\langle \Psi | : T_{\pm\pm}(x^{\pm}) : | \Psi \rangle = 0 . \qquad (6.210)$$

This assumes that the coordinates $x^{\pm}$ are indeed the Kruskal coordinates for the static black hole solutions. The solutions are of the form

$$\Omega = \chi = -\frac{\lambda^2 x^+ x^-}{\sqrt{\kappa}} + \frac{m}{\lambda\sqrt{\kappa}} . \qquad (6.211)$$

In order to show that this solution is a black hole we have to prove that it has a regular horizon. The location of the horizons is given, as in the classical solution, by

$$x^{\pm} = 0 , \qquad (6.212)$$

and it is easy to see that all terms in (6.207) are regular there. The fact that the Ω and χ are regular at these surfaces means that the coordinates $x^{\pm}$ used are also regular at the horizons and therefore they are of Kruskal-type. The location of the singularity, which is hidden inside the horizons, is modified with respect to to the classical one and is given by the condition

$$\Omega' = 0 , \qquad (6.213)$$

or, equivalently,

$$\phi = \phi_{cr} = -\frac{1}{2} \ln \frac{\kappa}{4} . \qquad (6.214)$$

Let us finally mention a special property of these thermal equilibrium solutions: they are stable against the addition of incoming classical fluxes. For instance, two such static solutions parametrized by m_1 and m_2 can be continuously matched across the line $x^+ = x_0^+$. This corresponds to a black hole in thermal equilibrium absorbing an incoming shock wave of energy $m_2 - m_1$. The state-dependent functions describing this physical scenario are

$$\langle \Psi | : T_{\pm\pm}(x^{\pm}) : | \Psi \rangle = \frac{m_2 - m_1}{\lambda x_0^+} \delta(x^+ - x_0^+) . \qquad (6.215)$$

The fact that before and after $x^+ = x_0^+$ the relation between the coordinates $x^{\pm}$ and the coordinates $\sigma^{\pm}$ of both static solutions remain the same, $\pm\lambda x^{\pm} = e^{\pm\lambda\sigma^{\pm}}$, indicates that the thermal incoming and outgoing fluxes are fixed and therefore described by the above state-dependent functions. We remark that this behavior is very different from that of the Schwarzschild black holes, which are unstable under perturbations of their thermal equilibrium configurations at the temperature $k_B T_H = \hbar/8\pi M$.

6.3.2.4 Backreaction in the Unruh state

We mention for completeness that one can also give the solution for the *Unruh state* bypassing the difficulties mentioned in Subsection 6.3.1. The state-dependent functions can be constructed by mixing the conditions for Boulware and Hartle–Hawking

$$\langle U | : T_{++}(\sigma^+) : |U\rangle = 0 \, ,$$
$$\langle U | : T_{--}(x^-) : |U\rangle = 0 \, , \qquad (6.216)$$

where it is *assumed* that σ^+ is the ingoing asymptotically flat null coordinate and x^- is the null outgoing Kruskal coordinate.

In the case of spherically reduced Einstein gravity the relation between asymptotically flat and Kruskal coordinates, involving the surface gravity at the horizon, depends on the mass, which is itself a dynamical variable changing during the evolution. In the present case the surface gravity, and consequently the Hawking temperature, is *constant*. This remarkable simplification with respect to the Schwarzschild case allows the state-dependent functions to be written down consistently. In particular, in full Kruskal coordinates, in which the semiclassical RST equations of motion can be easily solved, they read

$$\langle U | : T_{++}(x^+) : |U\rangle = -\frac{\kappa}{4x^{+2}} \, ,$$
$$\langle U | : T_{--}(x^-) : |U\rangle = 0 \, , \qquad (6.217)$$

leading to

$$\Omega = \chi = -\frac{\lambda^2 x^+ x^-}{\sqrt{\kappa}} - \frac{1}{4}\sqrt{\kappa} \ln \lambda x^+ + \frac{m}{\lambda\sqrt{\kappa}} \, . \qquad (6.218)$$

Notice that, despite the presence of the logarithmic term, the relation between $x^{\pm}$ and the asymptotically flat coordinates $\sigma^{\pm}$ remains the same as in the classical solutions: $\pm\lambda x^{\pm} = e^{\pm\lambda\sigma^{\pm}}$. This is not a trivial point, since it ensures the self-consistency of the conditions (6.216). Let us now analyse the spacetime structure of the above solution. The curvature singularity is given by

$$-\frac{\lambda^2 x^+ x^-}{\sqrt{\kappa}} - \frac{1}{4}\sqrt{\kappa}\ln \lambda x^+ + \frac{m}{\lambda\sqrt{\kappa}} = \frac{\sqrt{\kappa}}{4}\left(1 - \ln\frac{\kappa}{4}\right). \quad (6.219)$$

There is also a null singularity at $x^+ = 0$, which arises naturally due to the definition of the state-dependent function $\langle U| : T_{++}(x^+) : |U\rangle$ in Eq. (6.217). This singularity along the past horizon of the classical solution is not important, since in physically realistic situations (such as the "in" vacuum state construction that we will consider in the next subsection) the past horizon does not belong to the physical spacetime. We will be concerned here only with the future branch of the singularity and the future apparent horizon. The location of the apparent horizon, given by $\partial_+\phi = 0$, is represented by the curve

$$-\lambda^2 x^+ x^- = \frac{\kappa}{4}. \quad (6.220)$$

Let us now analyse the behavior of the above two curves. We know that for a static solution the spacelike singularity is always hidden inside the apparent (and event) horizon. This case is different, since the solution (6.218) is time dependent. As displayed in Fig. 6.6, the singularity is initially spacelike and hidden behind the apparent horizon. Unlike the static case, where the two curves "intersect" only at infinity, the dynamic makes singularity and apparent horizon meet at the intersection point (x_s^+, x_s^-) defined by

$$x_s^+ = \frac{\kappa}{4\lambda}e^{\frac{4m}{\kappa\lambda}},$$
$$x_s^- = -\frac{1}{\lambda}e^{-\frac{4m}{\kappa\lambda}}. \quad (6.221)$$

Beyond this point the singularity is timelike and naked. The evolution beyond x_s^- cannot be uniquely determined. This, in fact, signals the *breakdown of the semiclassical theory*. Some "extra" conditions are needed to deal with it. We expect that this signals the end of the evaporation. Our expectation is confirmed by realizing that the solution (6.218) naturally matches, along the null line $x^- = x_s^-$, the linear dilaton vacuum[16] since at $x^- = x_s^-$ we have

$$\chi = \Omega = -\frac{\lambda^2 x^+ x_s^-}{\sqrt{\kappa}} - \frac{\sqrt{\kappa}}{4}\ln(\lambda x^+) + \frac{m}{\lambda\sqrt{\kappa}} = -\frac{\lambda^2 x^+ x_s^-}{\sqrt{\kappa}} - \frac{\sqrt{\kappa}}{4}\ln(-\lambda^2 x^+ x_s^-). \quad (6.222)$$

---

[16]This prescription, however, is not without problems, as we will see in Subsection 6.3.3.1.

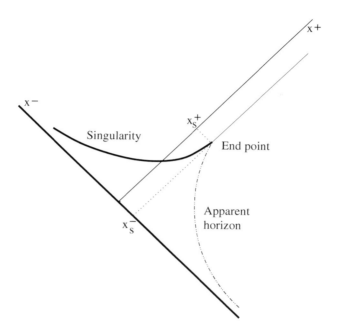

Fig. 6.6 Kruskal diagram of an evaporating RST black hole in the Unruh state.

This is a nice result, because it explicitly shows that at the finite time x_s^- the black hole has completely evaporated and the spacetime geometry becomes the Minkowski vacuum. In the higher dimensional interpretation, the near extremal dilaton black hole, due to the evaporation process, has evolved and turned into the extremal configuration.

Finally, we mention that the above matching with the linear dilaton vacuum is not smooth and requires the presence of a (small) emission of a negative energy shock wave, also called the thunderpop. The energy related to the thunderpop is determined in the following way. The full solution for $x^+ > x_s^+$ can be written as

$$\chi = \Omega = -\frac{\lambda^2 x^+ x^-}{\sqrt{\kappa}} - \frac{\sqrt{\kappa}}{4}\ln(\lambda x^+) + \frac{m}{\lambda\sqrt{\kappa}}\theta(x_s^- - x^-)$$
$$- \frac{\sqrt{\kappa}}{4}\ln(-\lambda x^-)\theta(x^- - x_s^-) \ . \tag{6.223}$$

In the gauge considered ($\chi = \Omega$) the constraint equations (6.198) become

$$\langle U| : T_{\pm\pm} : |U\rangle = -\sqrt{\kappa}\partial_\pm^2 \chi \ . \tag{6.224}$$

This, combined with (6.223), gives (for $x^+ > x_s^+$)

$$\langle U| : T_{++} : |U\rangle = -\frac{\kappa}{4x^{+2}} ,$$

$$\langle U| : T_{--} : |U\rangle = -\frac{\kappa}{4x^{-2}}\theta(x^- - x_s^-) - \frac{\kappa\lambda}{4}e^{\frac{4m}{\kappa\lambda}}\delta(x^- - x_s^-) . \quad (6.225)$$

The energy associated to the thunderpop is then

$$E_{thunderpop} = -\lambda \int dx^- x^- (-\frac{\kappa\lambda}{4}e^{\frac{4m}{\kappa\lambda}}\delta(x^- - x_s^-)) = -\frac{\kappa\lambda}{4} . \quad (6.226)$$

It is interesting to comment that this example, which is certainly a simplification of the full description of black hole formation plus evaporation to be considered in the next subsection, is nevertheless able to capture the main features of the evaporation process. This proves the usefulness of the Unruh state construction.

6.3.3 Black hole evaporation in the RST model

We now turn to the most interesting application, namely, the dynamical scenario describing black hole formation and subsequent evaporation.

The classical process was studied in Subsection 6.2.2. In terms of the classical asymptotically flat coordinates $\sigma_{cl}^+ (= \sigma_{in}^+)$ and σ_{cl}^-, the first approximation for the state-dependent functions reads

$$\langle \Psi| : T_{--}(\sigma_{cl}^-) : |\Psi\rangle = -\frac{\kappa}{2}\{\sigma_{in}^-, \sigma_{cl}^-\} ,$$

$$\langle \Psi| : T_{++}(\sigma_{cl}^+) : |\Psi\rangle = T_{++}^{cl}(\sigma_{cl}^+) , \quad (6.227)$$

where the relation between σ_{in}^- and σ_{cl}^- is given by

$$\sigma_{in}^- = -\frac{1}{\lambda}\ln(e^{-\lambda\sigma_{cl}^-} + \frac{P(x_f^+)}{\lambda}) . \quad (6.228)$$

In Kruskal coordinates, defined by

$$\pm\lambda x^\pm = e^{\sigma_{in}^\pm} , \quad (6.229)$$

the state-dependent functions become

$$\langle \Psi| : T_{++}(x^+) : |\Psi\rangle = T_{++}^{cl}(x^+) - \frac{\kappa}{4(x^+)^2} ,$$

$$\langle \Psi| : T_{--}(x^-) : |\Psi\rangle = -\frac{\kappa}{4(x^-)^2} . \quad (6.230)$$

The solution to the field equations plus constraints in Kruskal gauge then takes the form

$$\chi = \Omega = -\frac{\lambda^2 x^+}{\sqrt{\kappa}}\left(x^- + \frac{1}{\lambda^2}P(x^+)\right) - \frac{\sqrt{\kappa}}{4}\ln(-\lambda^2 x^+ x^-) + \frac{m(x^+)}{\lambda\sqrt{\kappa}}. \quad (6.231)$$

For simplicity we will restrict discussion to the case where the classical incoming matter distribution is concentrated in a single shock wave of mass m located at $x^+ = x_0^+$. We have therefore

$$\langle \Psi | : T_{++}(x^+) : | \Psi \rangle = \frac{m}{\lambda x_0^+}\delta(x^+ - x_0^+) - \frac{\kappa}{4(x^+)^2},$$

$$\langle \Psi | : T_{--}(x^-) : | \Psi \rangle = -\frac{\kappa}{4(x^-)^2} \quad (6.232)$$

leading to

$$\chi = \Omega = -\frac{\lambda^2 x^+ x^-}{\sqrt{\kappa}} - \frac{\sqrt{\kappa}}{4}\ln(-\lambda^2 x^+ x^-) - \frac{m}{\lambda\sqrt{\kappa}x_0^+}(x^+ - x_0^+)\theta(x^+ - x_0^+). \quad (6.233)$$

Before the shock wave $(x^+ < x_0^+)$ we have the Minkowski vacuum

$$\chi = \Omega = -\frac{\lambda^2 x^+ x^-}{\sqrt{\kappa}} - \frac{\sqrt{\kappa}}{4}\ln(-\lambda^2 x^+ x^-) \quad (6.234)$$

and after $(x^+ > x_0^+)$ we write the solution in the form

$$\chi = \Omega = -\frac{\lambda^2 x^+}{\sqrt{\kappa}}(x^- + \frac{m}{\lambda^3 x_0^+}) - \frac{\sqrt{\kappa}}{4}\ln(-\lambda^2 x^+ x^-) + \frac{m}{\lambda\sqrt{\kappa}}. \quad (6.235)$$

An important feature of this solution is that, despite the presence of the quantum corrections, the relation between the Kruskal coordinates and the asymptotically flat coordinates $\sigma^\pm$ is the same as the classical one, namely

$$\lambda x^+ = e^{\lambda \sigma^+},$$

$$-\lambda(x^- + \frac{m}{\lambda^3 x_0^+}) = e^{-\lambda \sigma^-}. \quad (6.236)$$

This ensures that $\sigma_{cl}^\pm = \sigma^\pm$ and, therefore, the choice of the state-dependent functions (6.232) is exact and does not receive additional quantum corrections. Therefore, the first order approximation turns out to be exact. We stress this fact because such a feature is not generic and is related with the fact that the Hawking temperature $k_B T_H = \frac{\hbar \lambda}{2\pi}$ is a constant. Later we will analyse a model, describing the evaporation of a near-extremal Reissner–Nordström black hole, where the first order approximation is not exact.

We will now consider the spacetime structure of the evaporating solution in the region $x^+ > x_0^+$, as depicted in Fig. 6.7. The curvature singularity

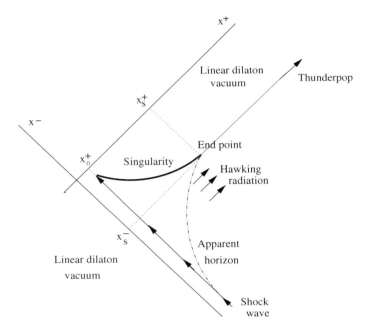

Fig. 6.7 Kruskal diagram for the evaporation of a black hole in the RST model.

is located along the spacelike line

$$-\frac{\lambda^2 x^+}{\sqrt{\kappa}}(x^- + \frac{m}{\lambda^3 x_0^+}) - \frac{\sqrt{\kappa}}{4}\ln(-\lambda^2 x^+ x^-) + \frac{m}{\lambda\sqrt{\kappa}} = \frac{\sqrt{\kappa}}{4}(1 - \ln\frac{\kappa}{4}) \,. \quad (6.237)$$

This is hidden inside the apparent horizon, given by the equation $\partial_+\phi = 0$, located along the timelike line

$$-\lambda^2 x^+ (x^- + \frac{m}{\lambda^3 x_0^+}) = \frac{\kappa}{4} \,. \quad (6.238)$$

As in the Unruh state construction of the previous subsection the evolution is such that the apparent horizon and the singularity intersect after a finite time at the point

$$x_s^+ = \frac{\kappa \lambda x_0^+}{4m}(e^{\frac{4m}{\kappa\lambda}} - 1),$$

$$x_s^- = -\frac{m}{\lambda^3 x_0^+}\frac{1}{(1 - e^{-\frac{4m}{\kappa\lambda}})}. \qquad (6.239)$$

The physical meaning of the intersection is that at this point the black hole has completely evaporated. This is confirmed by the fact that for $x^- = x_s^-$ the solution takes the form

$$\chi = \Omega = -\lambda^2 \frac{x^+}{\sqrt{\kappa}}(x_s^- + \frac{m}{\lambda^3 x_0^+}) - \frac{\sqrt{\kappa}}{4}\ln(-\lambda^2 x^+ x_s^-) + \frac{m}{\lambda\sqrt{\kappa}}$$

$$= -\frac{\lambda^2 x^+}{\sqrt{\kappa}}(x_s^- + \frac{m}{\lambda^3 x_0^+}) - \frac{\sqrt{\kappa}}{4}\ln(-\lambda^2 x^+ (x_s^- + \frac{m}{\lambda^3 x_0^+})). \quad (6.240)$$

This means that the evaporating solution naturally matches, along the outgoing null line $x^- = x_s^-$, the vacuum configuration

$$\chi = \Omega = -\frac{\lambda^2 x^+}{\sqrt{\kappa}}(x^- + \frac{m}{\lambda^3 x_0^+}) - \frac{\sqrt{\kappa}}{4}\ln(-\lambda^2 x^+ (x^- + \frac{m}{\lambda^3 x_0^+})). \quad (6.241)$$

Therefore the full solution in the region $x^+ > x_s^+$ can be written as

$$\chi = \Omega = -\frac{\lambda^2 x^+}{\sqrt{\kappa}}(x^- + \frac{m}{\lambda^3 x_0^+}) + \left[-\frac{\sqrt{\kappa}}{4}\ln(-\lambda^2 x^+ x^-) + \frac{m}{\lambda\sqrt{\kappa}}\right]\theta(x_s^- - x^-)$$

$$-\frac{\sqrt{\kappa}}{4}\ln(-\lambda^2 x^+ (x^- + \frac{m}{\lambda^3 x_0^+}))\theta(x^- - x_s^-). \qquad (6.242)$$

The appearance of the θ functions means that the matching is not smooth, but requires the presence of a δ function source along x_s^-. Evaluating the constraints

$$\langle\Psi|:T_{\pm\pm}:|\Psi\rangle = -\sqrt{\kappa}\partial_\pm^2 \chi \qquad (6.243)$$

one gets $(x^+ > x_s^+)$

$$\langle\Psi|:T_{++}:|\Psi\rangle = -\frac{\kappa}{4x^{+2}},$$

$$\langle\Psi|:T_{--}:|\Psi\rangle = -\frac{\kappa}{4x^{-2}}\theta(x_s^- - x^-) - \frac{\kappa}{4(x^- + \frac{m}{\lambda^3 x_0^+})^2}\theta(x^- - x_s^-)$$

$$+ \frac{\kappa}{4}\frac{(1 - e^{-\frac{4m}{\kappa\lambda}})}{x^- + \frac{m}{\lambda^3 x_0^+}}\delta(x^- - x_s^-). \qquad (6.244)$$

The energy associated to the thunderpop is

$$E_{thunderpop} = -\lambda \int dx^- (x^- + \frac{m}{\lambda^3 x_0^+})\frac{\kappa}{4}\frac{(1-e^{-\frac{4m}{\kappa\lambda}})}{(x^- + \frac{m}{\lambda^3 x_0^+})}\delta(x^- - x_s^-)$$

$$= -\frac{\kappa\lambda}{4}(1-e^{-\frac{4m}{\kappa\lambda}}) . \qquad (6.245)$$

The Hawking flux is the same in form as the one calculated in fixed background

$$\langle\Psi|T_{--}(\sigma^-)|\Psi\rangle|_{I^+} = \frac{\kappa\lambda^2}{4}\left[1 - \frac{1}{(1+\frac{m}{\lambda^2 x_0^+}e^{\lambda\sigma^-})^2}\right] . \qquad (6.246)$$

However, it must be emphasized the important difference that the radiation is suddenly turned off at x_s^- due to the matching with the final linear dilaton vacuum. Finally, it can be checked that the total energy radiated

$$E_{rad} = \int_{-\infty}^{\sigma_s^-} d\sigma^- \langle\Psi|T_{--}(\sigma^-)|\Psi\rangle_{I^+} \qquad (6.247)$$

plus the energy associated to the thunderpop $E_{thunderpop}$ equals the initial mass m, i.e., energy is conserved.

6.3.3.1 Information loss in the RST model

Let us now study a more involved situation than the one analysed in the previous subsection. We shall consider the general case where the black hole is formed by an influx of general incoming matter in the range $x_i^+ \leq x^+ \leq x_f^+$. The form of the solution in Kruskal gauge is the following:

$$\chi = \Omega = -\frac{\lambda^2 x^+}{\sqrt{\kappa}}\left(x^- + \frac{1}{\lambda^2}P(x^+)\right) - \frac{\sqrt{\kappa}}{4}\ln(-\lambda^2 x^+ x^-) + \frac{m(x^+)}{\lambda\sqrt{\kappa}} . \qquad (6.248)$$

In the region $x^+ < x_i^+$ we have the linear dilaton vacuum

$$\chi = \Omega = -\frac{\lambda^2 x^+ x^-}{\sqrt{\kappa}} - \frac{\sqrt{\kappa}}{4}\ln(-\lambda^2 x^+ x^-) , \qquad (6.249)$$

which defines the asymptotically flat coordinates $\sigma_{in}^{\pm}$ via

$$\pm\lambda x^{\pm} = e^{\pm\lambda\sigma_{in}^{\pm}} . \qquad (6.250)$$

The intermediate region $x_i^+ \leq x^+ \leq x_f^+$ is relevant for the following reason. The singularity curve

$$-\frac{\lambda^2 x^+}{\sqrt{\kappa}}(x^- + \frac{1}{\lambda^2}P(x^+)) - \frac{\sqrt{\kappa}}{4}\ln(-\lambda^2 x^+ x^-) + \frac{m(x^+)}{\lambda\sqrt{\kappa}} = \frac{\sqrt{\kappa}}{4}(1 - \ln\frac{\kappa}{4}) \tag{6.251}$$

is timelike for

$$\langle\Psi| : T_{++}(x^+) : |\Psi\rangle = T_{++}^{cl}(x^+) - \frac{\kappa}{4(x^+)^2} < 0 . \tag{6.252}$$

This is an interesting result because it tells us that as long as the ingoing "classical" stress tensor is less than the evaporation rate then the black hole does not form, i.e., the *singularity stays timelike and is naked*. We then identify a *subcritical regime* defined by configurations for which (6.252) [Russo et al. (1992c)] always holds for all x^+, see Fig. 6.8. For such

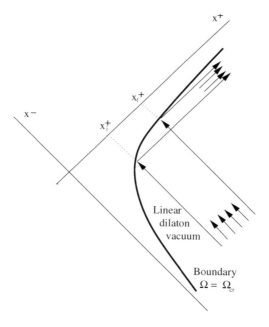

Fig. 6.8 Subcritical regime in the RST model. All incoming matter is reflected at the boundary line, which then behaves as a dynamical moving mirror.

configurations the singularity is located at those spacetime points where the field Ω reaches its critical (minimum) value $\Omega' = 0$. Beyond this critical value, the physical field ϕ will be complex. To maintain physical consistency one should then impose boundary conditions to the matter field to prevent Ω to get values less than Ω_{cr}, where

$$\Omega_{cr} = \frac{\sqrt{\kappa}}{4}(1 - \ln\frac{\kappa}{4}) . \tag{6.253}$$

The natural thing to do is to treat this critical line as a "mirror" and impose reflecting boundary conditions for the ingoing matter along the curve

$$\Omega = \Omega_{cr} . \tag{6.254}$$

Note that this is possible only when the curve is timelike. The boundary conditions to be imposed [Chung and Verlinde (1994); Das and Mukherji (1994); Strominger and Thorlacius (1994); Chung and Verlinde (1994)] are similar to those of the moving mirror case considered in Section 4.6

$$\langle\Psi|:T_{--}(x^-):|\Psi\rangle = \left(\frac{dp(x^-)}{dx^-}\right)^2 \langle\Psi|:T_{++}(x^+):|\Psi\rangle - \frac{\kappa}{2}\{p(x^-), x^-\} , \tag{6.255}$$

where $x^+ = p(x^-)$ is the trajectory of the boundary given by Eq. (6.254). The resulting model is *unitary* by construction, as all incoming quanta is reflected to future infinity. There is no information loss in the subcritical regime, but this is related to the fact that the black hole never forms.

The *supercritical* regime, characterized by

$$\langle\Psi|:T_{++}(x^+):|\Psi\rangle > 0 , \tag{6.256}$$

is different. In this case the critical line is the black hole spacelike singularity and it is hidden inside the apparent horizon, as in the case previously analysed. After x_f^+ the evolution is similar to the shock wave case. The form of the Hawking flux is

$$\langle\Psi|T_{--}(\sigma^-)|\Psi\rangle|_{I^+} = \frac{\kappa\lambda^2}{4}\left[1 - \frac{1}{(1+\frac{P(x_f^+)}{\lambda}e^{\lambda\sigma^-})^2}\right] . \tag{6.257}$$

As in the shock wave case the singularity and the apparent horizon intersect after some finite time x_s^-. This represents the end point of the evaporation. The linear dilaton vacuum can then be matched along x_s^-. In this view it is clear that the only type of radiation emitted is given by the Hawking flux plus the thunderpop. From Eq. (6.257) we cannot reconstruct the ingoing flux $\langle\Psi|:T_{++}(x^+):|\Psi\rangle$ because the only information that we can recover is $P(x_f^+)$. This is a clear example of a *breakdown of postdictability*.

In addition, we also have a *breakdown of predictability* since the quantum state of the radiation is not uniquely defined. Within this view, we have indeed a mixed state description for the radiation. In other words, there are correlations of the initial vacuum state that are lost for the external observer. The evolution is not unitary and the information is lost.

However the above picture is not without problems. As it was pointed out in [Strominger (1995)], the matching conditions imposed so far at the end point x_s^- are problematic. In Fig. 6.9 we represent the complete RST

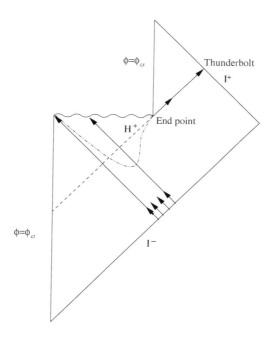

Fig. 6.9 Penrose diagram for the evaporation of a black hole in the RST model. It is similar to the Hawking picture for the evaporation of a Schwarzschild black hole. The boundary line $\phi = \phi_{cr}$ corresponds to the line $r = 0$ of four dimensional Minkowski space. There is a thunderbolt of infinite energy emanating from the end point of the evaporation (where the apparent horizon intersects the singularity).

picture of black hole evaporation, where the initial and final linear dilaton vacuum configurations are naturally restricted to the right regions of the critical lines

$$-\lambda^2 x^+ x^- = \frac{\kappa}{4} \tag{6.258}$$

$$-\lambda^2 x^+ (x^- + \frac{m}{\lambda^3 x_0^+})) = \frac{\kappa}{4}, \tag{6.259}$$

respectively.[17] These lines are analogous to the $r = 0$ boundaries of the four dimensional evaporating spacetime of Fig. 3.18. The two timelike lines are connected by a spacelike segment representing the black hole singularity. Let us evaluate the particle production across x_s^- due to the discontinuity in these boundaries. The formula is given in Section 4.6 and reads

$$\langle 0_{in} | N_k^{out} | 0_{in} \rangle = -\frac{1}{\pi} \int_{I^+} dx^- dx'^- u_k^{out}(x^-) u_k^{out*}(x'^-) \times \tag{6.260}$$

$$\left[\frac{dp}{dx^-}(x^-) \frac{dp}{dx^-}(x'^-) \frac{1}{(p(x^-) - p(x'^-))^2} - \frac{1}{(x^- - x'^-)^2} \right].$$

For $x^- > x_s^-$ it is

$$p(x^-) = -\frac{\kappa}{4\lambda^2} \frac{1}{x^- + \frac{m}{\lambda^3 x_0^+}} \tag{6.261}$$

and for $x^- < x_s^-$

$$p(x^-) = -\frac{\kappa}{4\lambda^2} \frac{1}{x^-}. \tag{6.262}$$

This gives

$$\langle 0_{in} | N_k^{out} | 0_{in} \rangle = -\frac{2}{\pi} \int_{-\infty}^{x_s^-} dx^- \int_{x_s^-}^{+\infty} dx'^- u_k^{out}(x^-) u_k^{out*}(x'^-) \times \tag{6.263}$$

$$\left[\left(\frac{\lambda^3 x_0^+}{m} \right)^2 - \frac{1}{(x^- - x'^-)^2} \right].$$

The discontinuity of the two timelike branches due to the presence of the black hole produces a nontrivial effect.[18] The above quantity strongly diverges and a similar thing happens for the stress tensor at x_s^- (similar considerations have been made at the end of Subsection 4.6.1). It gives an

---
[17] Note that the critical line (6.259) is the analytic continuation of the apparent horizon in the evaporating region.
[18] This was first pointed out in a different context by [Anderson and DeWitt (1986)].

infinite burst of outgoing energy at x_s^-, also called a *thunderbolt*,[19] in addition to the finite energy thunderpop encountered before. This strongly suggests that to treat the end point of the evaporation consistently, where the singularity is about to become naked, one needs to go beyond the semiclassical approximation. This prevents one to make any definitive conclusion about the information loss problem based on the pure RST model.

6.3.3.2 *Is the picture of RST black hole evaporation generic?*

One can ask at this point whether the main conclusions concerning the evaporation process drawn from the RST model depend on the particular local counterterm (6.187) added to the Polyakov action. Returning to the problem of solvability of the semiclassical theory, which motivated the RST model, it is important to remark that there are other ways to maintain the classical solvability at the quantum level [Bilal and Callan (1992); de Alwis (1993)]. A physically motivated proposal was given in [Bose et al. (1995)].[20] They added a different local term to the Polyakov action, namely

$$S_{bpp} = \frac{\hbar}{24\pi} \int d^2x \sqrt{-g}[(\nabla\phi)^2 - \phi R] \ . \qquad (6.264)$$

The new semiclassical action becomes $S_{cghs} + (S_P + S_{bpp})$.[21] There are two reasons for considering this action. The first one is that it respects the symmetry $\delta\phi = \epsilon e^{2\phi}, \delta g_{\mu\nu} = 2\epsilon e^{2\phi} g_{\mu\nu}$ already present in the classical theory and responsible for the existence of the free field. Moreover, in the Einstein frame this model takes a very simple form. It is the classical action

$$S = \frac{1}{2} \int d^2x \sqrt{-g} \left(e^{-2\phi} R + 4\lambda^2\right) \qquad (6.265)$$

coupled to the Polyakov action. Transforming back to the string frame (6.265) gives the CGHS action and the Polyakov action generates the above local counterterms.

Also in this case the exact solvability allows to follow the details of the evaporation analytically. The picture is qualitatively unchanged with respect to the RST model. Starting from the natural ground state solution of this model incoming matter will form a black hole, provided that the flux

---

[19] The presence of thunderbolts in the semiclassical CGHS model was also predicted by [Hawking and Stewart (1993)].
[20] Other solvable modifications of the RST model have been proposed in [Fabbri and Russo (1996); Cruz and Navarro-Salas (1996); Zaslavskii (1999)].
[21] This semiclassical theory can also be recovered within the reduced phase-space approach developed in [Mikovic (1995); Mikovic (1996)].

and the total energy are above the critical values.[22] The black hole will evaporate completely and at the end point one can again match the solution with the ground state configuration, with all the problems mentioned at the end of the previous subsection. Finally, below the threshold of black hole formation one has again a unitary evolution [Bose et al. (1996)].

6.4 The Semiclassical JT Model

In Section 4.4 we have seen how portions of the two-dimensional Anti-de Sitter space describe the near-horizon geometry of near-extremal Reissner–Nordström black holes and, in particular, their late-time radiation properties. This situation can be compared with the stringy black holes studied in the previous sections. The CGHS black holes describe the near-horizon geometry of the four-dimensional near-extremal dilatonic black holes. The role of the dilaton for the Reissner–Nordström black holes is given by the radial function, more precisely by its deviation from the extremal radius. In fact, as we have seen in Subsection 6.1.2 the two-dimensional action describing the near-horizon properties of these black holes is the JT model.[23]

Despite these similarities there is a big difference between these two situations. The stringy black holes have, near extremality, a *constant* (i.e., mass-independent) temperature $k_B T_H \approx \hbar/4\pi Q$. In contrast, the Reissner–Nordström black holes have a *non-constant* temperature: $k_B T_H \approx \hbar\sqrt{\Delta m/2\pi^2 Q^3}$, with a dependence on the mass. So we expect that the backreaction effects are more delicate to deal with than for the CGHS black holes. Related to this, the selection of the correct state-dependent functions in the semiclassical equations will be more involved. In particular, the very simple choice of the functions

$$\langle in| :T_{\pm\pm}(x^\pm): |in\rangle = -\frac{\hbar N}{48\pi x^{\pm 2}} , \qquad (6.266)$$

for the "in" vacuum is related to the fact that the relation between the Kruskal coordinates $x^\pm$ and the asymptotically flat coordinates $\sigma^\pm$ is always the same and does not change in the evolution. In the present case, instead, the relation between the Kruskal and the null Eddington–Finkelstein type coordinates involves the mass and therefore cannot be fixed *a priori*.

---

[22] The existence of a critical mass is absent in the RST model.
[23] The JT model can also be used to reproduce the deviation of the Bekenstein-Hawking entropy from extremality [Navarro-Salas and Navarro (2000)].

Let us now consider the near-horizon model coming from the Einstein–Maxwell theory minimally coupled to N scalar fields f_i. After the spherically symmetric reduction we get

$$S = \frac{1}{2}\int d^2x\sqrt{-g}\left(\tilde{\phi}R + \frac{2}{Q^2}\tilde{\phi} - \sum_{i=1}^{N}|\nabla f_i|^2\right), \qquad (6.267)$$

where, we remember, the radial function has been expanded as

$$e^{-\phi} = e^{-\phi_0}(1+\tilde{\phi}), \qquad (6.268)$$

and $le^{-\phi_0} = r_0 = Q$ is the extremal radius. We have also eliminated the overall coefficient Q^2 to agree with the convention used for the CGHS model.

Let us consider directly the semiclassical theory, which consists of the classical action plus the effective action for the matter fields

$$S_g(\tilde{\phi}) = \frac{1}{2}\int d^2x\sqrt{-g}\left(\tilde{\phi}R + \frac{2}{Q^2}\tilde{\phi}\right) - \frac{N\hbar}{96\pi}\int d^2x\sqrt{-g}(R\Box^{-1}R - \frac{8\xi}{2Q^2}). \qquad (6.269)$$

We have also included a "cosmological constant" term parametrized by ξ, which we can always add at the quantum level as it has been explained in Section 5.1. The usefulness of this will be seen shortly.

The semiclassical equations derived from the above action are

$$\partial_+\partial_-\rho + \frac{1}{4Q^2}e^{2\rho} = 0, \qquad (6.270)$$

$$\frac{1}{2}\partial_+\partial_-\tilde{\phi} + \frac{1}{4Q^2}\tilde{\phi}e^{2\rho} = -\frac{N\hbar}{24\pi}(\partial_+\partial_-\rho + \frac{\xi}{4Q^2}e^{2\rho}), \qquad (6.271)$$

$$-\partial_\pm^2\tilde{\phi} + 2\partial_\pm\rho\partial_\pm\tilde{\phi} = -\frac{N\hbar}{12\pi}((\partial_\pm\rho)^2 - \partial_\pm^2\rho) + \langle\Psi|:T_{\pm\pm}:|\Psi\rangle. \qquad (6.272)$$

The first equation is nothing else but the so called Liouville equation. It is just the constant curvature equation $R = -2/Q^2$ expressed in conformal gauge. Using this equation one can rewrite the equation for the field $\tilde{\phi}$ as follows

$$\frac{1}{2}\partial_+\partial_-\tilde{\phi} + \frac{1}{4Q^2}(\tilde{\phi} + (\xi-1)\frac{N\hbar}{24\pi})e^{2\rho} = 0. \qquad (6.273)$$

We observe that the semiclassical ρ and $\tilde{\phi}$ equations are then the same as the classical ones up to a redefinition of the field $\tilde{\phi}$

$$\tilde{\phi} \to \tilde{\phi} - (\xi-1)\frac{N\hbar}{24\pi}. \qquad (6.274)$$

6.4.1 Extremal solution

The classical extremal solution in the near-horizon approximation can be described, according to our analysis of Section 4.4, by the two-dimensional metric

$$ds^2 = -\frac{4Q^2}{(u_{in}-v)^2} du_{in} dv ,\qquad(6.275)$$

and the dilaton ($x \equiv Q\tilde\phi$)

$$\tilde\phi = \frac{2Q}{u_{in}-v}.\qquad(6.276)$$

The coordinates (u_{in}, v) are the near-horizon Eddington–Finkelstein null coordinates of the extremal Reissner–Nordström black hole. Moreover, the Kruskal coordinates (U, V) for the extremal black hole in the near-horizon limit are given by (see Section 4.5)

$$u_{in} = -\frac{2Q^2}{U},$$
$$v = -\frac{2Q^2}{V}.\qquad(6.277)$$

In terms of the Kruskal coordinates the extremal solution takes the form:

$$ds^2 = -\frac{4Q^2}{(U-V)^2} dU dV ,\qquad(6.278)$$

which just coincides with (6.275), but for $\tilde\phi$ we have

$$\tilde\phi = \frac{UV}{Q(U-V)}.\qquad(6.279)$$

At this point we use the freedom in the parameter ξ to fix it in such a way that the classical extremal configuration remains a solution of the backreaction equations. To do this we need an expression for the state-dependent functions. As already stressed in Section 4.5, in the natural vacuum state that we call $|in\rangle$ for later convenience, we have

$$\langle in|:T_{vv}(v):|in\rangle = 0 = \langle in|:T_{VV}(V):|in\rangle$$
$$\langle in|:T_{u_{in}u_{in}}(u_{in}):|in\rangle = 0 = \langle in|:T_{UU}(U):|in\rangle .\qquad(6.280)$$

Taking into account that the vacuum polarization terms also vanish in these coordinates for the solution (6.275) and (6.278)

$$(\partial_v \rho)^2 - \partial_v^2 \rho = 0 = (\partial_V \rho)^2 - \partial_V^2 \rho$$
$$(\partial_{u_{in}} \rho)^2 - \partial_{u_{in}}^2 \rho = 0 = (\partial_U \rho)^2 - \partial_U^2 \rho \,, \quad (6.281)$$

it is easy to see that the choice $\xi = 1$ allows the classical extremal solution to become an exact solution of the semiclassical equations. This reproduces the property of the RST model where the linear dilaton vacuum is also an exact solution of the semiclassical theory. We should remark, nevertheless, that the $|in\rangle$ state can be interpreted either as a Boulware state or as a Hartle–Hawking state. This is so because the corresponding state-dependent functions associated to each type of state (6.280) are equivalent, at the level of the dilaton-metric fields, for the extremal configuration in the near-horizon approximation. Going beyond the near-horizon approximation they become, obviously, different.

6.4.2 Semiclassical static solutions

For near-extremal black holes the near-horizon relation between the Eddington–Finkelstein and Kruskal coordinates is no longer given by a Möbius transformation as in Eq. (6.277), and this implies that the configurations corresponding to Boulware and Hartle–Hawking are, as expected, different.

We shall first analyse the semiclassical solutions with *Boulware* state-dependent functions

$$\langle B| : T_{vv}(v) : |B\rangle = 0$$
$$\langle B| : T_{uu}(u) : |B\rangle = 0 \quad (6.282)$$

in a static frame where $\rho = \rho(\frac{v-u}{2})$, $\tilde{\phi} = \tilde{\phi}(\frac{v-u}{2})$. In terms of the spatial coordinate $r^* = \frac{v-u}{2}$ the semiclassical equations become

$$\rho'' = \frac{1}{Q^2} e^{2\rho} \,, \quad (6.283)$$

$$\tilde{\phi}'' = \frac{2}{Q^2} \tilde{\phi} e^{2\rho} \,, \quad (6.284)$$

$$-2\tilde{\phi}'' + 4\rho'\tilde{\phi}' = -\frac{N\hbar}{6\pi}(\rho'^2 - \rho'') \,. \quad (6.285)$$

Because the first equation is unaffected by the quantum corrections the form of ρ coincides with its classical expression and is given by[24]

$$e^{2\rho} = \frac{Q^2 \kappa_+^2}{\sinh^2 \kappa_+ r^*} , \qquad (6.286)$$

where $\kappa_+ = \sqrt{\frac{2\Delta m}{Q^3}}$. Using this and combining the last two equations we obtain the following first order differential equation for $\tilde{\phi}$:

$$\tilde{\phi}' + \frac{\kappa_+}{\sinh \kappa_+ r^* \cosh \kappa_+ r^*} \tilde{\phi} - \frac{N\hbar}{24\pi} \kappa_+ \tanh \kappa_+ r^* = 0 . \qquad (6.287)$$

The solution of this equation is[25]

$$\tilde{\phi} = -Q\kappa_+ \tanh^{-1} \kappa_+ r^* + \frac{\hbar N}{24\pi} \left(\kappa_+ r^* \tanh^{-1} \kappa_+ r^* - 1 \right) . \qquad (6.288)$$

It is easy to see that in the limit $\Delta m \to 0$, which means $\kappa_+ \to 0$, we get back the extremal classical solution

$$ds^2 = -\frac{Q^2}{r^{*2}} du dv ,$$

$$\tilde{\phi} = -\frac{Q}{r^*} , \qquad (6.289)$$

which, as already explained above, does not receive quantum corrections. The case $\kappa_+ \neq 0$ is more interesting because it displays the quantum corrections typical of Boulware state. We have seen in the case of CGHS black holes that the typical feature of the backreaction solution in Boulware state is the appearance of a curvature singularity at the location of the classical horizon. Differently from that case the two-dimensional metric does not get quantum corrections, the dependence on $\hbar$ being all contained in $\tilde{\phi}$. In particular, at the classical horizon $r^* \to -\infty$ $\tilde{\phi}$ diverges and this signals the emergence of a singularity in the corresponding higher dimensional solution.

We shall now investigate a different set of exact solutions which we will relate to the *Hartle–Hawking* state. We shall use the fact that even for non-extremal black holes the near-horizon metric can always be written in the Poincaré form (see Section 4.5):

$$ds^2 = -\frac{4Q^2}{(x^- - x^+)^2} dx^+ dx^- . \qquad (6.290)$$

---

[24] See subsection 4.4.2.
[25] This solution has been given in a slightly different form in [Cadoni and Mignemi (1995b)].

In this Kruskal-type frame the vacuum polarization terms vanish.

A very simple way to generate exact solutions is to choose the state-dependent functions as follows:

$$\langle \Psi | : T_{\pm\pm}(x^{\pm}) : | \Psi \rangle = 0 \ . \tag{6.291}$$

Then the semiclassical solutions coincide exactly with the classical ones. It is easy to show that the classical solutions given in (u, v) coordinates

$$ds^2 = -\frac{Q^2 \kappa_+^2}{\sinh^2 \kappa_+ \frac{(v-u)}{2}} dv du \ ,$$

$$\tilde{\phi} = -Q\kappa_+ \tanh^{-1} \kappa_+ \frac{(v-u)}{2} \tag{6.292}$$

can be brought, in terms of the coordinates $x^{\pm}$ defined by

$$\frac{\kappa_+ x^+}{2} = \tanh \frac{\kappa_+ v}{2} \ ,$$

$$\frac{\kappa_+ x^-}{2} = \tanh \frac{\kappa_+ u}{2} \tag{6.293}$$

to the form

$$ds^2 = -\frac{4Q^2}{(x^+ - x^-)^2} dx^+ dx^- \ ,$$

$$\tilde{\phi} = 2Q \frac{1 - \frac{\Delta m}{2Q^3} x^+ x^-}{x^- - x^+} \ . \tag{6.294}$$

Let us interpret the result. As we have explained in detail in Section 4.5, the coordinates $x^{\pm}$ for which the metric takes the form (6.290), are related with the (non-extremal) Kruskal coordinates $U = -\frac{1}{\kappa_+} e^{-\kappa_+ u}$, $V = \frac{1}{\kappa_+} e^{\kappa_+ v}$ by Möbius transformations. Therefore, the state-dependent functions (6.291) are equivalent to the usual definition of the *Hartle–Hawking* state:

$$\langle H | : T_{VV}(V) : | H \rangle = 0$$
$$\langle H | : T_{UU}(U) : | H \rangle = 0 \ . \tag{6.295}$$

6.4.3 *General solutions*

Before turning to the most interesting, and also more complicated, dynamical solutions we shall present the generic solution to the semiclassical equations with unspecified state-dependent functions. The simplest way is to start solving the Liouville equation (6.270). It is well known that its general solution is given by[26]

$$ds^2 = -\frac{\partial_+ A_+ \partial_- A_-}{(1 + \frac{1}{4Q^2} A_+ A_-)^2} dx^+ dx^- , \qquad (6.296)$$

where $A_+ = A_+(x^+)$ and $A_- = A_-(x^-)$ are arbitrary chiral functions. These functions $A_\pm$ play a similar role to the arbitrary chiral functions $\omega_\pm$ that appear in the general solution of the CGHS model. They allow us to fix the conformal coordinates. It is convenient to select the following functions:

$$A_+ = x^+ \qquad A_- = -\frac{4Q^2}{x^-} . \qquad (6.297)$$

This way the form of the metric is fixed and takes the Poincaré form

$$ds^2 = -\frac{4Q^2}{(x^- - x^+)^2} dx^+ dx^- . \qquad (6.298)$$

In this gauge the solution for the dilaton $\tilde{\phi}$ is given by[27]

$$\frac{\tilde{\phi}}{Q} = \frac{1}{2}\partial_+ F(x^+) + \frac{F(x^+)}{x^- - x^+} + \frac{1}{2}\partial_- G(x^-) + \frac{G(x^-)}{x^+ - x^-} , \qquad (6.299)$$

where the two chiral functions $F(x^+)$ and $G(x^-)$ verify the constraint equations

$$-\frac{Q}{2}\partial_+^3 F = \langle \Psi | : T_{++}(x^+) : | \Psi \rangle$$

$$-\frac{Q}{2}\partial_-^3 G = \langle \Psi | : T_{--}(x^-) : | \Psi \rangle . \qquad (6.300)$$

The general solution for F, up to solutions of the homogeneous equation, can then be written as

$$F(x^+) = -\frac{2}{Q}\int\int\int \langle \Psi | : T_{++}(x^+) : | \Psi \rangle = -\frac{(x^+)^2}{Q}\int \langle \Psi | : T_{++}(x^+) : | \Psi \rangle$$

$$+ \frac{2x^+}{Q}\int x^+ \langle \Psi | : T_{++}(x^+) : | \Psi \rangle - \int \frac{(x^+)^2}{Q} \langle \Psi | : T_{++}(x^+) : | \Psi \rangle , \qquad (6.301)$$

and a similar expression for $G(x^-)$.

---
[26] See for instance [Jackiw (1984)].
[27] In an arbitrary gauge the generic solution can be found in [Filippov (1996)].

6.4.4 Black hole evaporation in the JT model

In this subsection we shall analyse the process of near-extremal black hole formation and subsequent evaporation. To simplify the mathematics we shall consider, as usual, the incoming matter represented by a single shock wave. The classical solution for this process can be easily derived from the general solution given in the previous subsection with the following classical expressions for the stress tensor:[28]

$$\langle in| : T_{++}(x^+) : |in\rangle = \frac{\Delta m}{Q^2}\delta(x^+ - x_0^+)$$
$$\langle in| : T_{--}(x^-) : |in\rangle = 0 , \qquad (6.302)$$

and the solution which describes this process is given by

$$\tilde{\phi} = 2Q\frac{1}{x^- - x^+} - 2Q\frac{\frac{\Delta m}{2Q^3}(x^+ - x_0^+)(x^- - x_0^+)}{x^- - x^+}\theta(x^+ - x_0^+) . \qquad (6.303)$$

Notice that the first term is the extremal configuration and the second one is obtained from the general formula for $F(x^+)$ given in Eq. (6.301) specialized to a shock wave. We shall now consider the quantum corrections to the above solution [Fabbri et al. (2000); Fabbri et al. (2001a)].[29]

6.4.4.1 First approximation for the state-dependent functions

The first thing to do is to give the quantum form for the state-dependent functions. For the initial extremal geometry they are exactly given by

$$\langle in| : T_{\pm\pm}(x^\pm) : |in\rangle = 0 , \qquad (6.304)$$

but once the classical matter is turned on the coordinate x^+ is no longer an Eddington–Finkelstein type null coordinate and therefore the state-dependent functions representing the absence of incoming radiation become different. If v is the Eddington–Finkelstein advanced null coordinate of the semiclassical solution, the absence of incoming radiation from past null infinity is expressed as

$$\langle in| : T_{vv}(v) : |in\rangle = 0 . \qquad (6.305)$$

But what is the precise way to define v in the near-horizon framework? Since we ignore, a priori, the relation between v and x^+ the first approximation to the problem, via a Hartree–Fock expansion, is to assume the

---
[28] The factor $1/Q^2$ arises because of the removal of the overall Q^2 factor in the action.
[29] Related work can be found in [Diba and Lowe (2002)].

classical relation. This implies that

$$\langle in| : T_{++}(x^+) : |in\rangle = T^{cl}_{++}(x^+) - \frac{\hbar N}{24\pi}\{v, x^+\} . \tag{6.306}$$

The relation between x^+ and v after x_0^+ is

$$\frac{\kappa_+(x^+ - x_0^+)}{2} = \tanh\frac{\kappa_+(v - v_0)}{2} . \tag{6.307}$$

The evaluation of the Schwarzian derivative gives

$$\langle in| : T_{++}(x^+) : |in\rangle = \frac{\Delta m}{Q^2}\delta(x^+ - x_0^+) - \frac{\hbar N}{12\pi}\frac{4\kappa_+^2 \theta(x^+ - x_0^+)}{(4 - \kappa_+^2(x^+ - x_0^+)^2)^2} . \tag{6.308}$$

We note that the second term, representing the absence of incoming quantum radiation at past null infinity in the higher dimensional picture, produces a quantum negative ingoing flux in the near-horizon region.

Before constructing the solution for $\tilde\phi$ we also need to determine $\langle in| : T_{--}(x^-) : |in\rangle$. This is easy since the form of the function $G(x^-)$ is constrained from the matching with the initial extremal solution. Indeed, to recover the extremal configuration at x_0^+ the function $G(x^-)$ can be at most a constant. Without loss of generality we can then take

$$G(x^-) = 0 . \tag{6.309}$$

This implies that

$$\langle in| : T_{--}(x^-) : |in\rangle = 0 . \tag{6.310}$$

This fact can be understood easily since the model has been constructed to describe the near-horizon behavior. When backreaction is included one expects that the outgoing radiation goes down at late times. This is indeed the meaning of Eq. (6.310).

We shall now analyse the physical properties of the solution in detail, mimicking the study performed in Subsection 6.3.3 for the RST model. The semiclassical solution with the above state-dependent functions is

$$\tilde\phi = 2Q\frac{1}{x^- - x^+} - 2Q\frac{\frac{\kappa_+^2}{4}(x^+ - x_0^+)(x^- - x_0^+)}{x^- - x^+}\theta(x^+ - x_0^+) + \frac{N\hbar}{\pi}\frac{(x^- - x^+)P'(x^+) + 2P(x^+)}{x^- - x^+}\theta(x^+ - x_0^+) , \tag{6.311}$$

where the function $P(x^+)$ is given by

$$P(x^+) = \frac{x^+ - x_0^+}{48} - \frac{4 - \kappa_+^2(x^+ - x_0^+)^2}{96\kappa_+} \tanh^{-1} \frac{\kappa_+(x^+ - x_0^+)}{2}, \quad (6.312)$$

and the prime denotes derivative with respect to x^+. Note that $P(x^+ = x_0^+) = 0 = P'(x^+ = x_0^+)$ and therefore the quantum contribution to $\tilde{\phi}$ vanishes at $x^+ = x_0^+$, allowing the matching with the initial extremal configuration. The curve $\tilde{\phi} = 0$ represents the location of the extremal radius $r = Q$.[30] Before x_0^+ it is located at $x^- = +\infty$ and after x_0^+ it is given by the curve

$$x^- = x_0^+ + \frac{4}{\kappa_+^2(x^+ - x_0^+)} + \frac{N\hbar}{12\pi\kappa_+^2 Q - \frac{N\hbar\kappa_+}{2}\tanh^{-1}\frac{\kappa_+(x^+-x_0^+)}{2}}. \quad (6.313)$$

The apparent horizon $\partial_+\tilde{\phi} = 0$ coincides, before the shock wave, with the extremal radius $x^- = +\infty$, and after the shock wave it is given by the following equation:

$$Q\left(1 - \frac{\kappa_+^2(x^- - x_0^+)^2}{4}\right) + \frac{N\hbar}{2\pi}\left(\frac{1}{2}(x^- - x^+)^2 P'' + (x^- - x^+)P' + P\right) = 0. \quad (6.314)$$

This curve has two branches, the outer and the inner apparent horizons. The main property of these curves is that, unlike the classical case, they intersect at $x^\pm = x_{int}^\pm$, where x_{int}^+ is given by

$$\tanh^{-1}\frac{(x_{int}^+ - x_0^+)\kappa_+}{2} = \frac{24\pi Q \kappa_+}{N\hbar} - \frac{\kappa_+(x_{int}^+ - x_0^+)}{\sqrt{4 - \kappa_+^2(x_{int}^+ - x_0^+)^2}}. \quad (6.315)$$

A graphic description of this process is represented in the Penrose diagram of Fig. 6.10. At the point $x_{int}^\pm$ the extremal radius curve $\tilde{\phi} = 0$ is null and both outer and inner horizons meet.[31] This means that we have arrived at the end point of the evaporation, in parallel to the result obtained within the RST model. We can also match the evaporating solution given in Eq.

---

[30]To understand the details of the evolution, the role played by the singularity in the RST model is here given by the extremal radius.
[31]These features were also found numerically in [Lowe and O'Loughlin (1993)].

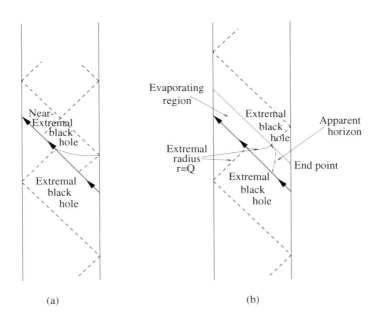

Fig. 6.10 (a) Penrose diagram for the formation of a near-extremal black hole by incoming matter, (b) Penrose diagram for the subsequent evaporation. The continuous line represents the extremal radius. The apparent horizon and the extremal radius intersect, in (b), at the finite point $x_{int}^{\pm}$. After x_{int}^{+} one can match the evaporating solution with the extremal configuration.

(6.311) with a static radiationless configuration along $x^+ = x_{int}^+$ [32]

$$\tilde{\phi} = -k\frac{(x^+ - x_{int}^-)(x^- - x_{int}^-)}{x^- - x^+}, \qquad (6.316)$$

where

$$k = \frac{Q\kappa_+^2(x_{int}^+ - x_0^+) - \frac{N\hbar}{\pi}P'(x_{int}^+)}{x_{int}^+ - x_{int}^-}. \qquad (6.317)$$

This solution turns out to be just the extremal black hole where $x^+ - x_{int}^-$ and $x^- - x_{int}^-$ are proportional to the near-horizon Kruskal coordinates (see Eq. (6.279)).

---

[32] Note that a matching along x_{int}^-, as in the RST model, is not possible. However, unlike in the RST model, where the matching involves a thunderpop, the matching along x_{int}^+ is smooth.

It is important to remark that the matching is possible if we impose that, after the end point, the state-dependent functions vanish

$$\langle in| : T_{++}(x^+) : |in\rangle = \frac{\Delta m}{Q^2}\delta(x^+ - x_0^+)$$
$$- \frac{\hbar N}{12\pi} \frac{4\kappa_+^2 \theta(x^+ - x_0^+)}{(4 - \kappa_+^2(x^+ - x_0^+)^2)^2}\theta(x_{int}^+ - x^+) \,. \quad (6.318)$$

This discontinuity of the state-dependent function at x_{int}^+ is an unpleasant fact and reflects the limitation of our approximation. How can we see that our initial assumption (6.306), (6.308) is indeed a first approximation? The answer will come by computing the new Eddington–Finkelstein advanced coordinate v_1 directly from the evaporating solution. To this end we have to rewrite the metric as

$$ds^2 = -\frac{x^2 - 2Q\tilde{m}_1(v_1)}{Q^2}dv_1^2 + 2dx dv_1 \,, \quad (6.319)$$

where $x = Q\tilde{\phi}$. Note that because the outgoing flux is zero (Eq. (6.310)), the solution can always be written in the above advanced Vaidya gauge. In the higher dimensional picture $v_{(1)}$ corresponds to the advanced Eddington–Finkelstein null coordinate. From (6.319) we can compute $v_1 = v_1(x^+)$ and also $m_1 = m_1(x^+)$. We get

$$v_1 = v_{cl} - \frac{N\hbar}{48\pi Q\kappa_+^2}\left(\frac{4}{4 - \kappa_+^2(x^+ - x_0^+)^2} - \left(\tanh^{-1}\frac{\kappa_+(x^+ - x_0^+)}{2}\right)^2\right),$$
$$(6.320)$$

where for brevity we have indicated with v_{cl} the classical v given in Eq. (6.307), and

$$\tilde{m}_1(x^+) = \Delta m - \frac{N\hbar\kappa_+}{24\pi}\tanh^{-1}\frac{\kappa_+(x^+ - x_0^+)}{2}$$
$$- \frac{N^2\hbar^2}{2304\pi^2 Q}\left[\frac{\kappa_+^2(x^+ - x_0^+)^2}{4 - \kappa_+^2(x^+ - x_0^+)^2} - \left(\tanh^{-1}\frac{\kappa_+(x^+ - x_0^+)}{2}\right)^2\right]. \quad (6.321)$$

The two main consequences of these results are:

- From Eq. (6.320), the relation between the Eddington–Finkelstein advanced coordinate v and x^+ gets quantum corrections. This means that the initial ansatz for the state-dependent function given in Eq. (6.308) is not exact.

- Although the mass function vanishes at the point $x^+ = x^+_{int}$, it continues decreasing after x^+_{int}. This indicates, in some sense, that the matching with the extremal solution is artificial.

It should be mentioned at this point that the next approximation consists in using the above relations, valid at $O(\hbar)$, to construct a new solution to $O(\hbar^2)$ along similar lines and so on. For the RST model this is not needed because the first order approximation is exact. As we will see, also in this case, fortunately, an exact treatment of the problem is possible.

6.4.4.2 Exact treatment for the state-dependent functions

Eq. (6.310) has two important consequences which allow an exact solution to our problem. First, it allows us to write the evaporating solution in the Poincaré coordinates:

$$ds^2 = -\frac{4Q^2}{(x^- - x^+)^2} dx^+ dx^-$$

$$\frac{\tilde{\phi}}{Q} = \frac{1}{2}\partial_+ F(x^+) + \frac{F(x^+)}{x^- - x^+} . \tag{6.322}$$

Second, we can also express it in the ingoing Vaidya form

$$ds^2 = -\frac{x^2 - 2Q\tilde{m}(v)}{Q^2} dv^2 + 2dxdv$$

$$\tilde{\phi} = \frac{x}{Q}, \tag{6.323}$$

where v is the exact Eddington–Finkelstein type coordinate for the evaporating solution. The compatibility between these two expressions leads to

$$\frac{dv}{dx^+} = \frac{2}{F}, \tag{6.324}$$

and

$$\tilde{m} = \frac{Q^3}{4}\left(-F''F + \frac{1}{2}F'^2\right), \tag{6.325}$$

where the prime means derivative with respect to x^+. Therefore, the solution to our problem is all contained in the function F, which gives the relation between v and x^+ and also the evaporating mass function $\tilde{m}$. The relation $v = v(x^+)$ is required to properly define the state-dependent function

$$\langle in | : T_{++}(x^+) : | in \rangle = T_{++}^{cl}(x^+) - \frac{\hbar N}{24\pi}\{v, x^+\} . \qquad (6.326)$$

Note that in our discussion we do not restrict to a shock wave, but rather consider a general situation.

Taking into account the constraint equation

$$-\frac{Q}{2}F''' = \langle \Psi | : T_{++}(x^+) : | \Psi \rangle , \qquad (6.327)$$

we finally derive the differential equation obeyed by the function F:

$$\frac{Q}{2}F''' = \frac{N\hbar}{24\pi}\left(-\frac{F''}{F} + \frac{1}{2}\left(\frac{F'}{F}\right)^2\right) - T_{++}^{cl} . \qquad (6.328)$$

It is important to insist that in this treatment the state-dependent function $\langle \Psi | : T_{++}(x^+) : | \Psi \rangle$, depending on F, has a dynamical character.

The derivative of the mass function with respect to x^+ is

$$\tilde{m}' = -\frac{Q^3}{4}F'''F , \qquad (6.329)$$

and using Eq. (6.328) we get a simple expression in terms of the v coordinate

$$\frac{d\tilde{m}(v)}{dv} = -\frac{N\hbar}{24\pi Q}\tilde{m}(v) + Q^2 T_{vv}^{cl} . \qquad (6.330)$$

If the incoming classical matter is turned off at some advanced time v_f the evaporating mass function $\tilde{m}(v)$ after v_f is given by

$$\tilde{m}(v) = \tilde{m}(v_f)e^{-\frac{N\hbar}{24\pi Q}(v-v_f)} , \qquad (6.331)$$

where the quantity $\tilde{m}(v_f)$ is given by the formal expression

$$\tilde{m}(v_f) = \sum_{n=0}^{\infty}\left(-\frac{N\hbar}{24\pi Q}\right)^n \int_{-\infty}^{v_f} dv_1 \int_{-\infty}^{v_1} dv_2 ... \int_{-\infty}^{v_n} dv_{n+1} T_{vv}^{cl}(v_{n+1}) . \qquad (6.332)$$

We observe that for $\hbar \to 0$, $\tilde{m}(v_f)$ is the total classical mass of the incoming matter. The most important fact is that the extremal configuration $\tilde{m} = 0$ is recovered at $v = +\infty$, leading to an infinite evaporation time. A similar conclusion using the adiabatic approximation was obtained in [Strominger and Trivedi (1993)].

To have a better picture of the evaporation process it is convenient to analyse the solution in the $x^{\pm}$ coordinates. The corresponding Penrose diagram is given in Fig. 6.11. The evolution of the "extremal radius" is

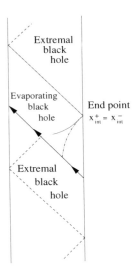

Fig. 6.11 Penrose diagram describing the near-horizon evaporation of a near-extremal Reissner–Nordström black hole in the exact treatment.

represented by the curve $\tilde{\phi} = 0$

$$x^- = x^+ - \frac{2F}{F'}, \qquad (6.333)$$

and the location of the inner and outer apparent horizons $\partial_+\tilde{\phi} = 0$ is given by

$$x^- = x^+ - \frac{F'}{F''} \pm \frac{\sqrt{F'^2 - 2FF''}}{F''}. \qquad (6.334)$$

It is easy to realize that these three curves[33] meet at the point x^+_{int} defined by

$$F'(x^+_{int})^2 - 2F(x^+_{int})F''(x^+_{int}) = 0. \qquad (6.335)$$

---

[33]For simplicity, in Fig. 6.11, as well as in Fig. 6.10, we only consider the extremal radius and outer apparent horizon curves.

This signals the end of the evaporation and is consistent with the fact that the mass function $\tilde{m}$, defined in Eq. (6.325), vanishes there:

$$\tilde{m}(x_{int}^+) = 0 . \tag{6.336}$$

Since in v coordinate the end of the evaporation takes place for $v = +\infty$, then the change of coordinates between v and x^+, displayed in (6.324), must be such that

$$F(x_{int}^+) = 0 . \tag{6.337}$$

From (6.335) this also implies that

$$F'(x_{int}^+) = 0 . \tag{6.338}$$

Moreover, the second derivative $F''(x_{int}^+)$ cannot vanish otherwise the solution to the differential equation (6.328) would be trivially $F = 0$. It follows immediately from (6.334) that

$$x_{int}^- = x_{int}^+ , \tag{6.339}$$

meaning that the intersection takes place at the AdS_2 boundary, as depicted in Fig. 6.11. One can then expand the function F around x_{int}^+ in the following way:

$$F(x^+) = \frac{F''(x_{int}^+)}{2}(x^+ - x_{int}^+)^2 + \cdots , \tag{6.340}$$

which implies the following Möbius-type relation between v and x^+ around x_{int}^+

$$v = -\frac{4}{F''(x_{int}^+)(x^+ - x_{int}^+)} + \cdots . \tag{6.341}$$

This means that the mass function $\tilde{m}$, that after v_f is given by

$$\tilde{m} = \frac{3\pi Q^4}{N\hbar} F''' F^2 , \tag{6.342}$$

behaves, close to the end point, as

$$\tilde{m} \sim \tilde{m}(v_f) e^{\frac{N\hbar}{6\pi Q} \frac{1}{F''(x_{int}^+)(x^+ - x_{int}^+)}} . \tag{6.343}$$

Therefore, we can estimate F''':

$$F'''(x^+) \sim \frac{4N\hbar \tilde{m}(v_f)}{3\pi Q^4 F''(x_{int}^+)^2 (x^+ - x_{int}^+)^4} e^{\frac{N\hbar}{6\pi Q} \frac{1}{F''(x_{int}^+)(x^+ - x_{int}^+)}} . \tag{6.344}$$

This implies that
$$F'''(x_{int}^+) = 0 , \qquad (6.345)$$
and also allows us to give the higher order correction to Eq. (6.340) in the form
$$F(x^+) \sim \frac{F''(x_{int}^+)}{2}(x^+ - x_{int}^+)^2 \left(1 - \frac{576\pi^2 \tilde{m}(v_f)}{QN^2\hbar^2} e^{\frac{N\hbar}{6\pi Q} \frac{1}{F''(x_{int}^+)(x^+ - x_{int}^+)}}\right) . \qquad (6.346)$$

From the above result for the function F one concludes that all the higher-order derivatives $F^{(n)}(x_{int}^+)$ for $n \geq 3$ vanish. The physical consequences are:

- The three curves (6.333), (6.334) become null at the end point, i.e.,
$$\frac{dx^-}{dx^+} \xrightarrow{x^+ \to x_{int}^+} 0 . \qquad (6.347)$$

- The fact that the mass function $\tilde{m}$ and all its derivatives vanish implies that the solution *becomes extremal* and no ad-hoc matching is required at the end point, in contrast with what happened in the first order approximation and also in RST.

The exact solvability of the model allows us to infer about the end point configuration. Close to the end point the dilaton function behaves as
$$\frac{\tilde{\phi}}{Q} = \frac{F''(x_{int}^+)}{2} \frac{(x^+ - x_{int}^+)(x^- - x_{int}^+)}{x^- - x^+} + O(e) , \qquad (6.348)$$
where $O(e)$ are exponentially small higher order terms. The leading order term is just the extremal configuration given in Eq. (6.279) expressed in Kruskal-type coordinates
$$U = \frac{Q^2 F''(x_{int}^+)}{2}(x^- - x_{int}^+)$$
$$V = \frac{Q^2 F''(x_{int}^+)}{2}(x^+ - x_{int}^+) . \qquad (6.349)$$
To recover the standard form
$$\tilde{\phi} = \frac{2Q}{u - v} , \qquad (6.350)$$

we have to introduce the Eddington–Finkelstein null coordinates

$$v = -\frac{4}{F''(x^+_{int})} \frac{1}{x^+ - x^+_{int}} \tag{6.351}$$

$$u = -\frac{4}{F''(x^+_{int})} \frac{1}{x^- - x^+_{int}} . \tag{6.352}$$

The near-horizon model also produces a non-trivial result concerning the Hawking radiation. The effect of the backreaction is to replace the fixed background relation

$$\frac{\kappa_+(x^- - x^+_0)}{2} = \tanh \frac{\kappa_+(u - v_0)}{2} , \tag{6.353}$$

producing the thermal radiation, with the pure Möbius transformation (6.352).

6.4.5 Beyond the near-horizon approximation and Hawking radiation

The non-trivial result of Eq. (6.352) is connected with the fact that at late times the evaporation flux vanishes (a natural condition, given that the end point of the evaporation is the extremal black hole). This contrasts with the calculations made in fixed background, which instead produce a late-time constant (thermal) flux. Unfortunately we do not have a solvable model from which to extract the exact Hawking radiation at all times. Nevertheless we can describe qualitatively the effects of the backreaction in the crucial relation $x^- \equiv u_{in} = u_{in}(u)$ from which one extracts the Hawking radiation. The relation $u_{in}(u)$ in fixed background can be easily determined in the usual scenario where a nonextremal black hole is created from extremality by means of a shock wave located at $v = v_0$. In the initial region we have

$$ds^2 = -(1 - \frac{Q}{r})^2 du_{in} dv \tag{6.354}$$

and after the shock wave

$$ds^2 = -\frac{(r - r_+)(r - r_-)}{r^2} du dv , \tag{6.355}$$

where $r_\pm = M \pm \sqrt{M^2 - Q^2}$ and $M = Q + \Delta m$. The relation between u_{in} and u can be worked out easily by demanding continuity of r at v_0. The result that we get is

$$\frac{du_{in}}{du} = \frac{(r-r_+)(r-r_-)}{(r-q)^2}, \qquad (6.356)$$

where $r = r(u, v_0)$ is given by inverting

$$\frac{v_0 - u}{2} = r + \frac{1}{r_+ - r_-}\left[r_+^2 \ln\frac{|r-r_+|}{r_+} - r_-^2 \ln\frac{|r-r_-|}{r_-}\right]. \qquad (6.357)$$

The graph of the function in Eq. (6.356) is given in Fig. 6.12(a).

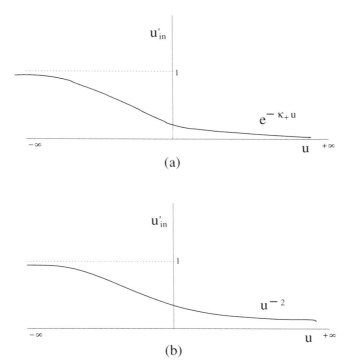

Fig. 6.12 (a) Plot of the function $u'_{in}(u)$ in the fixed background approximation: the late-time behavior is the usual exponential. In (b) the same function is plotted qualitatively by taking into account backreaction effects. The late-time behavior is modified to the power law $\sim u^{-2}$.

We now know from the analysis of the near-horizon model in the previous subsection that due to backreaction effects the relation between u_{in}

and u is largely modified at late times to a pure Möbius of the form

$$\frac{du_{in}}{du} \to \frac{4}{F''(x^+_{int})} \frac{1}{u^2} . \qquad (6.358)$$

At early times, instead, the backreaction effects are negligible and the function du_{in}/du is well approximated by the fixed background expression Eq. (6.356). Therefore, the qualitative behavior of this function should be of the form given in Fig. 6.12(b). Knowledge of this function allows to determine the quantum corrected Hawking flux

$$\langle in|T_{uu}(u)|in\rangle = -\frac{N\hbar}{24\pi}\{u_{in}, u\} , \qquad (6.359)$$

which of course has to be compatible with energy conservation[34]

$$\int_{-\infty}^{+\infty} du \langle in|T_{uu}(u)|in\rangle = \Delta m . \qquad (6.360)$$

6.4.6 Information loss in the JT model

Despite the fact that the Möbius transformation Eq. (6.352) involves no flux the existence of the event horizon, shifted with respect to the location of the initial extremal radius, involves the loss of some of the correlations of the initial vacuum state. A spacetime diagram representing these features is given in Fig. 6.13. Those correlations between the interior and exterior portion of the event horizon in the initial extremal black hole region are transferred, during the time evolution, into correlations between our universe and the interior region.[35] This produces information loss from the point of view of an external observer. Moreover, this information loss seems to take place at a zero energy cost in the interior region. This is because the JT model predicts that the solution along $x^+ = x^+_{int}$ is extremal and, therefore, radiationless. The only role of the interior region would be then to store the necessary correlations to ensure the purity of the initial state. However, also in this case the picture is not fully consistent. A similar analysis to that performed for the RST model in Subsection 6.3.3.1 unravels the possible presence of a thunderbolt along the prolongation of the initial extremal radius which reaches the Cauchy horizon. To see this,

---

[34] For a particular solution of this problem see [Fabbri et al. (2001b)].

[35] Note that the prolongation of the end point along the line $x^+ = x^+_{int}$ is the future Cauchy horizon and therefore the extension of the geometry across this surface is, in principle, problematic.

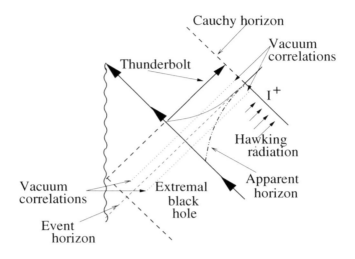

Fig. 6.13 Spacetime diagram representing the evaporation of a near-extremal Reissner–Nordström black hole. The Hawking radiation goes down at late times, as suggested by the JT model. Despite the absence of radiation beyond the event horizon and close to the future Cauchy horizon, there are non-vanishing correlations that are lost for an external observer.

consider the natural time coordinate u along the future Cauchy horizon defined, by analytic prolongation, in Eq. (6.352). The initial extremal radius is given by $x^- = +\infty$, which corresponds to $u = 0$. The natural vacuum state defined by the coordinate u has non-vanishing correlations between points $u < 0$ and $u > 0$, which are instead absent in the "in" vacuum state. Therefore the expectation value in the "in" vacuum state of the two-point function $\langle in| : \partial_- f(u < 0) \partial_- f(u > 0) : |in\rangle$ diverges in the limit $u \to 0$, producing the infinite energy thunderbolt given in Fig. 6.13. This again seems to indicate the limitations of the semiclassical theory to deal with the end point of the evaporation and, in general, with the information loss problem.

6.4.7 Unitarity in the semiclassical approximation?

It is natural to wonder, at this point, if there is any way to avoid the occurrence of information loss in the semiclassical approximation. For the problem under consideration we have seen, however, that there are inconsistencies in the emergence of a "thunderbolt" through the Cauchy horizon.[36] It turns out that information loss and thunderbolt are related one to the other. Indeed, the only possibility to avoid the thunderbolt is that the initial extremal horizon coincides with the event horizon.[37] The most important consequence of this would be that the late-time Möbius type (radiationless) relation (6.352) is replaced by the "inertial relation"

$$x^- \equiv u_{in} \approx u \ . \tag{6.361}$$

This is similar to the analogous condition imposed on the moving mirror trajectory in Subsection 4.6.3 making the function $p'(x^-)$ (here du_{in}/du) to come back to unity in the limit $x^- \to +\infty$, as displayed in Fig. 4.11. Under these conditions, for the same reasons explained in Subsection 4.6.3 there would be no information loss and unitarity would be recovered although at the price of having at least a small period of negative energy radiation.[38]

As a concluding remark, we want to stress that this possibility to reestablish unitarity (not realized in the semiclassical approximation) could be perhaps obtained within a proper quantum gravity approach. We mention that other possibilities were outlined at the end of Chapter 3. The challenge is to see whether one of them is realized physically and how the thermal spectrum at infinity is actually modified in a way consistent with energy conservation.

---

[36] We have seen a similar feature in the RST model, in Subsection 6.3.3.1, with the thunderbolt traveling instead to I^+.

[37] In the RST case an analogous requirement would prevent the formation of the event horizon itself.

[38] For restrictions on the possible amount of negative energy radiated see [Ford and Roman (1995)].

Bibliography

Adams, J.B., Harrison, B.K., Kaluder, L.T. Jnr, Mjolsness, R., Wakano, M., Wheeler, J.A. and Willey, R. (1958). Some implications of general relativity for the structure and evolution of the universe. *La Structure et L'Evolution de L'Univers*, Onzime Conseil de Physique, Institut International de Physique Solvay. R. Stoops, Brussels, pp. 97–146

Aharony, O., Gubser, S.S., Maldacena, J.M. and Ooguri, H. (2000). Large N field theories, string theory and gravity, *Phys. Rep.* **323**, pp. 183-386

Aldaya, V., Calixto, M. and Cerveró, J.M. (1999). Vacuum radiation and symmetry breaking in conformally invariant quantum field theory, *Commun. Math. Phys.* **200** pp. 325–354

de Alfaro, V., Nelson, J.E., Bandelloni, G., Blasi, A., Cavaglia, M. and Filippov, A.T., Eds. (1997). Proceedings of the 2nd meeting QG96 *Constrained Dynamics and Quantum Gravity*, S. Margherita Ligure (Italy), 1996. *Nucl. Phys. B Proc. Suppl.* **57**.

Alvarez-Gaumé, L., Sierra, G. and Gómez, C. (1989). Topics in conformal field theory, in *Physics and mathematics of strings*. Eds. L. Brink *et al.*, World Scientific, pp. 16–184

Alvarez, O. (1983). Theory of strings with boundaries: fluctuations, topology and quantum geometry, *Nucl.Phys. B* **216**, pp. 125–227

Amati, D. (2003). How strings solve the apparent contradiction between black holes and quantum coherence, *J. Phys. G* **29**, pp. 31–34

Anderson, A. and DeWitt, W. (1986). Does the topology of space fluctuate?, *Found. Phys.* **16**, pp. 91–105

Anderson, P.R., Hiscock, W.A. and Samuel, D.A. (1995). Stress-energy tensor of quantized scalar fields in static spherically symmetric spacetimes, *Phys. Rev. D* **51**, pp. 4337–4358

Ashtekar, A., Baez, J.C., Corichi, A. and Krasnov, K. (1998). Quantum geometry and black hole entropy, *Phys. Rev. Lett.* **80**, pp. 904–907

de Azcárraga, J.A. and Izquierdo, J.M. (1995). *Lie groups, Lie algebras, cohomology and some applications in physics*, Cambridge University Press, Cambridge, England

Balbinot, R. and Floreanini, R. (1985). Semiclassical two-dimensional gravity and

Liouville equation, *Phys. Lett. B* **151**, pp. 401–404

Balbinot, R. and Fabbri, A. (1999). Hawking radiation by effective two-dimensional theories, *Phys. Rev. D* **59**, 044031

Balbinot, R., Fabbri, A., Frolov, V.P., Nicolini, P., Sutton, P. and Zelnikov, A.I. (2001). Vacuum polarization in the Schwarzschild spacetime and dimensional reduction, *Phys. Rev. D* **63**, 084023

Balbinot, R. and Fabbri A. (2003). Two-dimensional black holes and effective actions, *Class. Quantum Grav.* **20**, pp. 5439–5454

Banks, T., Susskind, L. and Peskin, M.E. (1984). Difficulties for the evolution of pure states into mixed states, *Nucl. Phys. B* **244**, pp. 125–134

Banks, T., Dabholkar, A., Douglas, M.R. and O'Loughlin, M. (1992). Are horned particles the climax of Hawking evaporation?, *Phys. Rev. D* **45**, pp. 3607–3616

Banks, T. (1994). Lectures on black holes and information loss, *Nucl. Phys. Proc. Suppl.* **41**, pp. 21–65

Bañados, M., Teitelboim, C. and Zanelli, J. (1992). Black hole in three-dimensional spacetime, *Phys. Rev. Lett.* **69**, pp. 1849–1851

Bardeen, J.M., Carter, B. and Hawking, S.W. (1973). The four laws of black hole mechanics, *Commun. Math. Phys.* **31**, pp. 161–170

Bekenstein, J.D. (1972). Nonexistence of baryon number for static black holes, *Phys. Rev.* **D5**, pp. 1239–1246; and pp. 2403–2412

Bekenstein, J.D. (1973). Black holes and entropy, *Phys. Rev. D* **7**, pp. 2333–2346

Bekenstein, J.D. (1974). Generalized second law of thermodynamics in black hole physics, *Phys. Rev. D* **9**, pp. 3292–3300

Bekenstein, J.D. (1996). Black hole hair: twenty-five years after, gr-qc/9605059. 2nd International A.D. Sakharov Conference on Physics. pp. 216–219

Belinsky, V.A., Khalatnikov, I.M. and Lifshitz, E.M. (1970). Oscillatory approach to a singular point in the relativistic cosmology, *Advances in Physics* **19**, pp. 525–573

Benachenhou, F. (1994). Black hole evaporation. A survey. hep-th/9412189

Bertotti, B. (1959). Uniform electromagnetic field in the theory of general relativity, *Phys. Rev.* **116**, pp. 1331–1333

Bilal, A. and Callan, C. (1992). Liouville models of black hole evaporation, *Nucl. Phys. B* **394**, pp. 73–100

Birnir, B., Giddings, S.B., Harvey, J.A. and Strominger, A. (1992). Quantum black holes, *Phys. Rev. D* **46**, pp. 638–644

Birrel, N.D. and Davies, P.C.W. (1982). *Quantum fields in curved space*, Cambridge University Press, Cambridge, England.

Bose, S., Parker, L. and Peleg, Y. (1995). Semi-infinite throat as the end-state geometry of two-dimensional black hole evaporation, *Phys. Rev. D* **52**, pp. 3512–3517

Bose, S., Parker, L. and Peleg, Y. (1996). Hawking radiation and unitary evolution, *Phys. Rev. Lett.* **76**, pp. 861–864

Boulware, D. (1975). Quantum field theory in Schwarzschild and Rindler spaces, *Phys. Rev. D* **11**, pp. 1404–1423

Bousso, R. and Hawking, S.W. (1997). Trace anomaly of dilaton coupled scalars in two-dimensions, *Phys. Rev. D* **56**, pp. 7788–7791

Boyer, R.H. and Lindquist, R.W. (1967). Maximal analytic extension of the Kerr metric, *J. Math. Phys.* **8**, pp. 265–281

Brout, R., Massar, S., Parentani, R. and Spindel, P. (1995). A Primer for black hole quantum physics, *Phys. Rept.* **260**, pp. 329–454

Brown, J.D., Henneaux, M. and Teitelboim, C. (1986). Black holes in two space-time dimensions, *Phys. Rev. D* **33**, pp. 319–323

Brown, M.R., Ottewill, A.C. and Page, D.N. (1986). Conformally invariant quantum field theory in static Einstein spacetimes, *Phys. Rev. D* **33**, pp. 2840–2850

Bunch, T.S., Christensen, S.M. and Fulling, S.A. (1978). Massive quantum field theory in two-dimensional Robertson-Walker space-time, *Phys. Rev. D* **18**, pp. 4435–4459

Buric, M., Radovanovic, V. and Mikovic, A. (1999). One loop corrections for Schwarzschild black hole via 2D dilaton gravity, *Phys. Rev. D* **59**, 084002

Cadoni, M. and Mignemi, S. (1995a). On the conformal equivalence between 2D black holes and Rindler spacetime, *Phys. Lett. B* **358**, pp. 217–222

Cadoni, M. and Mignemi, S. (1995b). Non-singular four dimensional black holes and the Jackiw-Teitelboim model, *Phys. Rev. D* **51**, pp. 4319–4329

Callan, C.G., Giddings, S.B., Harvey, J.A. and Strominger, A. (1992). Evanescent black holes, *Phys. Rev. D* **45**, pp. 1005–1009

Candelas, P. (1980). Vacuum polarization in Schwarzschild spacetime, *Phys. Rev. D* **21**, pp. 2185–2202

Capper, D.M. and Duff, M.J. (1974). Trace anomalies and dimensional regularization, *Nuovo Cimento A* **23**, pp. 173–183.

Carlip, S. (2000). Black hole entropy from horizon conformal field theory, *Nucl. Phys. Proc. Suppl.* **88**, pp. 10–16

Carlitz, R.D. and Willey, R.S. (1987). Reflections on moving mirrors, *Phys. Rev. D* **36**, pp. 2327–2335. Lifetime of a black hole, *Phys. Rev. D* **36**, pp. 2336–2341.

Carter, B. (1966). The complete analytic extension of the Reissner–Nordström metric in the special case $e^2 = m^2$, *Phys. Lett.* **21**, pp. 423-424

Carter, B. (1971). An axisymmetric black hole has only two degrees of freedom, *Phys. Rev. Lett.* **26**, pp. 331–333

Carter, B. (1972). Unpublished notes.

Carter, B. (1973). Properties of the Kerr metric, in DeWitt, C. and DeWitt, B.S., *Black holes*, Proceedings of 1972 session of Ecole d'été de physique théorique, Gordon and Breach, New York

Chandrasekhar, S. (1931). The maximum mass of ideal white dwarfs, *Astrophysical Journal* **74**, pp. 81–82

Chandrasekhar, S. (1983). *The mathematical theory of black holes*, Oxford University Press, New York, USA

Christensen, S.M. and Fulling, S.A. (1977). Trace anomalies and the Hawking effect, *Phys. Rev. D* **15**, pp. 2088–2104

Christodoulou, D. (1970). Reversible and irreversible transformations in black hole physics, *Phys. Rev. Lett.* **25**, pp. 1596–1597

Christodoulou, D. and Ruffini, R. (1971). Reversible transformations of a charged black hole, *Phys. Rev. D* **4**, pp. 3552–3555

Chung, T.D. and Verlinde, H. (1994). Dynamical moving mirrors and black holes, *Nucl. Phys. B* **418**, pp. 305–336

Cruz, J. and Navarro-Salas, J. (1996). Solvable models for radiating black holes and area preserving diffeomorphisms, *Phys. Lett. B* **375**, pp. 47–53

Das, S.R. and Mukherji, S. (1994). Black hole formarion and space-time fluctuations in two-dimensional dilaton gravity, *Phys. Rev. D* **50**, pp. 930-940

Davies, P.C.W. and Fulling, S.A. (1977). Radiation from a moving mirror and from black holes, *Proc. R. Soc. London A* **356**, pp. 237–257

Davies, P.C.W. (1975). Scalar particle production in Schwarzschild and Rindler metrics, *J. Phys. A* **8**, pp. 609–616

Davies, P.C.W., Fulling, S.A. and Unruh, W.G. (1976). Energy momentum tensor near an evaporating black hole, *Phys. Rev. D* **13**, pp. 2720–2723

de Alwis, S.P. (1993). Black hole physics from Liouville theory, *Phys. Lett. B* **300**, pp. 330-335

D'Eath, P.D. (1996). *Black holes gravitational interactions*, Oxford University Press, New York, USA

Deser, S., Duff, M.J. and Isham, C.J. (1976). Nonlocal conformal anomalies, *Nucl. Phys. B* **111**, p. 45

DeWitt, B.S. (1975). Quantum field theory in curved spacetime, *Phys. Rep. C* **19**, pp. 295–357.

Diba, K. and Lowe, D.A. (2002). The stress energy tensor in soluble models of spherically symmetric charged black hole evaporation, *Phys. Rev. D* **65**, 024018

Doroshkevich, A.G., Zel'dovich Y.B. and Novikov, I.D. (1965). Gravitational collapse of nonsymmetric and rotating masses. *Zh. Eksp. @ Teor. Fiz.* **49**, pp. 170–181. English translation in *Sov. Phys. JETP* **22**, pp. 122–130

Elster, T. (1983). Vacuum polarization near a black hole creating particles, *Phys. Lett. A* **94**, pp. 205–209

Emparan, R., Fabbri, A. and Kaloper, N. (2002). Quantum black holes as holograms in AdS brane worlds, *J. High Energy Physics* **08**, 043

Epstein, H., Gaser, V. and Jaffe, A. (1965). Nonpositivity of the energy density in quantized field theories, *Nuovo Cimento* **36**, pp. 1016-1022.

Fabbri, A. and Russo, J.G. (1996). Soluble models in 2D dilaton gravity, *Phys. Rev. D* **53**, pp. 6995-7002

Fabbri, A., Navarro, D.J. and Navarro-Salas, J. (2000). Evaporation of near-extremal Reissner–Nordström black holes, *Phys. Rev. Lett.* **85**, pp. 2434-2437

Fabbri, A., Navarro, D.J. and Navarro-Salas, J. (2001a). Quantum evolution of near-extremal Reissner–Nordström black holes, *Nucl. Phys. B* **595**, pp. 381-401

Fabbri, A., Navarro, D.J. and Navarro-Salas, J. (2001b). A Planck-like problem for quantum charged black holes, *Gen. Rel. Grav.* **33**, pp. 2119-2124

Fabbri, A., Farese, S. and Navarro-Salas, J. (2003). Generalized Virasoro anomaly and stress tensor for dilaton-coupled theories, *Phys. Lett B* **574**, pp. 309–

318
Fabbri, A., Navarro-Salas, J. and Olmo, G.J. (2004). Particles and energy fluxes from a CFT perspective, *Phys. Rev. D* **70**, 064022

Filippov, A.T. (1996). Exact solutions of 1+1-dimensional dilaton gravity coupled to matter, *Mod. Phys. Lett. A* **11**, pp. 1691–1704

Finkelstein, D. (1958). Past-future asymmetry of the gravitational field of a point particle, *Phys. Rev.* **110**, pp. 965–967

Ford, L.H. and Roman, T.A. (1995). Averaged energy conditions and quantum inequalities, *Phys. Rev. D* **51**, pp. 4277-4286

Ford, L.H. (1997). Quantum field theory in curved spacetime, gr-qc/9707062.

Di Francesco, P., Mathieu, P. and Senechal, D. (1997). *Conformal field theory*, Springer–Verlag, New York, USA

Frolov, V.P. and Zelnikov, A. (1987). Killing approximation for vacuum and thermal stress-energy tensor in static spacetimes, *Phys. Rev. D* **35**, pp. 3031–3044

Frolov, V.P. and Novikov, I.D. (1998). *Black hole physics*, Kluwer Academic Publishers, Dordrecht, The Netherlands.

Frolov, V.P. (1992). Two-dimensional black hole physics, *Phys. Rev. D* **46**, pp. 5383–5394

Frolov, V.P., Sutton, P. and Zelnikov, A. (2000). The dimensional reduction anomaly, *Phys. Rev. D* **61**, 024021

Fulling, S.A. (1973). Nonuniqueness of canonical field quantization in Riemannian spacetime, *Phys. Rev. D* **7**, pp. 2850–2862

Fulling, S.A. and Davies, P.C.W. (1976), Radiation from a moving mirror in two-dimensional spacetime conformal anomaly, *Proc. R. Soc. London A* **348**, pp. 393–414

Fulling, S.A. (1989). *Aspects of quantum field theory in curved space-time*, Cambridge University Press, Cambridge, England

Gao, S. (2003). Late-time particle creation from gravitational collapse to an extremal Reissner–Nordström black hole, *Phys. Rev. D* **68**, 044028

Garay, L.J. and Mena Marugán, G.A. (2003). Immirzi ambiguity, boosts and conformal frames for black holes, *Class. Quant. Grav.* **20**, pp. L115–L121

Garfinkle, D., Horowitz, G.T. and Strominger, A. (1991). Charged black holes in string theory, *Phys. Rev. D* **43** pp. 3140-3143

Geroch, R.P. (1968). What is a singularity in general relativity?, *Ann. Phys.* **48**, pp. 526–540

Gibbons, G. (1975). Vacuum polarization and the spontaneous loss of charge by black holes, *Commun. Math. Phys.* **44**, pp. 245–264

Gibbons, G. (1979). Quantum field theory in curved space times, in *General Relativity, an Einstein centenary survey*, Eds. Hawking, S.W. and Israel, W. Cambridge University Press, Cambridge, England

Gibbons, G. and Maeda, K. (1988). Black holes and membranes in higher-dimensional theories with dilaton fields, *Nucl. Phys. B* **298**, pp. 741–775

Giddings, S.B. and Nelson, W.M. (1992). Quantum emission from two-dimensional black holes, *Phys. Rev. D* **46**, pp. 2486–2496

Giddings, S.B. (1994). Quantum mechanics of black holes, hep-th/9412138

Ginsparg P. (1990). Applied conformal field theory, Les Houches lectures, Eds. Brezin, E. and Zinn-Justin, J. North-Holland, Amsterdam, 1990, pp.1–168

Ginzburg, V.L. and Ozernoy, L.M. (1965). On gravitational collapse of magnetic stars, *Sov. Phys. JETP* **20**, p. 689

Grumiller, D., Kummer, W. and Vassilevich, D.V. (2002). Dilaton gravity in two-dimensions, *Phys. Rept.* **369**, pp. 327–430

Gullstrand, A. (1922). Allemegeine lösung des statischen Einkörper-problems in der Einsteinschen gravitations theorie, *Arkiv. Mat. Astron. Fys.* **16**, pp. 1–15

Harrison, B.K., Thorne K.S., Wakano, M. and Wheeler, J.A. (1965). *Gravitational theory and gravitational collapse.* University of Chicago Press, Chicago, USA.

Hartle, J.B. and Hawking, S.W. (1976). Path-integral derivation of black hole radiance, *Phys. Rev. D* **13**, pp. 2188–2203.

Harvey, J.A. and Strominger, A. (1992). Quantum aspects of black holes, hep-th/9209055

Hawking, S.W. (1972). Black holes in general relativity, *Commun. Math. Phys.* **25**, pp. 152–166.

Hawking, S.W. and Ellis, G.F.R. (1973). *The large scale structure of spacetime.* Cambridge University Press, Cambridge, England.

Hawking, S.W. (1974). Black hole explosions, *Nature* **248**, pp. 30–31

Hawking, S.W. (1975). Particle creation by black holes, *Commun. Math. Phys.* **43**, pp. 199–220.

Hawking, S.W. (1976). Breakdown of predictability in gravitational collapse, *Phys. Rev. D* **14**, pp. 2460–2473.

Hawking, S.W. (1982). The unpredictability of quantum gravity, *Commun. Math. Phys.* **87**, p. 395

Hawking, S.W. (1988). Wormholes in spacetime, *Phys. Rev. D* **37**, pp. 904-910

Hawking, S.W. (1992). Evaporation of two-dimensional black holes, *Phys. Rev. Lett.* **69**, pp. 406–409

Hawking, S.W. and Stewart, J.M. (1993). Naked and thunderbolt singularities in black hole evaporation, *Nucl. Phys. B* **400**, pp. 393–415

Hawking, S.W. (2004). The information paradox for black holes. Talk given at the GR17 Conference, Dublin, 18-24 July (2004)

Heusler, M. (1996). *Black hole uniqueness theorems.* Cambridge University Press, Cambridge, England.

Hiscock, W.A. (1981). Models of evaporating black holes I, *Phys. Rev. D* **23**, pp. 2813–2822

Hiscock, W.A. and Weems, D. (1990). Evolution of charged evaporating black holes, *Phys. Rev. D* **41**, pp. 1142–1151

Howard, K.W. (1984). Vacuum $< T_\mu^\nu >$ in Schwarzschild spacetime, *Phys. Rev. D* **30**, pp. 2532-2547

Immirzi, G. (1997). Quantum gravity and Regge calculus, *Nucl. Phys. Proc. Suppl.* **57**, pp. 65-72

Isham, C.J. (1977). Quantum field theory in curved space times: an overview, *Ann. N.Y. Acad. Sci.* **302**, pp. 114–157

Israel, W. (1967). Event horizons in static vacuum spacetimes, *Phys. Rev.* **164**, pp. 1776–1779

Israel, W. (1968). Event horizons in static electrovac spacetimes, *Commun. Math. Phys.* **8**, pp. 245–260

Israel, W. (1976). Thermo field dynamics of black holes, *Phys. Lett. A* **57**, pp. 107–110

Israel, W. (1986). Third law of black hole dynamics: a formulation and proof, *Phys. Rev. Lett.* **57**, pp. 397–399

Israel, W. (1987). Dark stars: the evolution of an idea, in Hawking, S.W. and Israel, W. Eds., *300 Years of Gravitation*, Cambridge University Press, Cambridge, England, pp. 199–276

Iyer, V. and Wald, R.M. (1994). Some properties of Noether charge and a proposal for dynamical black hole entropy, *Phys. Rev. D* **50**, pp. 846–864

Jackiw, R. (1984). Liouville field theory: a two-dimensional model for gravity?, in *Quantum Theory of Gravity*, Ed. Christensen, S.M. (Hilger, Bristol, 1984), pp. 403–420;

Jackiw, R. (1995). Another view on massless matter-gravity fields in two dimensions, hep-th/9501016.

Jacobson, T. (1991). Black hole evaporation and ultrashort distances, *Phys. Rev. D* **44**, pp. 1731–1739

Jacobson, T. (2003). Introduction to quantum fields in curved spacetime and the Hawking effect, gr-qc/0308048.

Karakhaniyan, D., Manvelyan, R. and Mkrtchyan, R. (1994). Area preserving structure of 2d gravity, *Phys. Lett. B* **329**, pp. 185–188

Kay, B.S. and Wald, R.M. (1991). Theorems on the uniqueness and thermal properties of stationary, nonsingular, quasifree states on spacetimes with a bifurcate Killing horizon, *Phys. Rept.* **207**, pp. 49–136

Kerr, R.P. (1963). Gravitational field of a spinning mass as an example of algebraically special metrics, *Phys. Rev. Lett.* **11**, pp. 237–238

Kiefer, C. (1999). Thermodynamics of black holes and Hawking radiation, in *Classical and quantum black holes*, Eds. Fre, P., Gorini, V., Magli, G. and Moschella, U. IOP Publishing, Bristol, England.

Kiem, Y., Verlinde, H. and Verlinde E. (1995). Black hole horizons and complementarity, *Phys. Rev. D* **52**, pp. 7053-7065

Klauder, J.R. and Skagerstam, B-S. (1985). *Coherent states*, World Scientific, Singapore

Klemm, D. and Vanzo, L. (1998). Quantum properties of topological black holes, *Phys. Rev. D* **58**, 104025

Klosch, T. and Strobl, T. (1996). Classical and quantum gravity in $1+1$-dimensions. Part I: a unifying approach, *Class. Quant. Grav.* **13**, pp. 965–984; Classical and quantum gravity in $1+1$-dimensions. Part II. The Universal coverings, *Class. Quant. Grav.* **13**, pp. 2395–2422

Kruskal, M.D. (1969). Maximal extension of Schwarzshild metric, *Phys. Rev.* **119**, pp. 1743–1745

Kummer, W. and Vassilevich D.V. (1999). Effective action and Hawking radiation for dilaton coupled scalars in two-dimensions, *Phys. Rev. D* **60**, 084021

Liberati, S., Rothman, T. and Sonego S. (2000). Nonthermal nature of incipient extremal black holes, *Phys. Rev. D* **62**, 024005

Lifschitz, E.M. and Khalatnikov, I.M. (1963). Investigations in relativistic cosmology, *Advances in Physics* **12**, pp. 185–249

Lowe, D.D. (1993). Semiclassical approach to black hole evaporation, *Phys. Rev. D* **47**, pp. 2446–2453

Lowe, D.A. and O'Loughlin, M. (1993). Nonsingular black hole evaporation and stable remnants, *Phys. Rev. D* **48**, pp. 3735–3742

Lowe, D.A. and Thorlacius, L. (1999). AdS/CFT and the information loss paradox, *Phys. Rev. D* **60**, 104012

Mann, R.B. (1992). Lower dimensional black holes, *Gen. Rel. Grav.* **24**, pp. 433–449

Mandal, G., Sengupta, A. and Wadia, S. (1991). Classical solutions of two-dimensional string theory, *Mod. Phys. Lett. A* **6**, pp. 1685–1692

Mikovic, A. (1995). Hawking radiation and backreaction in a unitary theory of 2D quantum gravity, *Phys. Lett. B* **355**, pp. 85–91

Mikovic, A. (1996). Unitary theory of evaporating 2D black holes, *Class. Quant. Grav.* **13**, pp. 209-220

Misner, C.W., Thorne, K.S. and Wheeler, J.A. (1973). *Gravitation.* Freeman, San Francisco, USA

Mukhanov, V., Wipf, A. and Zelnikov, A. (1994). On 4D Hawking radiation from effective action, *Phys. Lett. B* **332**, pp. 283–291

Myers, R.C. (1997). Pure states don't wear black, *Gen. Rel. Grav.* **29**, pp. 1217-1222

Navarro-Salas, J., Navarro, M. and Talavera, C.F. (1995). Weyl invariance and black hole evaporation, *Phys. Lett. B* **356**, pp. 217–222

Navarro-Salas, J. and Navarro, P. (2000). AdS_2/CFT_1 correspondence and near-extremal black hole entropy, *Nucl. Phys. B* **579**, pp. 250-266.

Newman, E.T., Couch, E., Chinnapared, K., Exton, A., Prakash, A. and Torrence, R. (1965). Metric of a rotating, charged mass, *J. Math. Phys.* **6** pp. 918–919

Nojiri, S. and Odintsov, S.D. (2001). Quantum dilatonic gravity in (D=2)-dimensions, (D=4)-dimensions and (D=5)-dimensions, *Int. J. Mod. Phys. A* **16**, pp. 1015-1108.

Nordström, G. (1918). On the energy of the gravitational field in Einstein's theory, *Proc. Kon. Ned. Akad. Wet.* **20**, pp. 1238–1245

Olmo, G.J. (2003). Quantum correlations in the evaporation of charged black holes, Master Thesis, University of Valencia

Oppenheimer, J.R. and Snyder, H. (1939). On continued gravitational contraction, *Phys. Rev.* **56**, pp. 455–459

Oppenheimer, J.R. and Volkoff, G.(1939). On massive neutron cores, *Phys. Rev.* **55**, pp. 374–381

Padmanabhan, T. (2001). *Theoretical astrophysics, II. Stars and stellar systems.* Cambridge University Press, Cambridge, England.

Page, D.N. (1976). Particle emission rates from a black hole: massless particles from an uncharged, nonrotating hole, *Phys. Rev. D* **13**, pp. 198–206. Particle emission rates from a black hole. II. Massless particles from a rotating hole, *Phys. Rev. D* **14**, pp. 3260–3273.

Page, D.N. (1980). Is black hole evaporation predictable?, *Phys. Rev. Lett.* **44**, pp. 301–304

Page, D.N. (1993). Black hole information, hep-th/9305040

Painlevé, P. (1921). La mécanique classique et la théorie de la relativité, *C.R. Acad. Sci. (Paris)* **173**, pp. 677–680

Parentani, R. (1996). The recoils of a dynamical mirror and the decoherence of its fluxes, *Nucl. Phys.* **B** 465, pp. 175–214

Parker, L. (1969). Quantized fields and particle creation in expanding universes. 1. *Phys. Rev.* **183**, pp. 1057–1068

Parker, L. (1975). Probability distribution of particles created by a black hole, *Phys. Rev. D* **12**, pp. 1519–1525

Parker, L. (1976). The production of elementary particles by strong gravitational fields, in *Asymptotic structure of spacetime*, Eds. Esposito, F.P. and Witten, L., New-York: Plenum.

Penrose, R. (1965). Gravitational collapse and spacetime singularities, *Phys. Rev. Lett.* **14**, pp. 57–59

Penrose, R. (1968). Structure of spacetime, in DeWitt, C.M. and Wheeler, J.A. eds. *Battelle Rencontres: 1967 Lectures in mathematics and physics* Benjamin, New York, p. 565

Penrose, R. (1969). Gravitational collapse: the role of general relativity, *Nuovo Cimento* **1**, pp. 252–276

Piqueras, C. (1999). Thermal aspects of Hawking radiation, Master Thesis, University of Valencia

Poisson, E. and Israel, W. (1990), Internal structure of black holes, *Phys. Rev. D* **41**, pp. 1796–1809

Polchinski, J. (1998). *String Theory*, Cambridge University Press, Cambridge, England

Polyakov, A.M. (1981). Quantum geometry of bosonic strings, *Phys. Lett. B* **103**, pp. 207–210

Preskill, J., Scharz, P., Shapere, A., Trivedi, S. and Wilczek, F. (1991). Limitations on the statistical description of black holes, *Mod. Phys. Lett. A* **6**, pp. 2353–2362

Preskill, J. (1992). Do black holes destroy information?, hep-th/9209058, in the Proceedings of the International Symposium on "Black holes, membranes, wormholes and superstrings", The Woodlands, Texas, p. 22

Price, R.H. (1972), Nonspherical perturbations of relativistic gravitational collapse, *Phys. Rev. D* **5**, pp. 2419–2438 and 2439–2454

Reissner, H. (1916). Uber die eigengravitation des elektrischen feldes nach der Einsteinschen theorie, *Ann. Phys.* **50**, pp. 106–120

Rindler, W. (1966). *Am. J. Phys.* **34**, p. 1174

Rindler, W. (2001). *Relativity*, Oxford University Press, Oxford, England

Robinson, D.C. (1975). Uniqueness of the Kerr black hole, *Phys. Rev. Lett.* **34**, pp. 905–906

Robinson, I. (1959). A solution of the Maxwell–Einstein equations, *Bull. Acad.*

Polon. Sci. **7**, pp. 351–352

Rovelli, C. (1996). Black hole entropy from loop quantum gravity, *Phys. Rev. Lett.* **77**, pp. 3288–3291

Russo, J.G., Susskind, L. and Thorlacius, L. (1992a). Black hole evaporation in $(1+1)$-dimensions, *Phys. Lett. B* **292**, pp. 13–18

Russo, J.G., Susskind, L. and Thorlacius, L. (1992b). End point of Hawking radiation, *Phys. Rev. D* **46**, pp. 3444–3449

Russo, J.G., Susskind, L. and Thorlacius, L. (1992c). Cosmic censhorship in two-dimensional gravity, *Phys. Rev. D* **47**, pp. 533–539

Schoutens, K., Verlinde, H. and Verlinde, E. (1993). Black hole evaporation and quantum gravity (hep-th/9401081), Proceedings of Trieste Spring School 1993

Schutz, B.F. (1985). *A first course in general relativity*. Cambridge University Press, Cambridge, England.

Sciama, D.W., Candelas, P. and Deutsch, D. (1981). Quantum field theory, horizons and thermodynamics, *Adv. in Phys.* **30**, pp. 327–366

Shapiro, S.L. and Teukolsky, S.A. (1983). *Black holes, white dwarfs and neutron stars*. Wiley, New York, USA.

Spradlin, M. and Strominger, A. (1999). Vacuum states for AdS_2 black holes, *J. High Energy Physics* **11**, 021

Starobinsky, A.A. (1973). Amplification of waves during reflection from a rotating black hole, *Sov. Phys. JETP* **37**, pp. 28–32

Stephani, H. (1982). *General relativity*. Cambridge University Press. Cambridge, England.

Stephens, C.R., 't Hooft, G. and Whiting, B.F. (1994). Black hole evaporation without information loss, *Class. Quant. Grav.* **11**, pp. 621-648

Strominger, A. and Trivedi, S. P. (1993). Information consumption by Reissner–Nordström black holes. *Phys. Rev. D* **48**, pp. 5778–5783

Strominger, A. and Thorlacius, L. (1994). Conformally invariant boundary conditions for dilaton gravity. *Phys. Rev. D* **50**, pp. 5177–5187

Strominger, A. (1995). Les Houches lectures on black holes, hep-th/9501071

Strominger, A. and Vafa, C. (1996). Microscopic origin of the Bekenstein–Hawking entropy, *Phys. Lett. B* **379**, pp. 99–104

Susskind, L. and Thorlacius, L. (1992). Hawking radiation and backreaction, *Nucl. Phys. B* **382**, pp. 123–147

Susskind, L. Thorlacius, L. and Uglum, J. (1993). The stretched horizon and black hole complementarity, *Phys. Rev. D* **48**, pp. 3743–3761

Susskind, L. (1995). The world as a hologram, *J. Math. Phys.* **36**, pp. 6377–6396

Susskind, L. and Uglum, J. (1995). String physics and black holes, hep-th/9511227

Szekeres, G. (1960). On the singularities of a Riemannian manifold, *Pub. Math. Debrecen* **7**, pp. 285–301

Tanaka, T. (2003). Classical black hole evaporation in Randall-Sundrum infinite brane world, *Prog. Theor. Phys. Suppl.* **148**, pp. 307–316

Teitelboim, C. (1984). The Hamiltonian structure of two-dimensional space-time and its relation with the conformal anomaly, in *Quantum Theory of Gravity*, Ed. Christensen, S.M. (Hilger, Bristol, 1984), pp. 327–344.

't Hooft, G. (1985). On the quantum structure of a black hole, *Nucl. Phys. B* **256**, pp. 727–745

't Hooft, G. (1990). The black hole interpretation of string theory, *Nucl. Phys. B* **335**, pp. 138–154

't Hooft, G. (1993). Dimensional reduction in quantum gravity theory, gr-qc/9310006

Thorlacius, L. (1994). Black hole evolution, hep-th/9411020

Thorne, K.S., Price, R.H. and Macdonald, A. (1986). *Black holes: the membrane paradigm*, Yale University Press, New Haven, USA

Thorne, K.S. (1994). *Black holes and time warps. Einstein's outrageous legacy.* W.W. Norton and Company, New York, USA

Tolman, R.C. (1939). Static solutions of Einstein's field equations for spheres of fluid, *Phys. Rev.* **55**, pp. 364–373

Townsend, P.K. (1998). Black Holes: lecture notes, hep-th/9707012

Unruh, W.G. (1974). Second quantization in the Kerr metric, *Phys. Rev. D* **10**, pp. 3194–3205

Unruh, W.G. (1976). Notes on black-hole evaporation, *Phys. Rev. D* **14**, pp. 870–892

Unruh, W.G. and Wald, R.M. (1984). What happens when an accelerating observer detects a Rindler particle, *Phys. Rev. D* **29**, pp. 1047–1056

Vaz, C. (1989). Approximate stress energy tensor for evaporating black holes, *Phys. Rev. D* **39**, pp. 1776–1779

Wald, R.M. (1974). Gendaken experiments to destroy a black hole, *Ann. Phys.* **45**, pp. 9–34.

Wald, R.M. (1975). On particle creation by black holes, *Commun. Math. Phys.* **82**, pp. 548–556.

Wald, R.M. (1984). *General Relativity*, Chicago University Press, Chicago, USA.

Wald, R.M. (1993). The first law of black hole mechanics, Hu, B.L., Ryan, M.P. and Vishveshwara, C.V. eds., *Directions in general relativity*, Vol.1,

Wald, R.M. (1994). *Quantum field theory in curved spacetime and black hole thermodinamics*, Chicago University Press, Chicago, USA.

Wald, R.M. (1998). Black holes and thermodynamics, in *Black holes and relativistic stars*, ed. Wald, R.M. Chicago University Press, Chicago, USA. pp. 155–176

Weinberg, S. (1972). *Gravitation and cosmology: principles and applications of the general theory of relativity*. Wiley, New York, USA

Wilzcek, F. (1993). Quantum purity at a small price: easing a black hole paradox, hep-th/9302096

Wilzcek, F. (1998). Lectures on black hole quantum mechanics, *Int. J. Mod. Phys. A* **13**, pp. 5279–5372

Wipf, A. (1998). Quantum fields near black holes (hep-th/9801025), in *Black holes: theory and observation*, Eds. Hehl, F.W., Kiefer, C. and Metzler, R. Springer, Berlin, Germany.

Witten, E. (1991). On string theory and black holes, *Phys. Rev. D* **44**, pp. 314–324

Zaslavskii, O.B. (1999). Exactly solvable models of dilaton gravity and quantum eternal black holes, *Phys. Rev. D* **59**, 084013

Zel'dovich, Ya B. (1972). Creation of particles and antiparticles in an electric and gravitational field, in *Magic without magic: John Archibald Wheeler*, ed. Klauder, J.R.; Freeman, San Francisco, USA, pp. 277-288

Index

AdS_2 space, 175–177

affine, 25–27, 36, 39, 50, 60, 124
angular momentum, 18, 19, 39, 51, 55, 56, 107, 108, 126, 130, 149, 196
anomalous transformation law, 154, 161, 206, 222, 224, 233
anomaly induced effective action, 225, 226, 229
Anti-de Sitter space, 145, 173, 174, 177, 178
apparent horizon, 3, 56–60, 257, 265, 276, 284, 288, 292–294, 305, 306, 310
area law, 59, 61, 62, 128, 129, 131

baby universe, 140
backreaction effects, 2, 136, 189, 233, 244, 296
backreaction problem, 4, 5, 195, 235, 236, 239, 243, 275
backscattering, 107, 108, 111, 112, 114, 117, 118, 125, 150, 161, 162, 215, 216, 229, 232, 241
Birkhoff's theorem, 7, 14, 40
black body radiation, 106
black hole thermodynamics, 130
blueshift, 38, 45, 52, 88, 112, 122, 123
Bogolubov coefficients, 78, 79, 86, 94, 98, 102, 107, 114, 126, 265, 268–270
Bogolubov transformation, 4, 76–78, 99, 100, 107, 265, 266, 268

Boulware, 173, 183, 209, 211, 230, 231, 239, 240, 271, 275, 281–283, 299, 300

Cauchy horizon, 45, 52, 315, 316
causal structure, 1, 26, 29, 40, 49, 133, 138, 139
charged black holes, 39, 66, 68, 69, 130, 244
complementarity, 142, 143
conformal compactification, 27
conformal symmetry, 4, 145, 215, 218, 221
conformal transformation, 145, 150, 152–156, 172, 184, 196, 206, 218, 220, 255, 260
conservations equations, 195, 238, 239
correlation functions, 4, 138, 145, 151, 152, 169, 171
critical line, 292, 294

degenecacy pressure, 10
density matrix, 114, 116, 135–137, 140, 213, 231, 233
dilaton black hole, 249, 251, 257, 259, 285
dilaton field, 66, 67, 71, 225, 229, 238, 245, 250, 251, 274

effective action, 66, 193, 203, 206, 207, 225–227, 275, 297
Einstein frame, 67, 251, 252, 254,

265, 295
end point, 243, 244, 276, 292, 293, 295, 296, 305, 312, 315, 316
entropy, 62–64, 66, 129–131, 139, 140
equivalence principle, 1, 3, 4, 32, 145, 148, 195, 197
ergosphere, 52, 54
eternal black hole, 24, 112, 263
extremal black hole, 40, 46, 47, 54, 69, 174, 178, 179, 244, 248, 266, 298, 306, 315

first law, 62, 63, 65, 128
Fock space, 74–76, 79, 80, 85, 86, 101, 104, 111, 142, 151, 163, 211, 231, 268
free field, 84, 107, 165, 254, 255, 262, 277–281, 295
future event horizon, 21–23, 31, 34–36, 41, 47, 48, 56, 59, 60, 62, 161, 215, 257, 259
future null infinity, 27, 56, 106, 184, 214, 274, 277

geometric optics, 84, 117, 122–126
gravitational collapse, 1–4, 7, 11, 16, 17, 19, 20, 22–25, 32, 40, 45, 50, 56, 76, 80–82, 97, 119, 128, 130, 132, 143, 145, 171, 185, 212, 215
grey-body factor, 114, 115

Hartle–Hawking, 173, 183, 210–212, 231, 233, 234, 239, 270, 272, 275, 282, 283, 299–301
Hawking temperature, 97, 107, 128, 212, 214, 231, 270, 272, 275, 283, 287
Hilbert space, 74, 142

information loss, 5, 131, 132, 137, 139, 140, 143, 276, 290, 292, 315, 316
inner horizon, 43, 44, 305
irreducible mass, 61

Kerr black hole, 52, 53

Killing horizon, 38, 62, 127
Klein–Gordon equation, 73, 83
Klein–Gordon product, 73, 75
Kruskal coordinates, 3, 23, 24, 30, 32, 36, 42, 147, 148, 182, 210, 211, 247, 260, 266, 272, 282, 283, 286, 287, 296, 298, 299, 301, 306

laws of black hole mechanics, 3, 62, 65, 66
linear dilaton vacuum, 71, 256, 257, 263, 265–267, 276, 281, 284, 285, 290, 292, 293, 299
Liouville equation, 297, 302
locally inertial coordinates, 3, 5, 32–37, 47, 48, 148, 195, 198, 200
Lorentz transformations, 33, 37, 155, 156
luminosity, 107, 108, 116–118, 131, 233

Möbius transformations, 154, 156, 171, 172, 181, 182, 301
Minkowski coordinates, 148
Minkowski space, 4, 27–29, 32, 57, 70, 73, 75, 76, 133, 145, 147, 166, 167, 169, 177, 193, 195–197, 201, 212, 217, 219, 246, 267, 293
mixed state, 135, 136, 141, 168, 171, 211, 213, 214, 293
moving-mirror, 146, 183

naked, 40, 59, 133
naked singularity, 54
near extremal, 270, 285, 296
near-horizon approximation, 5, 145, 148, 159, 165, 173, 177, 179, 183, 196, 209, 241, 243, 244, 248, 250, 251, 275, 298
negative energy, 214, 285
negative frequency, 73, 75, 76, 88, 102, 150, 184
neutron stars, 10, 11, 17
no-hair theorem, 18, 39, 40, 50, 63, 67, 130, 132
normal ordered stress tensor, 5, 161,

163, 164, 181, 193, 195, 196, 198, 199, 206, 213, 219, 221, 229, 232, 240
normal ordering, 153, 154, 165, 193, 195, 196, 198, 201, 208, 209, 221, 224, 227
null geodesic, 20, 21, 23, 24, 26, 29, 31, 36, 41, 56, 57, 60, 123, 124

particle detector, 90, 167
particle number, 78, 89, 127, 157, 162, 163, 171, 185
past event horizon, 23, 24, 35, 36, 42, 47, 48, 69
past null infinity, 27, 125, 133, 184, 276, 303
Planck distribution, 97
Planck mass, 132, 139
Planck scale, 2, 136, 137
Planck spectrum, 96, 107, 116
Polyakov effective action, 195, 203, 206–208, 225, 275
positive frequency, 74–77, 79, 84, 85, 87, 90, 151, 161, 169, 170, 172, 184, 196, 198, 219
potential barrier, 107, 108, 110, 112, 116, 139, 216, 233
pure state, 101, 135–141

quantum coherence, 135–137, 141, 142, 166
quantum hair, 140
quantum predictability, 4, 132, 137, 141
quantum stress tensor, 197, 201, 202, 204, 222, 229, 234, 274

redshift, 36–38, 88, 89, 107, 119, 122, 124, 142
Reissner–Nordström black hole, 5, 128, 175, 177, 179, 244, 245, 287, 296, 298, 310, 316
remnant, 139, 140
Rindler space, 4, 145–147, 159, 166, 177, 180, 181, 246

s-wave, 195, 216, 217, 233–235, 241
scattering, 107, 111, 113, 244, 266
Schwarzian derivative, 154, 156, 166, 172, 185, 187, 196, 199, 211, 272, 304
Schwarzschild geometry, 4, 20, 39, 108, 127, 145, 148, 166, 173, 209, 214
Schwarzschild radius, 10, 143
Schwarzschild spacetime, 3, 5, 7, 14, 24, 25, 29, 31, 57, 82, 108, 112, 210, 219, 231, 245
second law, 62–65, 129, 130
semiclassical approximation, 139, 144, 295
solvable, 5, 219, 244, 262, 282, 295
spacelike geodesic, 27, 29, 70
spacelike singularity, 284, 292
spherical reduction, 235, 250
spherical symmetry, 5, 7, 10, 18, 19, 40, 217, 237
state-dependent function, 195, 238–241, 243, 272, 273, 275, 276, 283, 284, 286, 296, 298, 301, 303, 308, 309
stationary black hole, 62, 63
string frame, 67, 69, 245, 250, 252, 257, 259, 275, 295
stringy black hole, 5, 66, 244, 245, 296
subcritical, 291, 292
superradiance, 56, 126, 127
surface gravity, 36–39, 41, 47, 54, 62, 66, 68, 71, 127–129, 146, 148, 270, 283

thermal bath, 64, 167, 196, 231, 272
thermal density matrix, 3, 101, 106, 163, 167, 171, 211, 231, 233
thermal radiation, 4, 64, 97, 98, 100, 119, 121, 126, 128, 131, 134, 135, 138, 163, 164, 185, 186, 215, 233, 273, 313
thermal state, 106, 165, 167, 186, 212, 232
thermodynamics, 2, 3, 62–65, 128–130
third law, 63

thunderbolt, 293, 295, 315–317
thunderpop, 285, 286, 290, 292, 295, 306
timelike geodesic, 14, 26, 29, 31
trace anomaly, 5, 202, 205, 208, 224, 225, 228, 230, 278
transplanckian, 142
trapped surface, 19, 56, 57, 257, 258

unitarity, 189, 191
Unruh effect, 4, 145, 159, 166, 167, 184, 196, 199
Unruh modes, 169–172
Unruh state, 173, 183, 214, 215, 226, 240, 241, 273, 283, 285, 286, 288
Unruh vacuum, 212, 232, 233

vacuum polarization, 210, 211, 213, 214, 234, 277, 299, 301
Vaidya, 81, 107, 126, 159, 178, 307, 308
Virasoro anomaly, 154, 155, 195, 198

wave packets, 90, 91, 94, 95, 98, 113, 153, 158
Weyl transformation, 26, 28, 31, 67, 155, 218
Weyl-invariant effective action, 206, 226–228
white dwarf, 7, 8, 10

zeroth law, 62